LTE BACKHAUL

LTE BACKHAUL

PLANNING AND OPTIMIZATION

Edited by

Esa Markus Metsälä

Juha T.T. Salmelin

Nokia Networks, Espoo, Finland

This edition first published 2016

Registered Office
John Wiley & Sons, Ltd, The Atrium, Southern Gate, Chichester, West Sussex, PO19 8SQ, United Kingdom

For details of our global editorial offices, for customer services and for information about how to apply for permission to reuse the copyright material in this book please see our website at www.wiley.com.

Library of Congress Cataloging-in-Publication Data

LTE backhaul : planning and optimization / edited by Esa Markus Metsälä and Juha T.T. Salmelin.
pages cm
Includes bibliographical references and index.
ISBN 978-1-118-92464-8 (hardback)
1. Long-Term Evolution (Telecommunications) 2. Telecommunication–Traffic. I. Metsälä, Esa Markus.
II. Salmelin, Juha T.T.
TK5103.48325.L7345 2015
621.3845′6–dc23

2015021968

A catalogue record for this book is available from the British Library.

Set in 10/12pt Times by SPi Global, Pondicherry, India
Printed and bound in Singapore by Markono Print Media Pte Ltd

1 2016

Contents

List of Contributors

Gerald Bedürftig
Nokia Networks
Berlin, Germany

Raimo Kangas
Nokia Networks
Tampere, Finland

Jouko Kapanen
Nokia Networks
Espoo, Finland

Raija Lilius
Nokia Networks
Espoo, Finland

Esa Markus Metsälä
Nokia Networks
Espoo, Finland

José Manuel Tapia Pérez
Nokia Networks
Espoo, Finland

Juha T.T. Salmelin
Nokia Networks
Espoo, Finland

Jari Salo
Nokia Networks
Doha, Qatar

Csaba Vulkán
Nokia Networks
Budapest, Hungary

Gabriel Waller
Nokia Networks
Espoo, Finland

Foreword

With LTE, the mobile network has evolved into a 150+ Mbps per user high-speed always-on packet network. Next we will see high-speed LTE networks becoming available for even larger populations, and solving capacity and speed bottlenecks that users currently experience. For many of us, mobile broadband is the preferred and primary access to the Internet.

The competition for the hearts and minds of LTE subscribers makes the user experience increasingly critical. Understanding the technology behind the service is the key to business success. Delving into the details of LTE technology soon reveals many items that affect performance, allowing room for optimization—and differentiation—in the market.

In general, operators today have more choice and support than ever in choosing their strategy for LTE planning and optimization tasks, including IP and backhaul tasks. The myriad challenges operators face can be addressed by specific professional services, purchased from an expert organization, or issues can be solved by in-house professionals. Many large networks are operated as a service, and a continuum of possibilities exists, from traditional in-house operation to fully managed service operations, and everything in between.

Whatever the technology and business strategy of the operator, high-bandwidth LTE radio needs to be reflected in the IP backhaul. For the LTE backhaul, a number of new areas call for special attention, namely security, synchronization, availability, end-user QoS and dimensioning, to name a few.

LTE IP planning professionals depend on both LTE and IP knowledge, and greatly benefit from realistic guidance for their projects. This book is of great help when assessing technical and economical alternatives and when creating solid and reliable real-life backhaul designs for LTE success.

Igor Leprince
Executive Vice President, Global Services
Nokia

Acknowledgments

The editors would first like to acknowledge the contributing authors, who are our colleagues at Nokia Networks: Gerald Bedürftig, Raimo Kangas, Jouko Kapanen, Raija Lilius, José Manuel Tapia Pérez, Jari Salo, Csaba Vulkán and Gabriel Waller. Your knowledge has been the most essential ingredient in this project.

For specific review comments and for bigger and smaller suggestions and contributions we would like to thank: Heikki Almay, Antti Pietiläinen, Jyri Putkonen, Eugen Wallmeier, Raimo Karhula, Pekka Koivistoinen, Zoltán Vincze, Péter Szilágyi, Balázs Héder, Attila Rákos, Gábor Horváth, Lajos Bajzik, Dominik Dulas, Michal Malcher, Puripong Thepchatri, Lasse Oka, Steve Sleiman, Taufik Siswanto, Matti Manninen (Elisa), Timo Liuska (Juniper Networks) and Mika Kivimäki.

Also, we would like to thank the team at John Wiley & Sons for very good cooperation and an easy editing process, especially Mark Hammond, Tiina Wigley, Sandra Grayson, Teresa Netzler, Tim Bettsworth and Victoria Taylor.

We appreciate the patience and support of our families and our authors' families during the writing period.

We are grateful for comments and suggestions for improvements or changes that could be implemented in forthcoming editions of this book. This feedback can be sent to the editors' email addresses: esa.metsala@nokia.com and juha.salmelin@nokia.com.

Esa Markus Metsälä and Juha T.T. Salmelin
Espoo, Finland

List of Abbreviations

2G	second generation (mobile system)
3G	third generation (mobile system)
3GPP	Third Generation Partnership Program
ASN.1	Abstract Syntax Notation One
ABS	almost blank subframe
ACK	acknowledgement signal
AFxx	assured forwarding behavior group xx
AH	Authentication Header
AM	acknowledged mode
AMBR	Aggregate Maximum Bit Rate
AMR	adaptive multi-rate coding
ANR	automatic neighbor relation
AOM	administration of measurements
AP	Application Protocol
APN-AMBR	Access Point Name–Aggregate Maximum Bit Rate
AQM	active queue management
ARP	Address Resolution Protocol
ATM	asynchronous transfer mode
BBF	Broadband Forum
BC	boundary clock
BCMP	Baskett, Chandy, Muntz and Palacios
BE	best effort
BFD	bidirectional forwarding detection
BH	Backhaul, Busy Hour
BITW	bump in the wire
BMAP	batch Markovian arrival processes
BMCA	best master clock algorithm
BSC	base station controller
BSHR	bidirectional self-healing ring

BTS	base station
CA	carrier aggregation, certification authority
CAC	connection admission control
capex	capital expenditure
CBS	committed burst size
CDF	cumulative distribution function
CDMA	Code Division Multiple Access
CE	customer equipment
CET	carrier Ethernet
CIR	committed information rate
CLI	command line interface
CM	configuration management
CMP	Certificate Management Protocol
CoDel	controlled delay
CoMP	coordinated multi-point
CORBA	Common Object Request Broker Architecture
CoS	class of service
C-plane	control plane
CPE	customer premises equipment
CPU	central processing unit
CRC	cyclic redundancy check
CRL	certificate revocation list
CRS	common reference signals
CSFB	Compact Small Form-factor Pluggable
CSV	comma-separated values
CUBIC	TCP with cubic window increases function
CWDM	coarse wavelength division multiplexing
DC	dual connectivity
DCF	discounted cash flows
DCH	dedicated channel
DCN	data communications network
DHCP	Dynamic Host Configuration Protocol
DL	downlink
DNS	domain name system
DNU	do not use
DOCSIS	data over cable service interface specification
DoS	denial of service
DPD	dead peer detection
DSCP	differentiated services code point
DSL	digital subscriber line
DWDM	dense wavelength division multiplexing
DWRR	deficit weighted round robin
EAPS	Ethernet Automatic Protection Switching
EBS	excess burst size
ECMP	equal cost multipath
eCoMP	enhanced CoMP

EDGE	enhanced data rates for GSM evolution
EF	expedited forwarding
eICIC	enhanced inter-cell interference coordination
EIR	excess information rate
E-LAN	Ethernet service, multipoint-to-multipoint
E-line	Ethernet service, point-to-point
EMS	element management system
eNB	evolved NodeB
e2e	end-to-end
EPC	evolved packet core
ERP	Ethernet ring protection
ESM	EPS session management
ESP	Encapsulating Security Payload
E-tree	Ethernet service, point-to-multipoint
E-UTRAN	Evolved Universal Terrestrial Radio Access Network
EXP	experimental bits
FCAPS	fault, configuration, accounting, performance, security
FCFS	first come, first served
FDD	frequency division duplex
FD-LTE	full duplex LTE
FeICIC	further enhanced inter-cell interference coordination
FIFO	first in, first out
FTP	File Transfer Protocol
GbE	gigabit Ethernet
GBR	guaranteed bit rate
GE	gigabit Ethernet
G.Fast	up to Gigabit/s fast short distance digital subscriber line
GLONASS	Global Navigation Satellite System, Russia
GNSS	global navigation satellite system
GPON	gigabit-capable passive optical network
GPRS	general packet radio service
GPS	Global Positioning System
GSM	Global System for Mobile communications
GTP	general packet radio service Tunneling Protocol
GTP-U	general packet radio service Tunneling Protocol user
HARQ	hybrid automatic repeat request
HetNet	heterogeneous networks
HRM	hypothetical reference model
HSPA	high-speed packed access
HSRP	Hot Standby Router Protocol
HTML	Hypertext Markup Language
HTTP	Hypertext Transfer Protocol
ICIC	inter-cell interference coordination
ICMP	Internet Control Message Protocol
IEEE	Institute of Electrical and Electronics Engineers
IETF	Internet Engineering Task Force

IKE	Internet key exchange
IMS	IP Multimedia Subsystem
IMT-A	international mobile telecommunications advanced
impex	implementation expenditure
IP	Internet protocol
IPsec	Internet Protocol Security architecture
IRC	interference rejection combining
IRR	internal rate of return
ISD	inter-site distance
itag	video parameter classification
ITU	International Telecommunication Union
ITU-T	ITU Telecommunication Standardization Sector
IU	indoor unit
KPI	key performance indicator
L1	Layer 1 in Open Systems Interconnection data link layer
L2	Layer 2 in Open Systems Interconnection data link layer
L2 VPN	Layer 2 virtual private network
L3 VPN	Layer 3 virtual private network
LAG	link aggregation group
LAN	local area network
LDF	load distribution factor
LFA	loop-free alternate
LOS	line of sight
LSP	label switched path
LTE	long-term evolution
LTE-A	long term evolution advanced
M/G/R-PS	M/G/R Processor Sharing model
MAC	media access control
MAN	metropolitan area network
MAP	Markovian arrival processes
MBH	mobile backhaul
MBMS	Multimedia Broadcast Multicast Service
MEF	Metro Ethernet Forum
MeNB	master eNB
MGW	media gateway
MIB	management information base
MIMO	multiple input, multiple output
MLO	multilayer optimization
ML-PPP	multilayer point-to-point protocol
MME	mobile management entity
MPEG4	Moving Pictures Experts Group
M-plane	management plane
MPLS	multiprotocol label switching
MPLS TC	multiprotocol label switching traffic class
MPLS-TP	multiprotocol label switching traffic profile
MSP	multiplex section protection

MS-SPRING	multiplex section protection ring
MSTP	Multiple Spanning Tree Protocol
MTBF	mean time between failures
MTTR	mean time to repair
MTU	maximum transfer unit
MVI	multi-vendor interface
MWR	microwave radio
NaaS	network management system as a service
NAS	network application server
NETCONF	Network Configuration Protocol
NGMN	Next Generation Mobile Network
NG-SDH	Next Generation Synchronous Digital Hierarchy
nLOS	near line of sight
NLOS	non line of sight
NMS	network management system
non-GBR	non-guaranteed bit rate
NP	non-protected
NPV	net present value
NTP	Network Time Protocol
O&M	operation and maintenance
OAM	operations administration and maintenance
OC-3	optical carrier level 3
ODU	outdoor unit
OID	object identifier
opex	operational expenditure
OSPF	Open Shortest Path First
OSS	operation support system
OTDOA	observed time difference of arrival
OTT	over the top
OU	outdoor unit
P	protected (IPsec)
PDF	probability distribution function
PDH	plesiochronous digital hierarchy
PDN	public data network
PDP	packet data protocol
PDU	protocol data unit
PE	provider edge
PE–PE	provider edge to provider edge
P-GW	packet data network gateway
PHB	per-hop behaviors
PHY	physical layer
PKI	public key infrastructure
PLMN	public land mobile network
PM	performance monitoring
PON	passive optical network
ppb	parts per billion

PPP	point-to-point protocol
ppm	pulse per minute
pps	pulse per second
PRC	primary reference clock
PS HO	packet service handover
PSK	pre-shared key
PTP	Precision Time Protocol
QCI	quality of service class indicator
QNA	queuing network analyzer
QoE	quality of experience
QoS	quality of service
RA	radio access
RBID	radio bearer identification
RC	resource coordination
RE	range extension
RED	random early detection
RF	radio frequency
RFCs	request for comments
RLC	radio link control
RN	relay node
RNC	radio network controller
ROI	return on investment
RRC	radio resource control
RRH	remote radio head
RRM	radio resource management
RSTP	Rapid Spanning Tree Protocol
RTO	retransmission timeout timer
RTP	Real-time Transport Protocol
RTT	round trip time
RX	receive, receiver
S1	Interface between eNB and MME/S-GW
S1-AP	S1 Application Protocol
S1-MME	interface between eNB and MME
S1-U	interface between eNB and S-GW
SA	security association
SACK	selective acknowledgment
SCEP	Simple Certificate Enrollment Protocol
SCF	Small Cell Forum
SCTP	Stream Control Transmission Protocol
SDH	synchronous digital hierarchy
SEG	security gateway
SeNB	slave eNB
S-GW	serving gateway
SLA	service level agreement
SMS	short message service

SMTP	Simple Mail Transfer Protocol
SNMP	Single Network Management Protocol
SOA	service-oriented architecture
SOAP	Simple Object Oriented Access Protocol
SON	self-organizing network
SONET	synchronous optical network
SP	strict priority scheduling
SP-GW	combined node of S-GW and P-GW
S-plane	synchronization plane
SPQ	strict priority queuing
SRLG	shared risk link group
SRVCC	single radio-voice call continuity
SS7	signaling system 7
SSH	secure shell
SSL	secure sockets layer
SSM	synchronization status messages
STM	synchronous transport module
STP	Spanning Tree Protocol
SyncE	Synchronous Ethernet
TCP	Transmission Control Protocol
TDD	time division duplex
TD-LTE	time division duplex LTE
TDM eICIC	time domain enhanced inter-cell interference coordination
TFRC	Transmission Control Protocol-friendly rate control
TLS	Transport Layer Security protocol
TMN	Telecom Management Network
TTI	transmission time interval
TTL	Time to Live
TWAMP	Two-Way Active Measurement Protocol
TX	transmit, transmitter
UDP	User Datagram Protocol
UE	user equipment
UL	Uplink
U-plane	user plane
USB	universal serial bus
VDSL	very high bit rate digital subscriber line
VLAN	virtual local area network
VLL	virtual leased line
VoIP	voice over Internet protocol
VoLTE	voice over long-term evolution
VPLS	virtual private local area network service
VPN	virtual private network
VPWS	virtual private wire service
VRF	virtual routing and forwarding
VRRP	Virtual Router Redundancy Protocol

WACC	weighted average cost of capital
W-CDMA	Wideband Code Division Multiple Access
WDM	wavelength division multiplexing
WFQ	weighted fair queuing
WRR	weighted round robin
X2-AP	X2 application protocol
X2-U	interface between eNB and eNB
xDSL	"any kind of" digital subscriber line
XG-PON	10 gigabits/passive optical network
XML	Extensible Markup Language
XPIC	cross-polarization interference cancellation

1

Introduction

Esa Markus Metsälä and Juha T.T. Salmelin
Nokia Networks, Espoo, Finland

1.1 To the Reader

This book intends to offer guidelines and insight for long-term evolution (LTE) backhaul planning and optimization tasks and is aimed at technical professionals working in the field of network planning and operations. With LTE backhaul, several functional areas like synchronization, Quality of Service (QoS) and security, to name a few, require major new analysis, when designing for a high-performing and well-protected network. And in addition, the capacity needs of the LTE and LTE Advanced (LTE-A) radio typically mandate a major upgrade to the currently supported backhaul capacity, which often means introducing new backhaul links and technologies.

As with any network design project, several feasible and technically sound approaches exist. Many of the examples given in this text highlight topics that the authors find especially important. For every design, high-level goals are unique, as are the boundaries set for the project, and the examples should be tailored where necessary to match the individual design target. All of the views presented reflect the authors' personal opinions and are not necessarily that of their employers.

The book aims to give an objective, standards-based view of the topics covered. Many of the LTE backhaul related aspects are, however, not written as binding standards. As such, there is room for different implementations, dependent on the capabilities of mobile network elements such as evolved NodeB (eNB), security gateways, backhaul elements and related management systems.

A basic command of LTE and Internet Protocol (IP) networking is useful for getting the most out of this book. Mobile backhaul, and its key services and functions, is discussed in greater detail in Metsälä and Salmelin (2012), which can be used as a reading companion.

LTE Backhaul: Planning and Optimization, First Edition. Edited by Esa Markus Metsälä and Juha T.T. Salmelin.

The book's chapters approach each major topic of LTE with illustrations, complementing these with both short examples, including questions with model answers, and a few longer case studies.

Network designs are also influenced by non-technical drivers, such as the budget available for the project. Strategic input and comparing alternative designs from the financial side is important, which is why these topics are covered in a separate chapter.

1.2 Content

The book is divided into eight chapters. Chapter 2 is a bird's-eye view of LTE backhaul: what is it all about? While the book's focus is on technical matters and planning advice, the financial modeling of LTE backhaul is discussed in Chapter 3. Backhaul dimensioning is a challenge, and the theoretical basis for backhaul dimensioning and end user application behavior is covered in Chapter 4. Chapter 5 covers planning advice in the form of guidelines and examples, while Chapter 6 focuses on two bigger walk-through network design cases. Network management of the backhaul network and its relation to the LTE radio network management is the topic of Chapter 7 and the book is summarized in Chapter 8.

1.3 Scope

The essential scope of the book is the planning of the LTE IP backhaul, with focus on the LTE-specific design requirements and how to meet these requirements using Ethernet, IP and other packet protocols, with security of the backhaul taken into account in all phases of the design.

In order to help dimensioning LTE backhaul, the theoretical basis for analyzing backhaul capacity needs is given. As well, several end user aspects related to Transmission Control Protocol (TCP) behavior over the LTE network are investigated, since these may heavily affect end user perception of the LTE service.

Network management systems are traditionally separate for the backhaul and for the LTE radio network; however, several benefits can be exploited from the integration of these tools, as discussed in a Chapter 7.

Detailed planning of backhaul physical layer technologies—like optical links, wavelength division multiplexing and wireless (microwave) links—would all need a book of their own to be properly covered, and such reference books exist, and those should be used as additional sources of knowledge for detailed planning with those technologies.

The standards for the radio technologies discussed in this book in relation to the evolution of LTE-A are those finalized by the Third Generation Partnership Program (3GPP) at the spring of 2015. Constructing design guidelines for functionalities where standardization is in progress is difficult; however, key LTE advanced functions and their foreseen impact to backhaul are included in section 2.7.

Reference

Metsälä E. and Salmelin J. (eds) (2012) *Mobile Backhaul*. John Wiley & Sons, Ltd, Chichester, UK, doi: 10.1002/9781119941019.

2

LTE Backhaul

Gerald Bedürftig[1], Jouko Kapanen[2], Esa Markus Metsälä[2] and Juha T.T. Salmelin[2]
[1] *Nokia Networks, Berlin, Germany*
[2] *Nokia Networks, Espoo, Finland*

2.1 Introduction

This chapter gives an overview of different aspects of LTE backhaul. An introduction about the different elements of a backhaul network, the different end-to-end (e2e) services and the requirements of the LTE Mobile Access network are given. A short explanation of the different L1 possibilities for the defined network areas is also included. In addition, a prospective of future requirements and an overview of the relevant standards are provided. A general understanding of packet-based backhaul networks is a prerequisite for an understanding of this chapter. More detailed information can be found in Metsälä and Salmelin (2012).

With LTE, the backhaul network will be further extended toward the core network. For second-generation (2G) networks the backhaul consisted of an access part and in most cases one level of aggregation until the base station controller (BSC) was reached. Third generation (3G) further concentrated the radio network controller (RNC) site locations and so additional aggregation layers needed to be considered. With the elimination of the RNC in the LTE architecture, a further concentration of the mobility management entity (MME), serving gateway (S-GW) and packet data network gateway (P-GW)—which in most cases were combined as a serving and packet data network gateway (SP-GW)—was the consequence. In addition to multiple aggregation levels, parts of the IP backbone network may be passed to connect the eNBs with the core network elements. Figure 2.1 gives an overview of the nomenclature and Figure 2.2 shows the different network areas used in this book.

As seen in Figure 2.2 the network that is the focus of this book will be the network between eNB and the respective core network elements. Backhaul elements are the network elements which are located between eNB and core network elements. Their purpose is simply to provide the relevant e2e services (see Section 2.2) by closing the geographic distance between the mobile nodes in a secure and cost-efficient manner.

LTE Backhaul: Planning and Optimization, First Edition. Edited by Esa Markus Metsälä and Juha T.T. Salmelin.

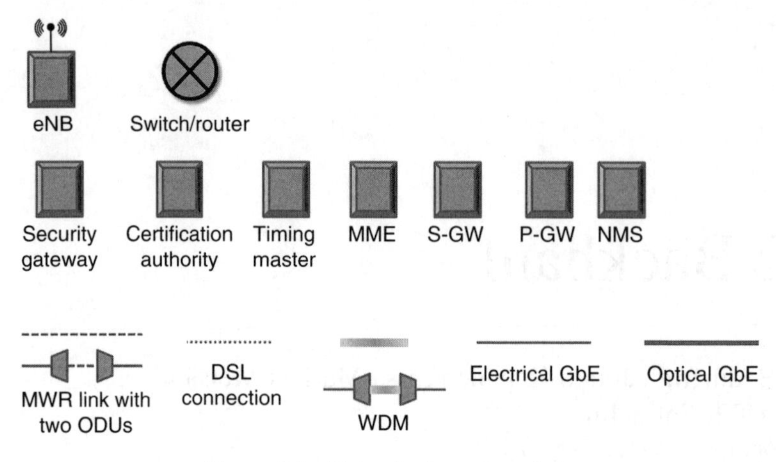

Figure 2.1 Network element symbols

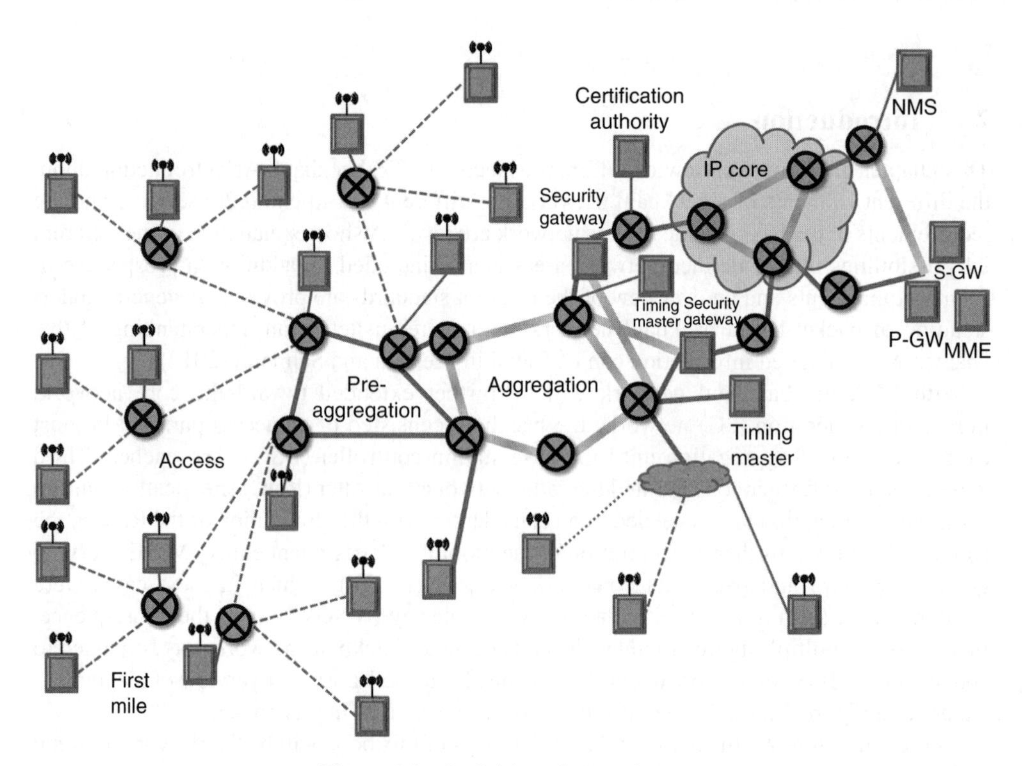

Figure 2.2 Definition of network areas

In the future, "fronthaul" will be relevant for remote radio heads (RRH) or distributed antenna systems and "small cell" backhaul, which is about the backhauling of high-density base stations in urban areas. Currently, fronthaul still requires dedicated fibers. Other wireless technologies or even fronthaul via a shared switching network are being considered but are not yet ready. Section 2.8 gives a short overview of small cell backhaul planning. Additionally, in the future functionality like caching and distributed security features may become relevant.

2.2 LTE Backhaul Planes

Different kinds of user and control planes (C-planes) have to be transported across the backhaul network.

Figure 2.3 gives an overview of the logical planes which are needed. This section gives a first overview and all different planes will be described in detail in subsequent sections.

Most important are the user traffic S1 user plane (U-plane), which has to be transported from eNB to S-GW and P-GW (or combined SP-GW), and the control traffic S1 C-plane between eNB and MME. In addition, the management traffic which consists of all the data needed for FCAPS (fault, configuration, accounting, performance, security) support of the network elements and the optional X2 traffic consisting of user and control plane between neighboring eNBs has to be considered. The special role of the synchronization plane can be realized in various ways. Finally, additional traffic which is due to active measurements of either probes or inbuilt functionality to monitor the e2e performance of the backhaul network is relevant.

2.2.1 3GPP Planes and Protocol Stacks

This section gives a short overview of the different planes and protocol stacks. Respective security considerations are provided in Section 2.4.6.

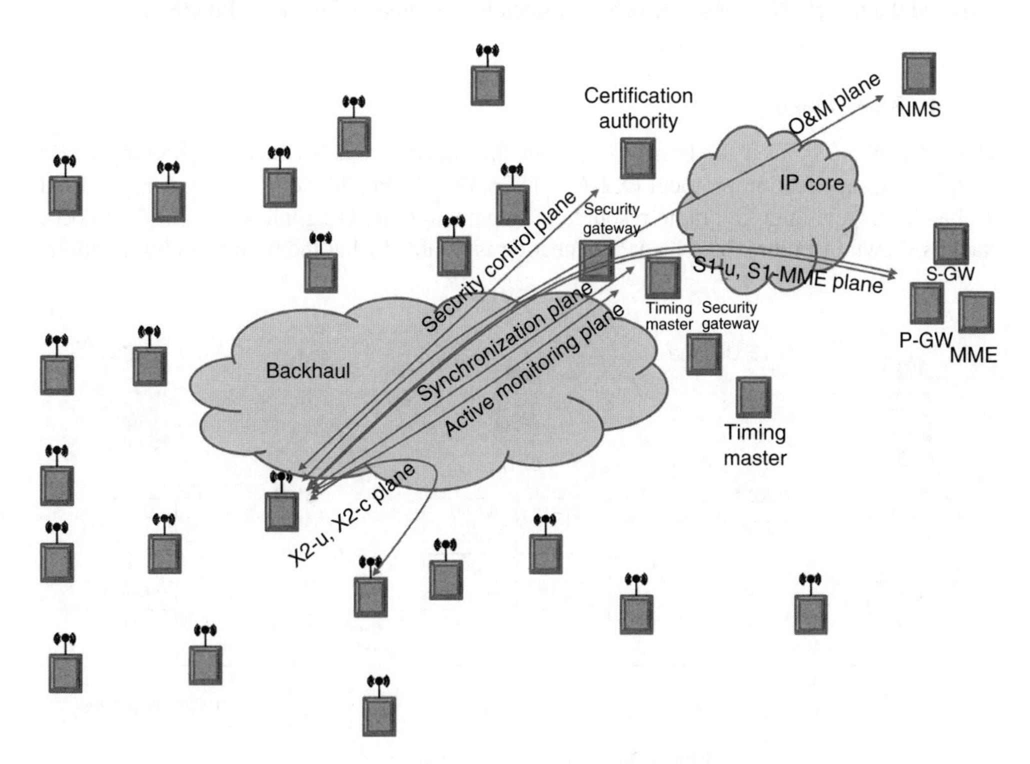

Figure 2.3 LTE related e2e planes of backhaul network

2.2.1.1 S1-U Plane

The S1-U plane is used to transport user data between the eNB and the S- and P-GW using a general packet radio service tunneling protocol user (GTP-U). Each S1 bearer consists of a pair of GTP-U tunnels (one for the uplink [UL] and one for the downlink [DL]). The eNB performs mapping between radio bearer IDs (RBID) and GTP-U tunnel endpoints. The GTP-U protocol is defined in TS29.060 (3GPP, 2015a) and its position in the protocol stack is shown in Figure 2.4.

2.2.1.2 S1-MME (S1-C Plane)

The S1-MME plane is used to transfer signaling information between the eNB and MME using S1-AP protocol TS36.413 (3GPP, 2015b). It is used for S1 bearer management, mobility and security handling and for the transport of network application server (NAS) signaling messages between the user equipment (UE) and MME. S1-MME protocol stack is shown in Figure 2.5.

2.2.1.3 X2-U-Plane

The X2-U-plane is used for forwarding user data between the source eNB and target eNB during inter-eNB handovers. A GTP-U tunnel is established across the X2 between the source eNB and the target eNB. Thus the protocol stack is the same as for S1-U (Figure 2.4).

2.2.1.4 X2-C-Plane

The X2-C-plane is used for transferring signaling information between neighboring eNBs using the X2-application protocol (X2-AP) TS36.423 (3GPP, 2015c). This signaling is used for handovers and inter-cell radio resource management (RRM) signaling. An X2-C protocol stack is shown in Figure 2.5. X2-AP connection is established at the time neighbor relations

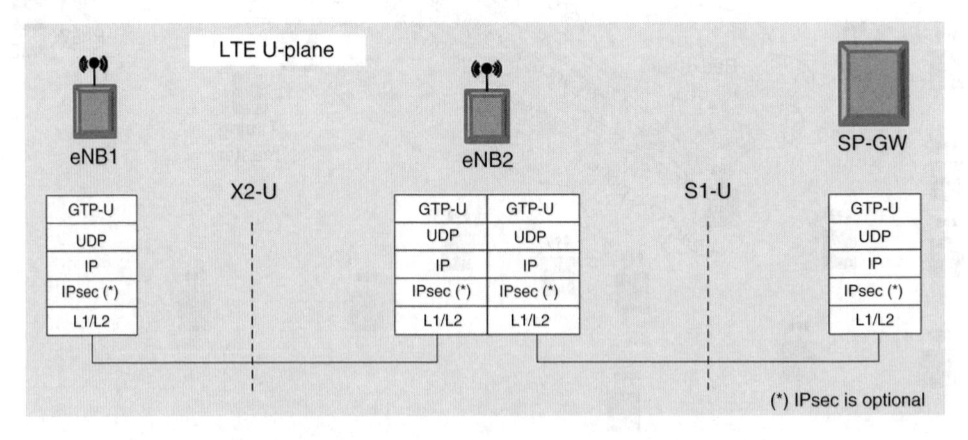

Figure 2.4 LTE U-plane protocol stack

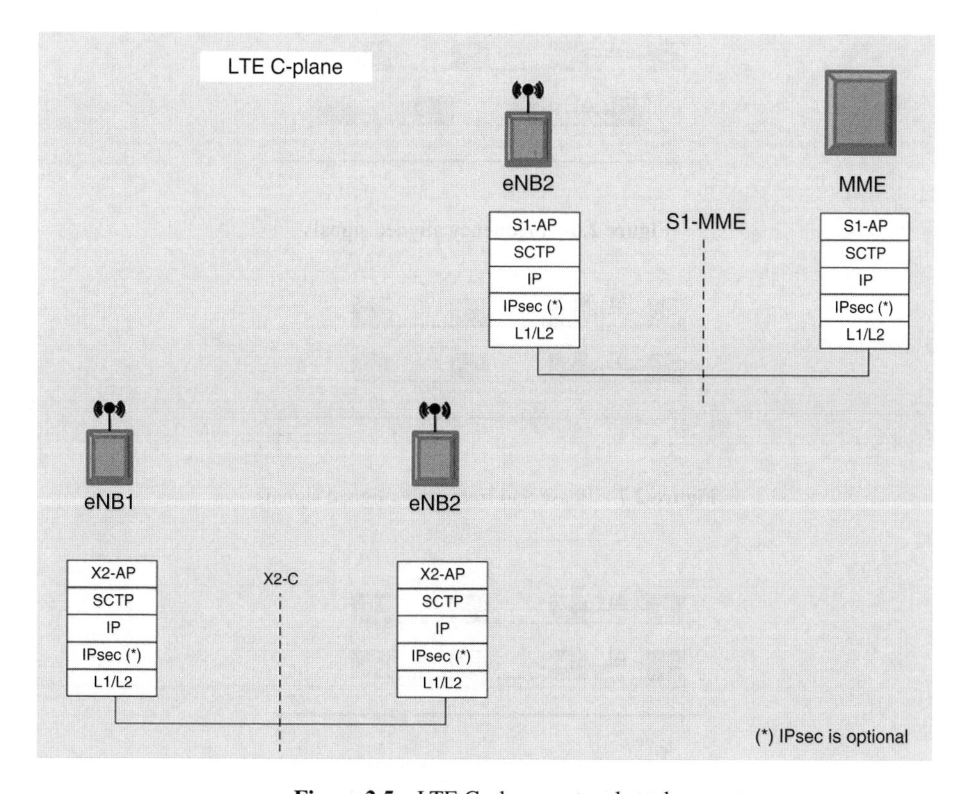

Figure 2.5 LTE C-plane protocol stack

are formed, i.e. at eNB startup or when neighbors are added by manual operation and maintenance (O&M) intervention, or by automatic neighbor relations. Prior to X2-AP setup, a Stream Control Transmission Protocol (SCTP) connection is initialized via the usual SCTP four-way handshake.

It is important to note that all the protocol stacks include IP. There is no other possibility given by the standard.

2.2.2 Synchronization Plane

Synchronization is important to assure proper functionality of the eNBs. In section 2.4.4 it is shown that different radio technologies have different synchronization requirements. Frequency synchronization has to be distinguished from phase and time synchronization. For frequency synchronization it is sufficient to transport information about the frequency across the backhaul network. This can be achieved by physical mechanisms (synchronous Ethernet, synchronous digital hierarchy [SDH]) or by Precision Time Protocol (PTP) based mechanisms. Another opportunity is to decouple this completely from the backhaul network and to use GNSS (global navigation satellite system) solutions. Figure 2.6 shows schematically two frequency-aligned signals.

In addition to frequency, the phase information includes additional information about the point in time a signal occurs. This is shown in Figure 2.7.

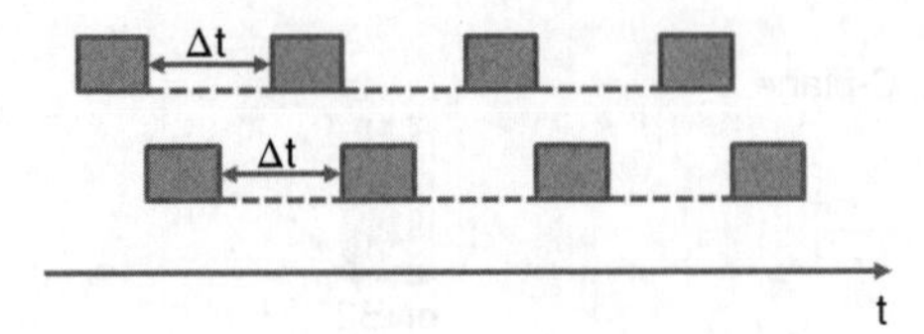

Figure 2.6 Frequency aligned signals

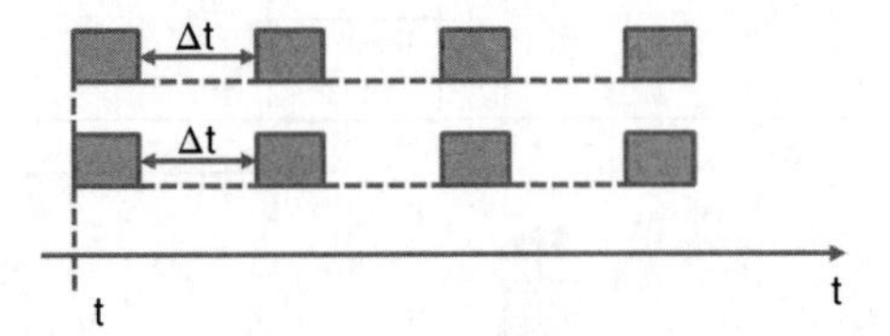

Figure 2.7 Phase and frequency aligned signals

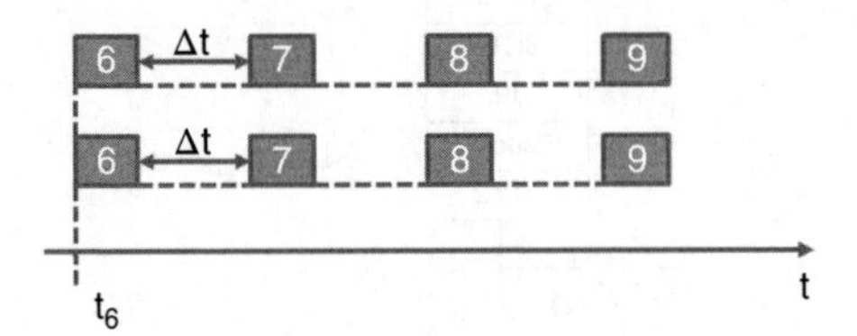

Figure 2.8 Time, phase and frequency aligned signals

The next complexity step would be the time alignment. A simplified picture is shown in Figure 2.8.

Chapter 6 of Metsälä and Salmelin (2012) gives a much more detailed overview of the different possibilities to transport synchronization signals across the network and the respective standards and challenges. For a planner it is important to understand the fundamental difference between frequency and time and phase synchronization. Frequency synchronization is rather simple compared to phase and time synchronization, in that it is simple to plan the network and the impact of frequencies being out of specification is not as dramatic as when time synchronization is out of specification. Three possibilities are distinguished to achieve frequency synchronization:

- physically based like plesiochronous digital hierarchy (PDH), SDH or synchronous Ethernet
- algorithm-based PTP solutions
- GNSS-based systems.

At first glance, the reliability of the physically based seems to be higher than for PTP solutions, as they do not depend on network performance. But this benefit can only be achieved if these networks are planned properly and all configuration and topology changes are well considered and updated in the synchronization plan. As many operators do not spend this effort once the network is deployed, their networks have synchronization problems. In many cases this does not affect one but a couple of base stations and these problems are static problems, i.e. they remain and will not be solved automatically. The quality of PTP solutions depends on the network performance. This increases the fluctuation of the signal and on a short timescale the frequency is not as stable as for the other solutions. On the other hand, the benefit is that all base stations are tuned independently and network problems typically only occur for a certain period. The algorithm will tune all the frequencies back to target value once the problem is resolved. Thus the PTP solution is much more robust and the effort which has to be spent to plan and maintain this solution is lower compared to any physical solution (in a volatile network environment with many changes happening in the network). In addition it has to be considered that base stations have very good oscillators. They can stay in holdover for several days without being out of specification.

For network-based phase and time synchronization the following challenges occur:

- planning and maintenance challenges like for physically based solutions
- network performance dependencies like for frequency PTP solutions
- much shorter holdover times for base station oscillators
- a significant impact on the overall performance of the base station in case they are out of specification.

It is obvious that the need for time and phase synchronization leads to the most challenging network requirements and, in the event of failure, has the biggest impact on the radio network performance. This combination of challenging requirements and huge impact makes it difficult to deploy the e2e synchronization solution and additional effort has to be spent to plan, implement and maintain the network.

Section 2.4.4 specifies the exact synchronization requirements for different LTE features.

2.2.3 Management Plane

The management plane (M-plane) is the interface between eNB and the O&M system. This interface is not specified in detail by 3GPP. Typically, the transport layer uses TCP. NTP (Network Time Protocol) uses User Datagram Protocol (UDP) instead of TCP. NTP is typically used to define the time of the base station which is used for time stamping alarms, performance counter, certification expiry and other notifications. This has nothing to do with the time synchronization mentioned in Section 2.2.2.

The M-plane consists of all data needed to manage and monitor the status of the network elements. From a planning perspective especially, the performance monitoring (PM) data are relevant. More and more data from network elements are collected to monitor the network status and to initiate proactive analysis. The volume of this traffic has to be estimated and the right point in time has to be defined to collect these data. A stable management connection is essential to avoid any unnecessary site visits and to assure continuous counter collections. The M-plane protocol stack is shown in Figure 2.9.

2.2.4 Active Monitoring Plane

The e2e performance of the backhaul network is important for the e2e quality of experience (QoE) of mobile services. Thus it is very important to monitor different e2e key performance

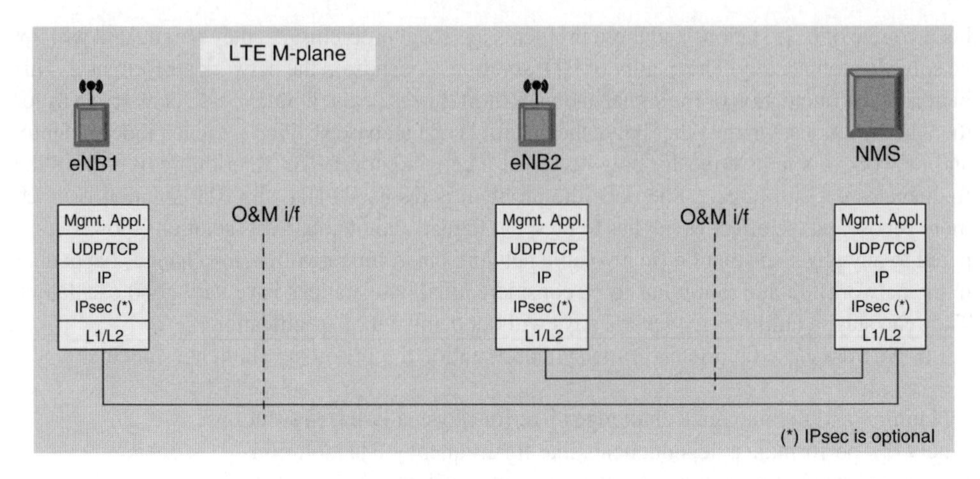

Figure 2.9 Management plane protocol stack

indicators (KPIs). The most cost-efficient solution for this is to initiate active measurement traffic in different class of service (CoS) categories and to analyze the collected data. Although the amount of traffic should be small compared to user and control traffic, it is important to consider this properly. In addition to the measurement traffic itself, the M-plane traffic (especially collection of PM data) of dedicated probes has to be considered. The frequency and phase of active measurements may also have an impact on the performance of PTP algorithms in the network. PTP algorithms do not like signals that are sent with a correlated phase. For example, many dedicated messages which are sent phase synchronous with one pulse per second (pps) may fill certain buffers and this may have an impact on the performance of the PTP messages.

2.2.5 Security Control Plane

Setting up secured communications with Internet Protocol Security architecture (IPsec) tunnels requires control protocols in addition to the tunnels themselves. With public key infrastructure (PKI) architecture, certificates need to be fetched from a certification authority (CA). The protocol may be based on the Certificate Management Protocol (CMP) or Simple Certificate Enrollment Protocol (SCEP).

For IPsec, the IKE (Internet key exchange) protocol has to be used.

2.2.6 Control and User Plane of Additional Proprietary Applications

Vendors are working on additional proprietary functionality in the eNBs, for example the storage of local content for low-latency applications or the caching of frequently used information. These applications cause additional control and user traffic and have to be considered for proper dimensioning. On the other hand, they may also reduce the user traffic, for example in the case of caching.

Example 2.1

Question: What are the key traffic types in the LTE backhaul?

Answer: U-plane traffic carries the mobile user packets encapsulated in GTP-U tunneling protocol for both the S1 and the X2 interfaces. U-plane bearers at S1 and X2 are managed by related signaling (C-plane) with S1-AP (S1 application protocol) and X2-AP. Both the U-plane and the C-plane may be protected by IPsec.

When synchronization is arranged by backhaul network, for example PTP, these packets need to be delivered to the eNB from the timing server. The packet timing mechanism is not specified by 3GPP so different implementations exist.

Network management traffic (O&M connectivity) allows remote management of the eNBs. As with synchronization, O&M channel implementation may vary but often relies on TCP. O&M traffic clearly needs to be cryptographically protected. Related to management, different measurement protocols typically need to be supported for monitoring and troubleshooting purposes.

To support the identification of nodes, and the security and protection of the network, PKI relies on certificates and on the use of protocols like CMP and SCEP. Additionally, IPsec is mandated by 3GPP for protection of the S1 and X2 interfaces and relies on the use of IKE as the IPsec C-plane protocol. Furthermore, there may exist vendor- or operator-specific applications.

2.3 Radio Features of LTE and LTE-A

2.3.1 LTE

The specifications for the initial LTE system that were defined in 3GPP's Release 8 were intended to optimize the system for increased IP and data traffic compared to 2G and 3G. New air interface technology with single-carrier bandwidths up to 20 MHz enables data rates up to 50 Mbps in the UL and 150 Mbps in the DL (assuming category 4 UE).

Compared to 3G and high-speed packed access (HSPA) systems, the network architecture is simplified so that no separate radio network controller is needed. Instead most of the functionality is embedded in the base station, which is the only type of network element within the radio network. The base station eNB then interfaces directly with the core network and other eNBs, both of which features are different from 2G and 3G.

U-plane and C-plane functionality are clearly separated in the core network by having MME as the C-plane element and S-GW as the U-plane element. All this supports better scalability of the network: the radio network is flat, consisting only of eNBs, and the U-plane and C-plane capacities in the core network can be scaled up independently.

Scalability is also embedded in the air interface, since now carrier bandwidth may range from 1.4 MHz to 20 MHz.

The circuit switched core network is abandoned and, instead, the core network is completely IP based. And the SS7 (signaling system 7) signaling stack has also been retired. As the LTE radio does not support circuit switched services, the voice services need to be implemented with IP-based voice over long-term evolution (VoLTE) or, alternatively, circuit switched fallback to 2G/3G is possible.

For backhaul, all interfaces are by 3GPP definition IP protocol based and there are no existing alternatives. Because, unlike the controller (like the RNC in 3G), the air interface encryption has been terminated in the eNB and so now part of the mobile system traffic path (between the base station/eNB site and the core) is unsecure and needs some other means of encryption (e.g. IPsec). The X2 interface between eNBs means horizontal traffic streams between eNBs. Due to topological and other restrictions in practice, it is often arranged via a central point higher in the network. Nevertheless, the X2 type of traffic between neighbor base stations is new compared to 3G or 2G.

2.3.2 LTE-A

The drivers of 3GPP for LTE advanced (LTE-A) has been to further develop LTE to fulfill the International Telecommunication Union (ITU) requirements set for international mobile telecommunications advanced (IMT-A):

- increased number of simultaneously active users
- increased peak data rate
- higher spectral efficiency
- improved performance at cell edges.

Further drivers on mobile operator point of view may include aspects such as:

- increased adoption of mobile broadband, and greater availability and choice in terms of devices
- enhanced coverage (spreading across more locations), and increase in usage intensity
- machine-to-machine communications.

LTE-A is enabled by new technologies and features, and enhancements to existing technologies, like carrier aggregation, MIMO (multiple input, multiple output), Coordinated Multi-Point (CoMP), Inband Relaying (relay nodes), and HetNets (heterogeneous networks).

2.3.2.1 Carrier Aggregation

Carrier aggregation (CA) allows up to five LTE Release 8 compatible component carriers being combined, each having a bandwidth of 1.4 MHz to 20 MHz and providing a maximum of 100 MHz aggregated bandwidth, and provides almost as high a spectrum efficiency and peak rates as single allocation does. The principle of CA is presented in Figure 2.10.

CA can be implemented either intra-band, meaning the component carriers belong to the same operating frequency band, or inter-band, meaning the component carriers belong to different operating frequency bands.

The capability of CA to improve single user throughput depends on the number of users in a cell. If the number of users is low, scheduling over multiple carriers provides significant throughput gain since the available radio resources can be allocated to the user(s) with the most favorable radio conditions (e.g. taking into account different propagation characteristics of different frequency bands). On the other hand, scheduling a high number of users over multiple carriers provides only marginal gain. Illustrations of CA impact, as a function of the cell load (offered load), both in UL and DL direction, is presented in Figure 2.11.

CA Impacts on Backhaul Transport

Improving average cell throughput both in UL and DL practically means needing a higher backhaul bandwidth. With current approaches CA does not affect the synchronization and backhaul latency requirements; however, lower RTT (round trip time) is preferred to be able to fully utilize the higher data rates for single users.

2.3.2.2 Multiple Input, Multiple Output (MIMO)

MIMO technology is used to increase the peak data rate when two or more (2, 4, 8, etc.) parallel data streams are transmitted and received with multiple antennas by using the same resources in both frequency and time. The principle of MIMO (2 × 2 MIMO) has been illustrated in Figure 2.12.

Transmission peak data rates depend on the number of antennas on the transmitter and the receiver, the used bandwidth and the configuration of radio parameters. With MIMO, up to 25% average DL cell throughput gain is achievable. Examples of MIMO peak data rates with different antenna configurations are presented in Figure 2.13.

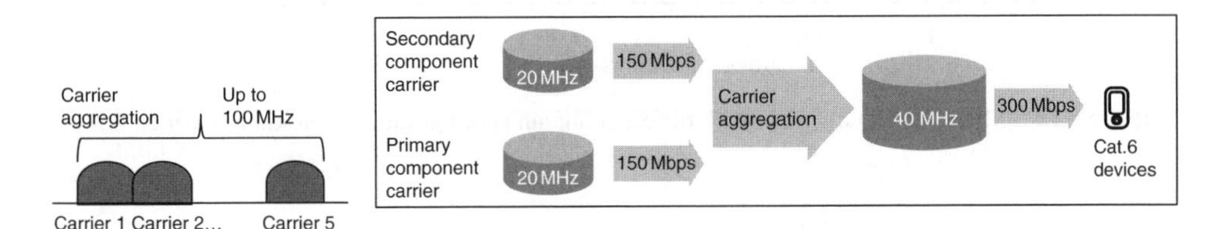

Figure 2.10 Principle of carrier aggregation

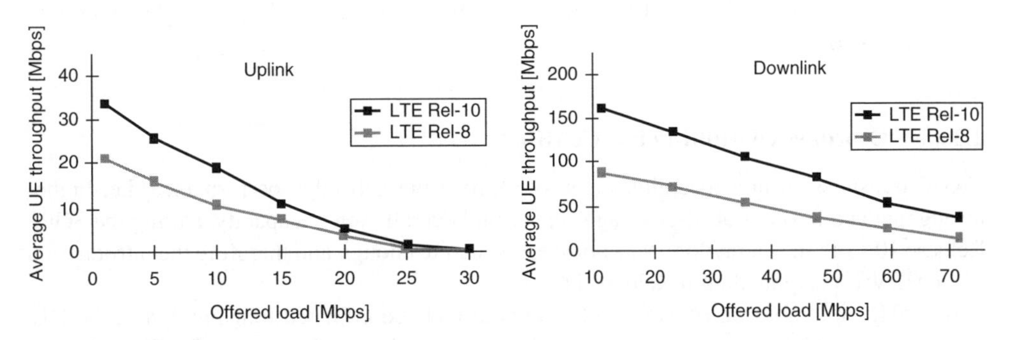

Figure 2.11 Carrier aggregation improves average cell throughput both in UL and DL (Nokia Networks, 2014)

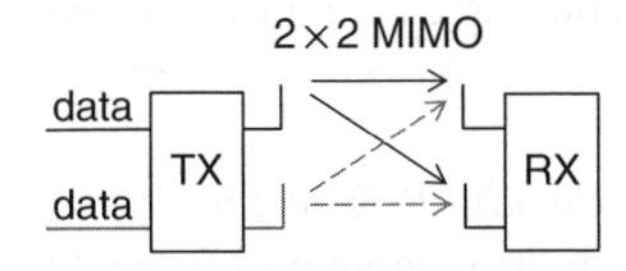

Figure 2.12 Principle of MIMO

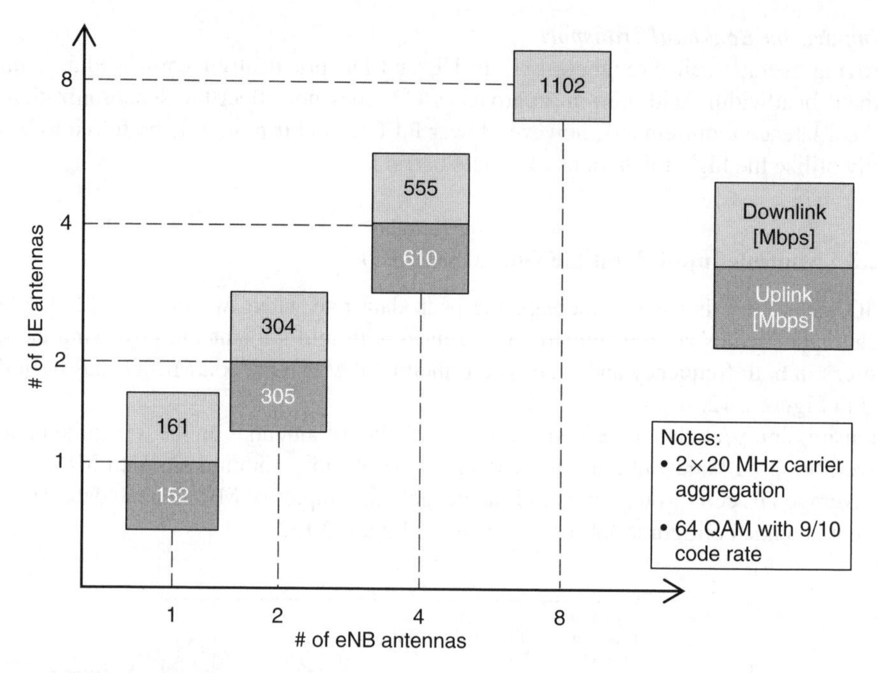

Figure 2.13 MIMO peak data rates with different antenna configurations (Nokia Networks, 2014)

MIMO Impacts on Backhaul Transport
Improvement in average cell throughput (DL) and higher peak rates naturally affects the required backhaul capacity, but will have no effect on synchronization or backhaul latency requirements. However, lower RTT is preferred to be able to fully utilize the higher data rates for single users.

2.3.2.3 Coordinated Multi-Point (CoMP)

CoMP transmission and reception aims to improve the cell edge performance, i.e. higher throughput and fewer "out-of-coverage" users, and overall system capacity. During the 3GPP Release 10 work the technology was not seen as mature enough and therefore the introduction of CoMP was postponed until Release 11.

In CoMP, a number of TX (transmit) points provide coordinated transmission in the DL, and a number of RX (receive) points provide coordinated reception in the UL. The principle of CoMP transmission and reception is presented in Figure 2.14.

The critical system deployment issue is the communication between the TX/RX points (cells), and therefore intra-site CoMP technologies, requiring communication just between the sectors of a single eNB, have been the focus of Release 11 work.

CoMP (Release 11) Impacts on Backhaul Transport
Slight improvement in average cell throughput may affect the required backhaul capacity. For almost every kind of CoMP, time domain synchronization is needed (for both frequency division duplex [FDD] and time division duplex [TDD]) for the involved cells, but in the case

of intra-eNB CoMP that is an internal eNB requirement and does not affect, and is not visible on, the backhaul. There are no specific requirements for backhaul latency.

2.3.2.4 Relay Nodes (RN)

Self-backhauling and RNs provide enhanced coverage and capacity at cell edges and improve the average spectral efficiency of the whole system. RNs use the LTE-A air-interface for self-backhauling, and the radio resources are shared between RN(s) and UEs connected directly to the so-called donor eNB (macro eNB). The principle of RNs is described in Figure 2.15.

The highest gain from RNs is achieved in coverage limited scenarios (e.g. in large cells) that serve multiple RNs. Examples of achievable performance gains in some RN configurations as a function of the number of RNs and inter-site distance (ISD) are presented in Figure 2.16.

RN Impacts on Backhaul Transport

Higher overall system performance (average cell throughput gain, see Figure 2.16) means that higher bandwidth is needed from the backhaul in the same proportion. There is no effect on macro eNB synchronization; however, the RN needs to able to synchronize the UE (like the donor/macro eNB). Relaying increases the total RTT for the UEs under the RN, but not significantly.

2.3.2.5 HetNets, TDM eICIC and FeICIC

Release 10 time domain enhanced inter-cell interference coordination (TDM eICIC) reduces the interference in HetNets by controlling the transmissions of overlapping macro and (low power) small cells (micro eNB) in the time domain. Macro eNB provides the low power nodes with an "almost blank subframe (ABS) muting pattern" that indicates the time slots (subframes)

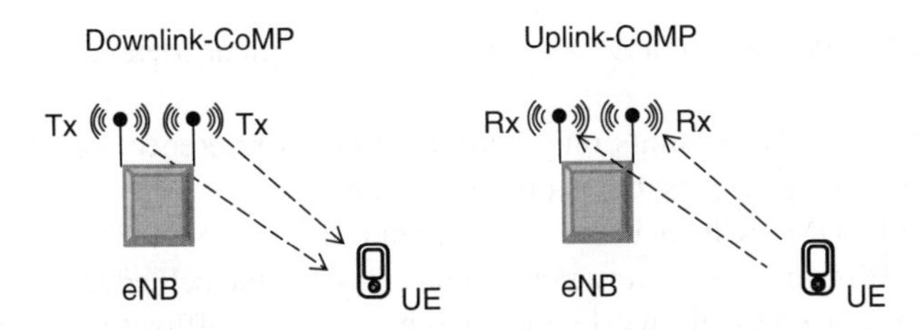

Figure 2.14 Principle of CoMP (intra-eNB)

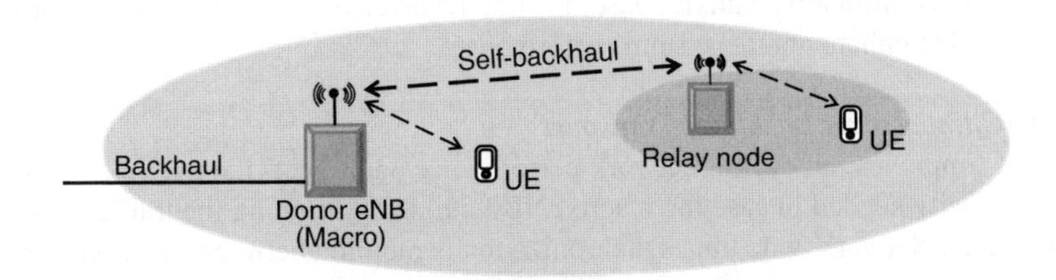

Figure 2.15 Principle of relay nodes

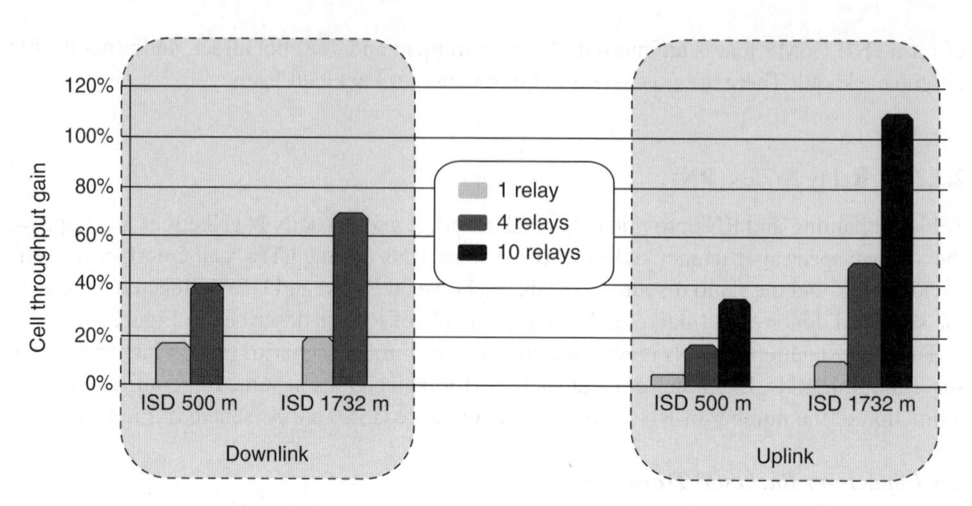

Figure 2.16 Examples of achievable cell throughput gains in some relay node scenarios (Nokia Networks, 2014)

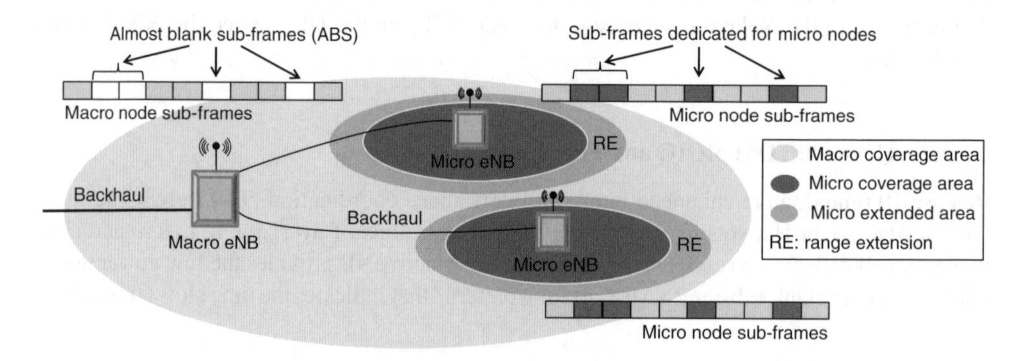

Figure 2.17 Principle of time domain enhanced inter-cell interference coordination

during which macro eNB transmits only certain common reference signals (CRS) that are mandatory for the legacy devices, but is otherwise muted.

The principle of TDM eICIC is illustrated in Figure 2.17.

During these ABSs, the low power micro eNBs can schedule cell edge UEs, allowing them to transmit with lower inter-cell interference, i.e. it allows the offloading of more UEs to the small cell(s) by using larger range extension (RE).

Further enhanced inter-cell interference coordination (FeICIC) combines Release 10 network-based coordination and Release 11 UE receiver-based interference coordination (CRS IC), further enhancing the system performance.

TDM eICIC Impacts on Backhaul Transport

TDM eICIC improves the overall network performance and affects the macro eNB backhaul transport capacity needed in case the macro eNB (site) is aggregating the traffic of the micro eNB backhauls. Network wide time synchronization is required between the macro and small cells. Current Release 11 specifications allow ABS pattern change the fastest every 40 ms, requiring less than 10 ms X2 latency. It is, however, likely that in Release 10/11 implementations the ABS pattern updates are made semi-statically on a rate of tens of seconds to minutes, which would not set any specific requirements for the X2 latency.

Example 2.2

Question: LTE-A (up to Release 11) introduces a number of new functionalities that may affect backhaul. What are these and what would their impact be?

Answer: One can classify the new functions into two categories: those that mainly affect network dimensioning, requiring more capacity from the backhaul, and those that additionally may require new functionality from the backhaul or tighter service level agreements (SLAs), in terms of, for example, delay and loss.

CA and MIMO are examples of the first category; these features greatly enhance air interface throughput but as such do not require anything else than adequate bandwidth on the LTE backhaul, to be able to fully exploit the radio interface.

CoMP of Release 11 focuses on intra-eNB CoMP, where the coordination occurs within a single eNB, and thus there is no impact on the backhaul in terms of functionality.

With RN, the eNB is using its own air interface as a backhaul (self-backhaul). As such, RN can be considered a separate backhaul mechanism. To the traditional macro eNB backhaul (donor eNB backhaul) there is no impact other than additional capacity needed.

Heterogeneous networks and different interference cancellation mechanisms (ICIC, TDM eICIC, FeICIC, etc.) target coordination between macro and small cells. The coordination mechanism defined up to Release 11 likely do not require very fast coordination signaling between cells; however, items in the standardization also discuss faster control messaging, which would have the impact of reducing the maximum latency allowed between coordinated cells. Time domain synchronization is needed for all coordinated cells.

2.4 Requirements for LTE Backhaul (SLAs)

Mandatory and optional services which have to be provided or transported by the backhaul network are defined in section 2.2. Now the more precise requirements have to be described in detail. These requirements are not completely strict. In some cases they are defined by standards. Some depend on the expected traffic load and provided e2e user applications. Some even depend on the specific implementation of the different equipment vendors. These requirements will be the starting point of planning and optimization, which are described in Chapter 5.

2.4.1 Capacity

Compared to 2G and 3G, LTE base stations are able to provide a massive growth of possible bandwidth. Figure 2.18 and Figure 2.19 give an overview about the technical possibilities for peak and average rates. Numbers based on results shown in Holma and Toskala (2011).

Peak rates can only be achieved under ideal air interface conditions. More detailed traffic models are considered in Chapter 4 and Chapter 5. Additional capacity is needed for the control, management and synchronization traffic. Compared to the user traffic, this traffic is quite small. Further details are provided in Chapter 5.

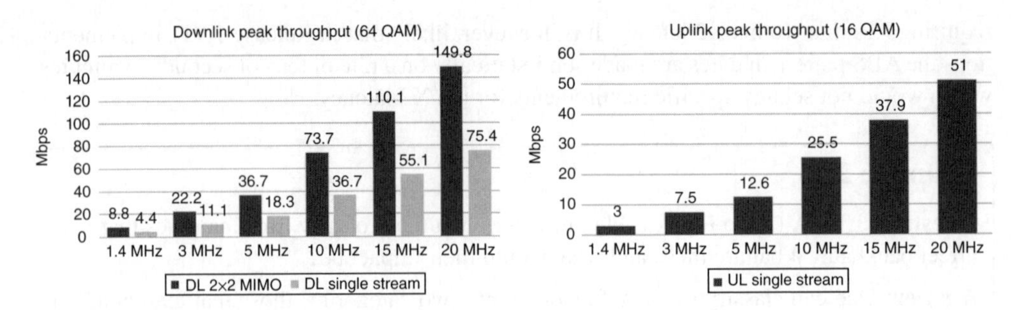

Figure 2.18 User plane peak capacities of LTE base station for different frequency bands

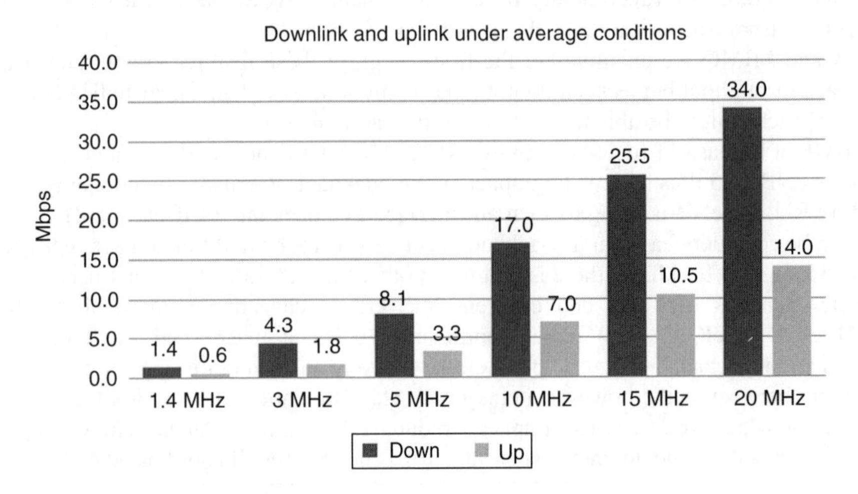

Figure 2.19 Average capacity of LTE base station for different frequency bands

2.4.2 Latency and Loss

2.4.2.1 Latency and Loss for S1 and Management Plane

In addition to capacity increase, latency reduction is one of the key goals of the LTE deployments. Compared to 2G and 3G networks, the contribution of the backhaul to the e2e performance is significant. No strict latency requirements exist. The delay requirements for the backhaul depend on the complete e2e delay of end customer applications (user entity to user entity or user entity to application server) and on the delay budget which is given to the backhaul. Thus the starting point to define the backhaul requirements is the 3GPP e2e specification TS23.203 (3GPP, 2015d).

Table 2.1 gives an overview of the e2e requirements and the typical recommendations for the backhaul network (including the IP core). The U- and C-plane e2e requirements are derived from 3GPP TS23.203. The M-plane requirements are based on typical network management system requirements.

Some additional remarks are necessary for Table 2.1. First of all the values shown are upper bounds. If an operator wants to differentiate in this area, smaller delay values for the backhaul are always helpful. The best achievable round trip time (RTT) for LTE which can be achieved between the user's handset and the operator's Internet gateway is roughly 10–20 ms. This delay does not consider any backhaul or transport delay. Assuming an additional 20 ms one-way delay for backhaul, this time will be increased to 50–60 ms, which is of course still

Table 2.1 Delay and loss requirements for different planes (upper bound)

Plane	One way packet delay		One way packet delay variation		Packet loss ratio	
	e2e requirement	Backhaul recommendation[a]	e2e requirement	Backhaul recommendation	e2e requirement	Backhaul recommendation
U-plane real time	50 ms	20 ms	±10 ms	±10 ms	10^{-3}	10^{-4}
U-plane non-real time	300 ms	20 ms	none	none	10^{-6}	10^{-7}
C-plane	100 ms	20 ms	none	none	10^{-6}	10^{-7}
M-plane (of eNB)		100 ms	none	none	10^{-3}	10^{-6}

[a] See additional textual remarks.

sufficient for most applications, but there is room to differentiate. To achieve 20 ms RTT the one-way delay of the backhaul should be below 5 ms.

The second issue concerns the C-plane. The C-plane mentioned in Table 2.1 is the control traffic of any e2e application (like VoLTE). From an eNB perspective this control traffic is part of the S1-U traffic. A typical requirement for S1-C is in the same magnitude of 20 ms as for the U-plane discussed earlier.

2.4.2.2 X2 Delay Requirements

Figure 2.20 shows inter-eNB handover signaling for X2 handover—eNB1 is the source BTS and eNB2 is the target eNB.

The X2 transport can affect handover U-plane interruption due to one-way delay of messages 5 and 6 (Figure 2.20). UE random access and first UL RRC message (message 7) in the target cell will typically take 20–30 ms from the reception of the handover command in the source cell, based on drive test measurements. From a transport delay point of view, the X2 transport network should not contribute more than this value as a one-way delay; otherwise, the U-plane interruption time will be increased.

Note: having an X2 delay > 30 ms does not necessarily introduce noticeable service degradation for the end user (voice over Internet protocol [VoIP] quality loss or disturbance throughput degradation for Web browsing). Hence the value of 30 ms is a very strict requirement from an e2e quality point of view. This is especially so in low-mobility (universal serial bus [USB] stick dominant) networks.

To summarize, as a rule of thumb, the X2 one-way delay between the eNBs should be less than 30 ms in order for the transport network to not introduce *additional* delay to the inter-eNB handover procedure (on top of the air interface X2 delay). As mentioned, however, X2 delays of > 30 ms do not as such cause handover call drops or e2e quality degradation from an end-user perspective.

The X2 delay requirement raises questions mainly concerning the location of access routers and, in the case of IPsec, the location of the security gateway (SEG).

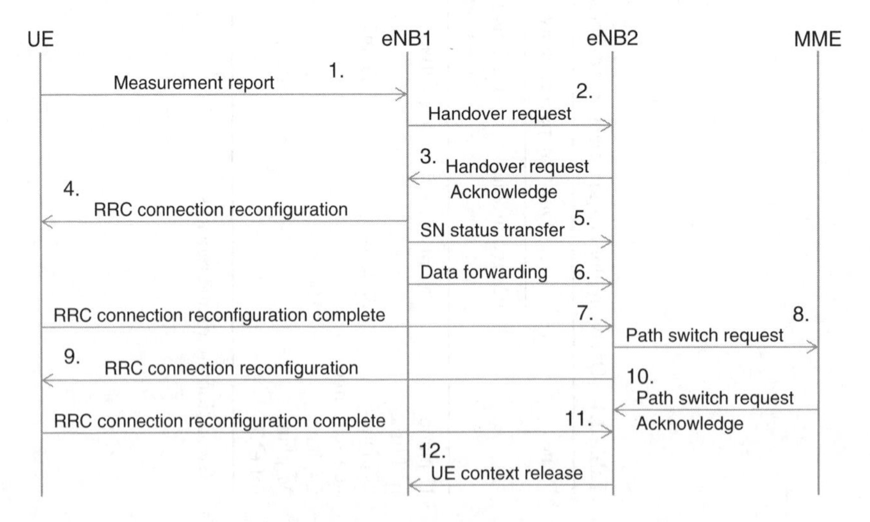

Figure 2.20 X2 handover signaling

2.4.3 QoS Capabilities

Quality of service (QoS) separation is essential to distribute limited resources between different traffic classes. As the mobile backhaul network may also be the limiting factor for certain scenarios, the backhaul network needs to offer certain capabilities to assure QoS. In detail these are:

- CoS separation (including scheduling)
- shaping
- policing

These functionalities are needed at different points in the network. The base stations need to be able to map the radio domain classes to the respective transport mechanisms. More recommendations are given in Chapter 5.

2.4.4 Synchronization

Table 2.2 gives an overview about the synchronization requirements for LTE-FDD and LTE-TDD.

The synchronization input requirement in Table 2.2 depends on the vendor specific implementation. The 16 parts per billion (ppb) mentioned give a rough guide as to which has to be cross-checked with the specific vendor requirements. More detailed explanation of input requirements can be found in Metsälä and Salmelin (2012).

Some more advanced LTE features will require phase synchronization. Table 2.3 gives an overview of the impact of each feature. Some of the requirements are still not completely defined and may vary according to the values mentioned.

2.4.4.1 Synchronization Plane Requirements for PTP

In case PTP mechanisms are used for frequency synchronization of the eNBs the backhaul network has to fulfill certain delay requirements. These requirements are vendor specific, as all the PTP algorithms are proprietary. In most cases the delay and delay variation has to be limited and the number of delay jumps (a sudden change of the average delay) per day should be below a certain threshold. Typical requirements are specified by certain delay masks. As these masks are not really useful in daily operation, some vendors have translated these requirements into engineering guidelines. A "natural" network fulfilling these requirements

Table 2.2 Synchronization requirements for LTE mobile standards (see Chapter 6 of Metsälä and Salmelin, 2012)

Mobile technology	Air interface frequency requirement	Synchronization input requirement[a]	Time/phase requirement
LTE-FDD	±50 ppb	±16 ppb,	
LTE-TDD	±50 ppb	±16 ppb,	±1.5 μs for cell radius < 3 km ± 5 μs for cell radius > 3 km

[a] May vary from vendor to vendor, but gives a good typical requirement.

Table 2.3 LTE feature-dependent phase synchronization requirements

3GPP release	Feature	Time/Phase requirement	Comment
8	LTE-CDMA interworking	±10 μs	
	Interference rejection combining (IRC)	±10 μs	depending on gain requirement
9	MBMS	±1 μs (G.8271.1), ± 5 μs, simulations	
	Observed time difference of arrival (OTDOA)	±1 μs	depending on location accuracy requirement
10	eICIC	±2,5 μs to ± 10 μs	depending on cell radius

will be sufficient to assure proper functionality of the PTP algorithm. "Natural" means a network that is not artificially simulated by delay pattern or load generators which do not consider certain statistical behavior. Below is an example of Nokia engineering rules for frequency synchronization:

- Maximum one-way delay should be <100 ms.
- Jitter <±5 ms.
- Clock packet stream should have highest priority or at least same priority as the real-time traffic.
- High-priority traffic share of bandwidth should be <60%.
- Maximum number of hops: 20.
- Maximum number of microwave hops: 10.
- Fewer than six delay jumps per day.
- Packet loss <2%.

2.4.5 *Availability*

The availability requirements of the backhaul network are derived from the availability requirements of the end-user service. A typical requirement could be an availability value of 99.95%. In most cases the availability is dominated by the last access hop. Chapter 5 looks at how the e2e availability is mapped to different subavailabilities.

2.4.6 *Security*

Due to the fact that the RNC has been removed from the network architecture, it is crucial to encrypt the traffic between eNBs and the evolved packet core (EPC). Otherwise, the plain-user traffic would be visible on the S1 interface. In addition, attacks could be possible by feeding traffic into the network. Access points for this could be Ethernet ports of eNBs or Ethernet connections of small cells.

To achieve this, the functionality of an SEG and a CA is needed. Whether these are dedicated network elements or included in other network elements depends on the specific implementation. Figure 2.21 shows a dedicated security gateway and certification authority.

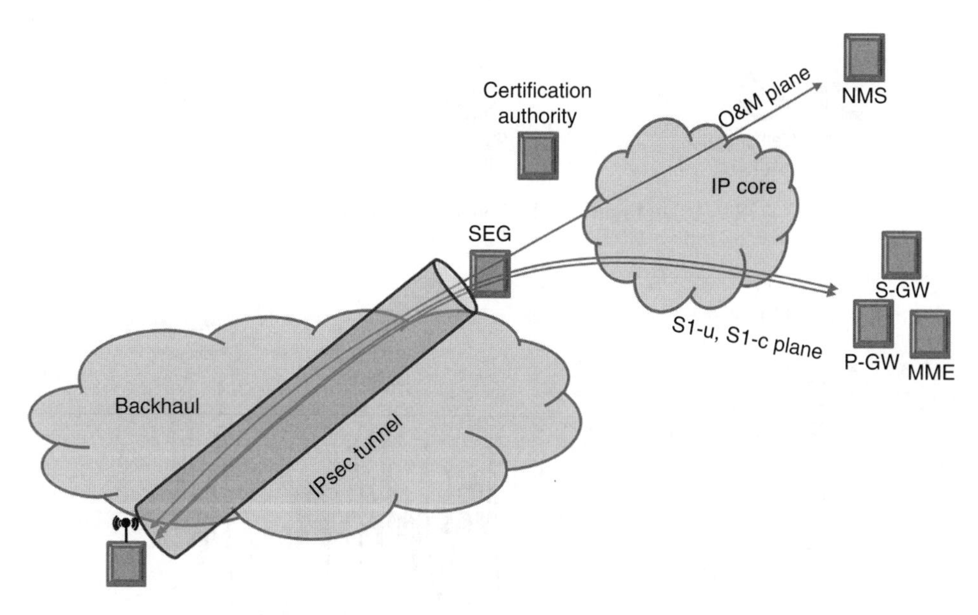

Figure 2.21 Security architecture

The additional control traffic for the authorization of certificates is negligible, but of course the availability of the control connection is essential because the complete e2e connectivity depends on available certificates. The impact of IPsec is more significant on the U-plane, as an additional overhead will be introduced. For a traffic mixture of different packet sizes the additional overhead due to IPsec is about 10%. Chapter 5 considers the security aspects involved in the planning process. As a consequence of the encryption some tracing functionality for fault analysis and special compression technologies might be impacted.

Figure 2.21 shows the IPsec tunnel for the S1 and M-plane. The M-plane may also be protected by transport layer security (TLS) mechanisms. Different X2 security scenarios are discussed in Chapter 5. In case synchronization is provided with PTP technologies, some 1588v2 inbuilt security mechanisms can be used. It is not recommended to transport the PTP packets in an IPsec tunnel, as this has some impact on the delay variation. For optional proprietary planes additional IPsec tunnels may be required.

2.4.7 *Examples*

There follow two examples of what happens if the requirements are not properly met by the backhaul networks.

2.4.7.1 Impact of Shaping on FTP Performance

The following example shows the possible impact of missing shaping on the e2e performance of the File Transmit Protocol (FTP) service. The network topology is given in Figure 2.22. The microwave radio (MWR) link next to the base station is critical.

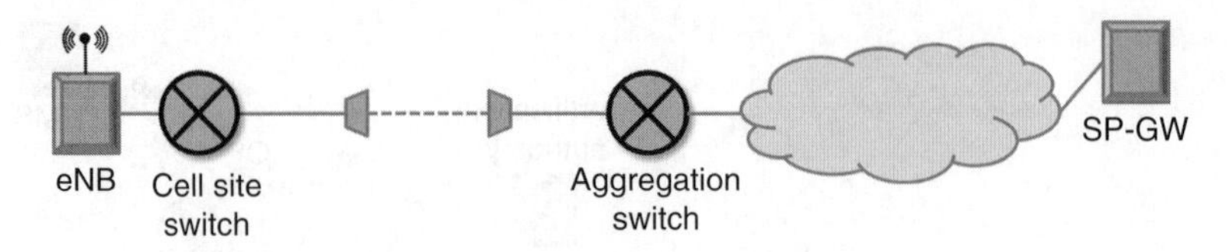

Figure 2.22 Topology example

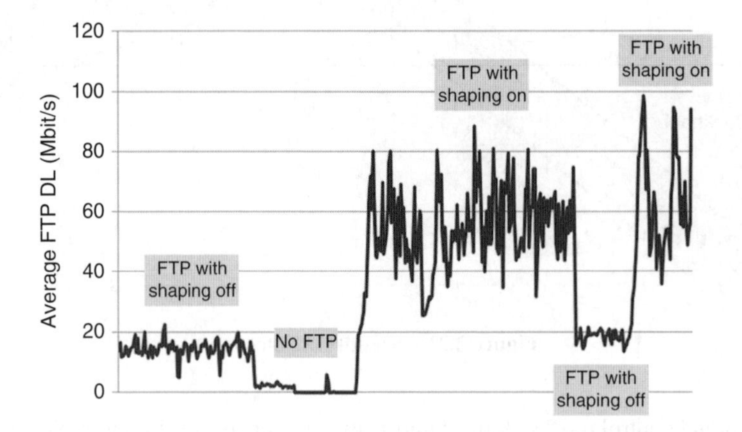

Figure 2.23 FTP throughput for different shaping configurations

The queues of the MWR outdoor unit (ODU) equipment are typically quite small. Thus shaping has to be enabled in the switches connected to the ODU units. Figure 2.23 shows the impact of different shaping configurations on the e2e FTP throughput.

In summary, the average throughput of 55 Mbps with shaping is much higher than the throughput of 18 Mbps without shaping.

Example 2.3

Question: What is the reason for the throughput increase with "shaping on" configuration?

Answer: Without the "shaping on" configuration, the site switch, having a high-capacity Ethernet port (such as 1 Gbps), may temporarily transmit packets at the full peak rate, even though the average rate of packets per second would be far less. Assume that the microwave hop supports a capacity of 100 Mbps. Due to an incoming burst, some packets need to be stored in an internal buffer until the output queue to the microwave link hop can consume them.

When this internal buffer cannot hold all packets during the burst, some packets have to be discarded. This causes serious degradation of TCP throughput. (See further discussion in Chapter 4.) To smooth the incoming packet flow and avoid loss of packets, the shaping of the traffic at the egress of the switch is configured.

2.4.7.2 Impact of Synchronization on Radio KPIs

Synchronization requirements are well specified for all kinds of base stations. Nevertheless, the real impact of a base station being out of sync is not investigated in detail in the literature. Some real measurements have been done in 2G handover scenarios (Bregni and Barbieri, 2003). A general impact on handover failure rate and dropped call rate is expected. This will be noticed first by fast-moving subscribers. Figures 2.24–2.26 show examples of the impact of missing synchronization on different radio KPIs for 3G systems. The deployment scenario of the first example is shown in Figure 2.24, where three base stations are located near a highway. A fourth has no direct neighborhood of any fast-moving user. Due to configuration mistakes in the backhaul network, base stations B and D have been out of sync for a longer period. Base stations A and C are still synchronized properly.

A synchronization assessment has been performed in the network and the problem of B and D has been fixed. Figures 2.25 and 2.26 show different radio KPIs before and after fixing

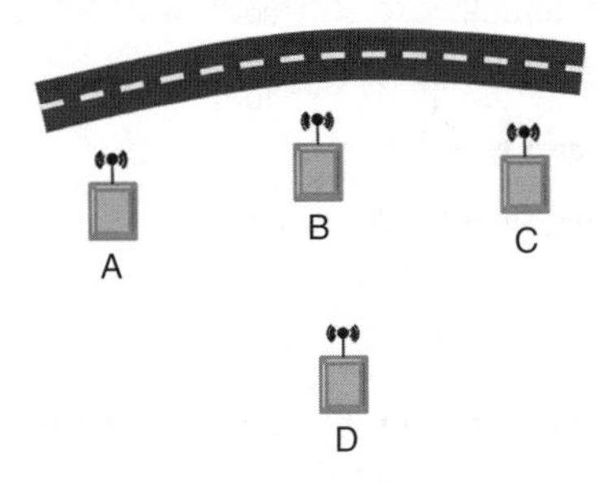

Figure 2.24 Simplified network picture

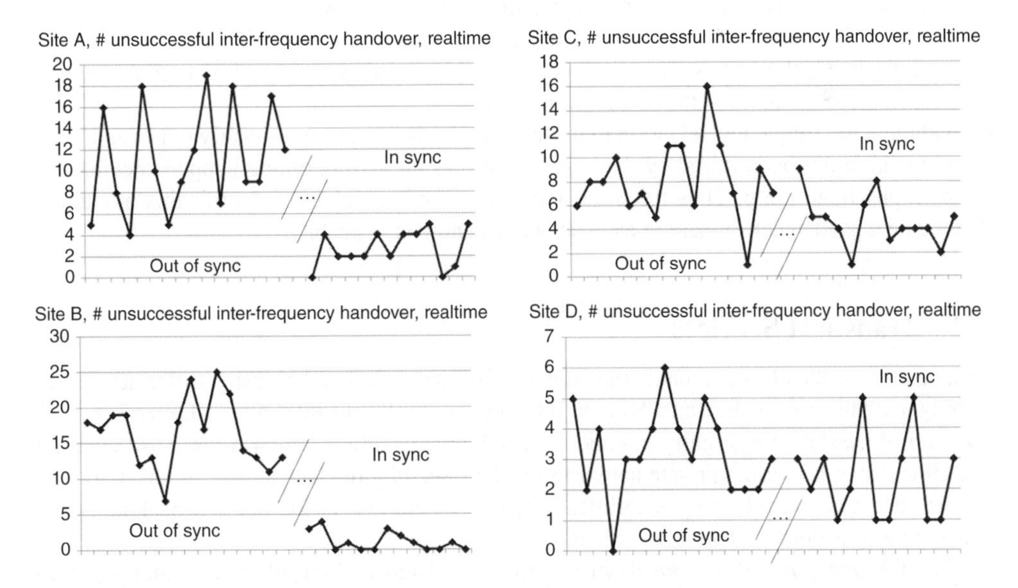

Figure 2.25 Unsuccessful inter-frequency handover, real time per day for the base stations shown in Figure 2.24

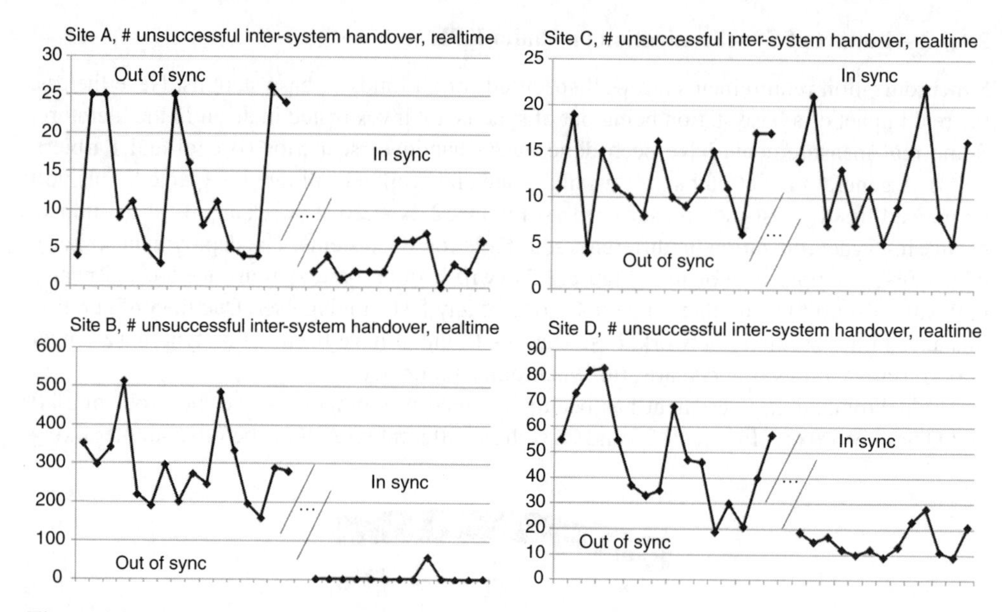

Figure 2.26 Unsuccessful inter-system handover, real time per day for the base stations shown in Figure 2.24

the problem. It can be seen that the sync problem of base station B affected not only the radio KPIs of B itself but also base stations A and C.

Due to the fact that base station D has no fast-moving users, the impact of missing synchronization is smaller. The biggest impact is visible for base station B.

Figure 2.27 gives another example. In this case the base station first has been in sync. Due to a wrong input, the base station has been smoothly tuned out of specification showing an impact on the radio KPIs within over several days.

Although the frequency deviation in these examples has been significantly out of specification, no alarm has been raised by the base stations, as the wrong tuning happened constantly without any huge jump. Thus the base station has not been able to identify this mistuning. Only a dedicated synchronization assessment has shown the problem.

2.5 Transport Services

Independent of the physical media the connectivity can be realized by using different e2e service types. This means the Open System Interconnection Data Link Layer 1/Layer 2 (L1/L2) protocol stack shown in Figures 2.4 and 2.5 can be realized in different ways. Most common are either L2 or L3 virtual private networks (VPNs); see Figures 2.28 and 2.29. There are a lot of pros and cons with these possibilities. Legacy constraints for 2G and 3G backhaul often have to be considered. Chapter 5 gives more details.

The different possibilities are described in more detail in Metsälä and Salmelin (2012). Other alternatives (such as involvement of point-to-point [PPP] and multilayer point-to-point [ML-PPP] transport layer protocols) are also explained there.

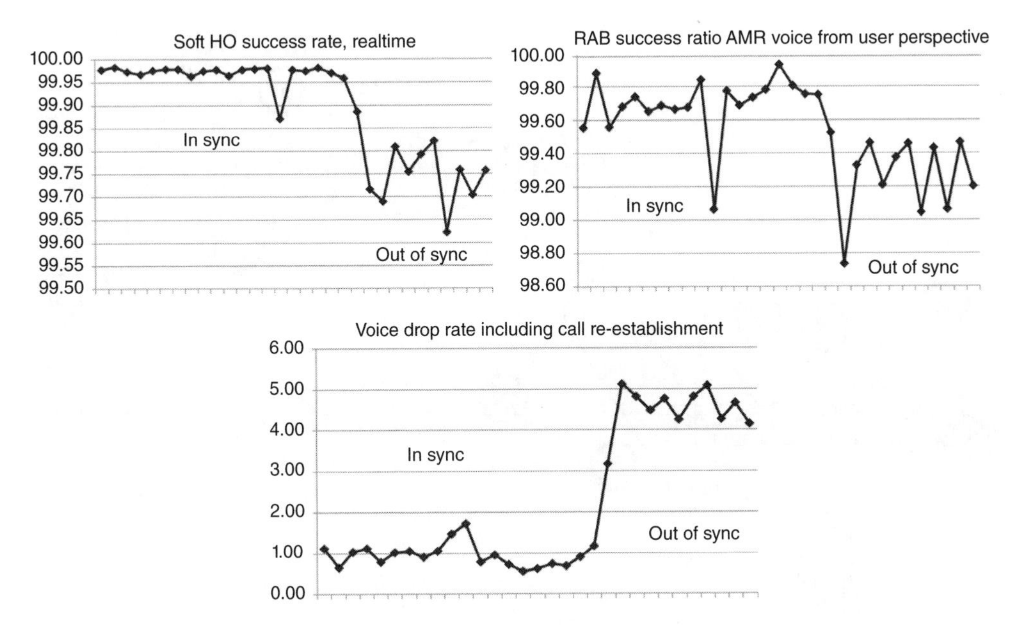

Figure 2.27 Different radio KPIs per day for another base station which changed from in sync to out of sync

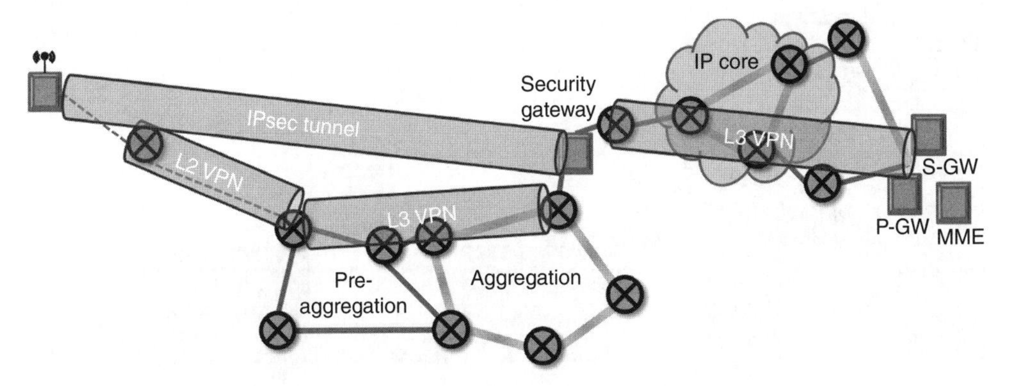

Figure 2.28 Chain of L2 and L3 VPNs in case of microwave backhaul

2.6 Planning Problems

After the requirements of the LTE backhaul network have been defined, a more precise definition of the planning problem which is to be solved shall be given in this section. The general planning process of a backhaul network is visualized in Figure 2.30. As there are many different understandings and definitions of planning, this section also gives an idea of the wording used in this book.

First of all, strategic planning is about the big picture. It has to determine which backhaul solutions should be in the toolbox for a specific operator. Should the operator go for leased lines or an own backhaul network? With one's own network, what should be the high-level architecture and which technologies should be considered for the first mile in the next five years? This planning has to consider business and technology aspects. As part of this strategic

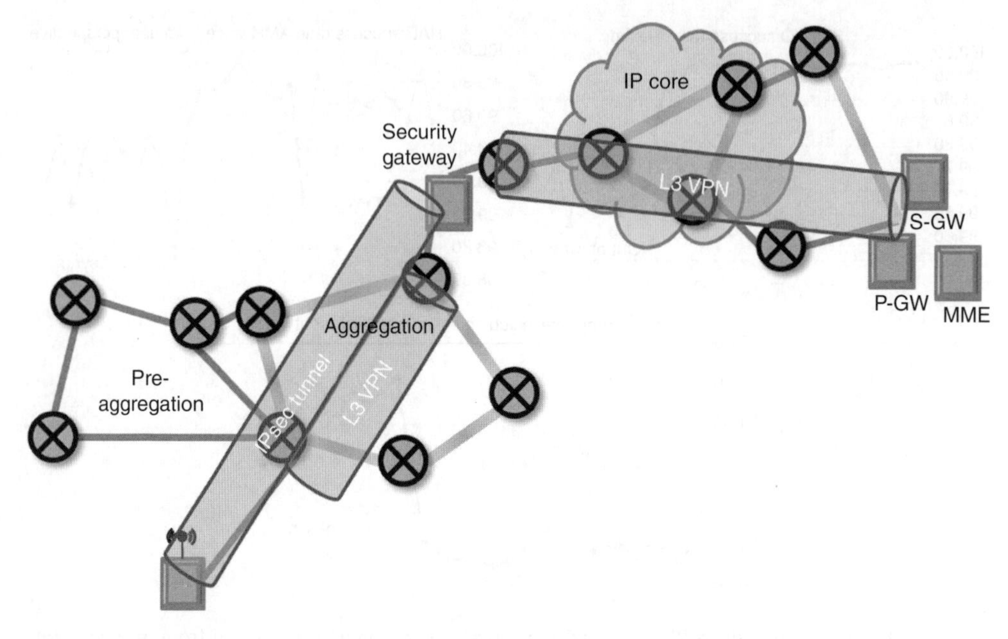

Figure 2.29 L3 VPN in case of optical backhaul

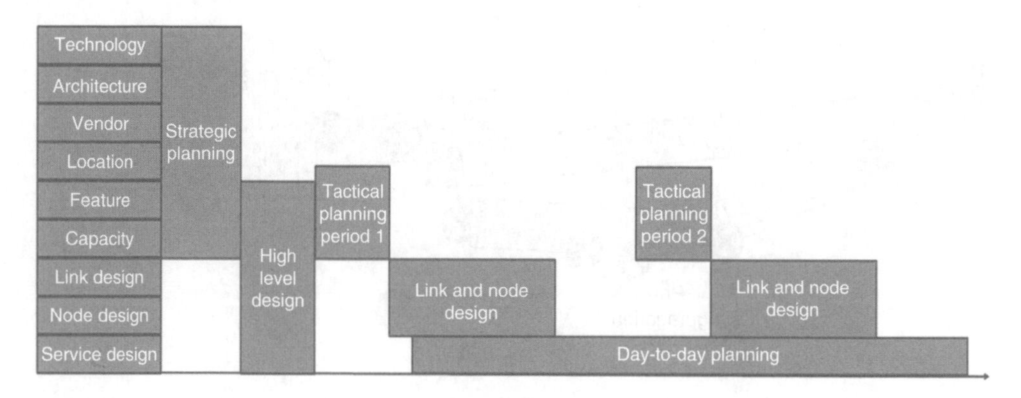

Figure 2.30 High-level planning process

planning, first capacity planning has to be done to get an idea of the necessary link and node dimensioning of the future network.

Strategic planning is followed by the high-level design, which defines in more detail which features and configuration patterns will be used in the network. Once this is finished, the stepwise implementation of the strategy has to be planned. This step is called "tactical planning." Which part of the network will be changed in the next six months and what will be the detailed parameterization? The frequency of the tactical planning varies from operator to operator. Some have a yearly timeslot; some on a quarterly base.

Tactical planning triggers rollout projects to realize the respective network changes. The network planning tasks in these projects are link designs (in most cases fiber or MWR links) or node designs (in most cases switches or routers). These detailed design activities follow the guidelines which have been elaborated in the high-level design. Once the network is rolled out, the network will be used to realize new e2e transport services. Whether these new e2e services are planned by an operational team or a planning team varies from operator to operator.

The most important part of the tactical planning stage is capacity management. The planning team has to decide when additional capacity is needed to fulfill the forecasted traffic demand. For legacy technologies, like PDH and SDH, the capacity management dominates planning tasks.

Performance issues of the network have been in most cases a pure operational issue due to node or link failures. The only exceptions have been, for example, link performance problems of MWR links or synchronization issues.

In new packet-based networks this has changed. Performance monitoring of the network is essential also for capacity management and performance optimization. A purely theoretical plan cannot simulate all the issues which may occur in the real packet network. Indications from performance monitoring may trigger tactical planning or the need to optimize the performance by re-parameterization. In this sense there is a certain shift from a pure proactive planning to a hybrid planning considering proactive and reactive planning steps.

Strategic planning has to consider business and technical issues. Combining the business modeling of Chapter 3, the dimensioning in Chapter 4 and the planning advice in Chapter 5 will enable the reader to achieve better results in strategic planning. The high-level design is discussed in detail in Chapter 5. The tactical planning is dominated by capacity aspects. Again Chapter 4 and Chapter 5 will help to dimension networks according to the respective traffic requirements. The parts of the node design for switches and routers are covered in Chapter 5. For MWR or fiber design aspects, the reader is referred to existing literature. Day-to-day planning and performance monitoring are covered in Chapter 5.

2.7 LTE Backhaul Technologies

As shown in Figure 2.4, Figure 2.5 and Figure 2.9, the IP layer is mandatory in all the relevant protocol stacks. Thus there is no freedom to choose between IP, asynchronous transfer mode (ATM) or other layers. Of course, the L1/L2 layers can still be chosen. In most cases, the eNB will have optical or electrical Ethernet interfaces to be connected to the backhaul. Depending on the vendor, other inbuilt functionality may provide all kinds of legacy interface, like PDH level E1 (PDH E1), SDH synchronous transport module level 1 (STM-1) or synchronous optical network (SONET) optical carrier level 3 (OC-3). Mapping the IP layer to these interfaces can be realized with different methodologies (see Metsälä and Salmelin, 2012). Once the traffic requirements on the IP layer are known, it is straightforward to calculate the necessary bandwidth considering the mapping overhead and to do the planning with the available legacy methodologies for PDH/SDH/SONET networks.

In this book it is assumed that an Ethernet interface is used to connect to the backhaul network. In any case the legacy interfaces at eNB will only be needed at the beginning of the rollout to cope with certain constraints during the rollout of an "all-IP"[1] network.

In this section a short overview about the different L1/L2 technologies used in the different backhaul network areas is given. Backhaul technologies which are relevant for macro eNBs

[1] "all IP" is used for any mixture of Ethernet/IP/MPLS technologies.

will be mentioned. Technologies for small cells may be different, as some small cells may have less stringent requirements. In this section, focus will be given to the impact of the technologies on the later planning and optimization tasks.

2.7.1 Access

As defined in Figure 2.2, access mostly consists of the first[2] mile from the base station and a maximum of one additional hop. As mentioned above, it is assumed that the first backhaul equipment is connected via an Ethernet connection to the eNB. The following technologies are the technologies that are suitable to reach the next logical level in the backhaul network (see Figure 2.2). Figure 2.31 gives an overview of different media used for the first mile.

2.7.1.1 Wireline Technologies

Native Ethernet

Native Ethernet (electrical or optical Ethernet) may be available in some locations, minimum rate used should be 100 Mbps, but 1 Gbps is recommended. The distance which can be spanned by native Ethernet is limited. In most cases Ethernet will only be the local medium to connect to other equipment on the same site.

DSL

Digital subscriber line (DSL) technologies reuse the existing copper infrastructure. Different standards support different bandwidth. The achievable performance and capacity strongly depend on the cable lengths of the final copper connection. Table 2.4 gives an overview of the most important characteristics of the different standards. In case DSL is used for mobile backhaul, this will most probably be a combined leased line, which starts with a CPE (customer premises equipment) located at the eNB and ends at a handover point in the aggregation

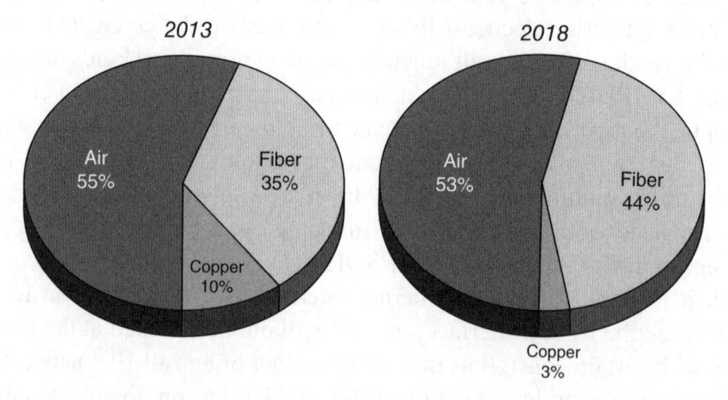

Figure 2.31 Physical media used in the backhaul for 2013 and 2018 (Infonetics Research, 2014)

[2] We start to count at the base station.

Table 2.4 DSL standards suitable for LTE backhaul

Name	Reference standard	Maximum speed down/up [Mbps]	Symm/Asymm	Loop length
VDSL[a]	ITU G.993.1	52/16	A	$l < 1200$ m
VDSL2	ITU G.993.2	250…4	S+A	50 m $< l <$ 5 km
VDLS2 Vectoring[b]	ITU G.993.5	250…4	S+A	50 m $< l <$ 5 km
G.Fast	draft ITU G.9700 ITU G.9701	1000…150	S+A	50 m $< l <$ 250 m

[a] VDSL: very high bit rate digital subscriber line.
[b] Coordination of line signals leads to reduction of crosstalk in real field deployments.

network. Many different protocol stacks are possible and shall not be discussed in detail in this book. Essential for the further planning steps are the following capabilities of the DSL line:

- monthly cost
- bandwidth in UL and DL
- availability
- delay/delay variation
- asymmetry of delay.

On the other hand, some DSL technology causes additional challenges for some PTP implementations. Most "any kind of" digital subscriber lines (xDSL) do not provide the bit rates required by LTE large bandwidth deployments, a notable exception being low-bandwidth deployments at 10 MHz FDD bandwidth or less. Reliability of xDSL may also be a concern for the operator hence making it a niche technique used for otherwise hard-to-reach locations.

Fiber

A fiber connection to the eNB is the most favorable solution as it solves all capacity and performance problems of current backhaul networks. Nevertheless, the percentage of base stations which are connected via fiber is still below 35% (2013). Of course, there are many differences between urban and rural areas as well as between countries. For example, China, South Korea and Japan have a fiber connectivity of nearly 100%. In many countries, the first mile is the problem and the effort which has to be spent for the connection of an available fiber access point to the base station is too high in many areas. The physical aspects of fiber planning will not be considered in this book. For the strategic planning it is of course necessary to model the cost of fiber deployment. Once the fiber is deployed the impact on the next planning steps is quite small, as a high performance and capacity are given. Two kinds of fiber deployments have to be distinguished: active and passive. In the active area two kinds of fiber scenarios are relevant:

- grey, which means usage of the fiber with a single frequency
- colored, which means usage of the fiber with multi frequencies in coarse wavelength division multiplexing (CWDM) and dense wavelength division multiplexing (DWDM) systems.

The wavelength division multiplexing (WDM) systems are necessary to use fiber deployments more efficiently. A single fiber can be used with several frequencies (colors) at the same time. They are mostly used in backbone systems but also first deployments in the aggregation part of backhaul networks come up. The single mode usage of fiber is mostly used for gigabit Ethernet (GbE) connections. From a planning perspective, the capacity of a fiber connection is constant and does not depend on any environmental constraint. For availability aspects, it is necessary to keep track of the conduct usage to avoid the usage of redundant fibers in the same conduct.

On the other hand, there are passive optical networks (PONs). The difference to "active" optical networks is that the connection uses passive components such as optical splitters to distribute the signal in point-to-multipoint connectivity, the benefit here being a potential cost saving. The capacity between leaf nodes is shared by TDM. The distance between the root node and the leaf nodes can be up to 20 km. The capacity provided may be symmetrical or asymmetrical, with leaf node bit rates up to 2.5 Gbps (gigabit-capable passive optical network [GPON]) or even 10 Gbps for 10 gigabits/passive optical network [XG-PON]) in downstream and 1.25 Gbps (2.5 Gbps for XG-PON) in upstream.

DOCSIS

Data over cable service interface specification (DOCSIS) enables digital data transport using cable-TV network. With a 6 MHz RF channel a DL bandwidth of 38 Mbps can be achieved. Compared to DSL, a higher UL performance of up to 27 Mbps can be achieved. Although a further bundling of channels is possible, to achieve higher bandwidths the technology is not often used for the backhaul of macro sites. However, with the release of DOCSIS 3.1, it will be possible to offer even 10 Gbps DL and 1 Gbps UL capacity in certain circumstances.

2.7.1.2 Wireless Technologies

Point-to-Point Microwave Radio

Although the bandwidth requirements of eNBs increased significantly, MWRs are still a viable solution for many base stations. This is due to new technologies like higher modulation schemes, dual polarization and MIMO technologies.

Modern equipment completely includes all the dedicated functionality of the microwave into the ODU (outdoor unit) and is connected to any kind of indoor part via the Ethernet (providing power in some cases). Any kind of L2/L3 functionality is provided by the indoor unit, which might be a switch or router. The physical aspects of the MWR planning are not the focus of this book. (See e.g. Donnevert, 2013.)

Relevant aspects for the planning and optimization in this book are link capacity and link availability. Due to adaptive modulation mechanisms the availability may not be on/off, but depending on the weather conditions the link capacity may be varying. In addition, shaping as well as QoS mechanisms of the equipment have to be considered.

Figure 2.32 gives an idea of the possible capacity for different modulation schemes and relative thresholds (dB).

Cross-polarization interference cancellation (XPIC) allows horizontal and vertical channels to be transmitted over the same channel bandwidth; load balancing can also be applied, hence for dual-polarized link hops the bit rates are doubled. Alignment of antennas for dual-polarized link hops can be very resource-consuming, but this should be done carefully to guarantee proper operation of the XPIC.

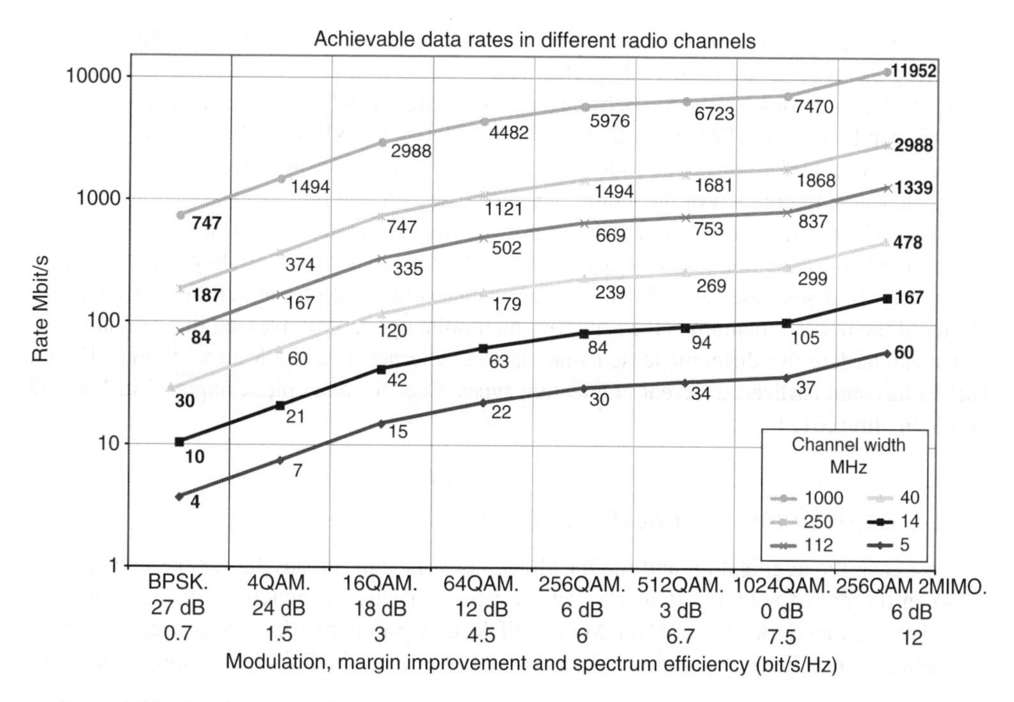

Figure 2.32 Radio data rate as a function of modulation with different radio channel bandwidths

E-band (70/80 GHz) MWRs are available and emerging as a viable option for dense urban links; maximum hop capacity is up to 3–4 Gbps. Due to very narrow-beam antennas (about 1 degree half-power beam width), there are special requirements for antenna alignment and resistance to sway, for example from wind.

In many cases a problem in increasing MWR transport network capacity for LTE is the increased RF interference. This is made worse by two factors: the larger channel bandwidth per link hop and high increased interference margin required by higher-order modulations. The cumulative design impact from these factors is that channel reuse distance is increased, while simultaneously the spectrum needs to be more efficiently harvested. If the interference analysis is not carried out during design phase, this can afterwards cause serious problems after MWR roll-out. In addition to meticulous frequency planning and interference analysis, performance monitoring of the MWR transport network should be planned ahead (and carried out during operation).

Other Wireless Technologies

There are many other wireless technologies which are used for the backhaul of base stations. As they have limited bandwidth, high delay or low availability they are not relevant for the backhaul of macro base stations. Examples are point-to-multipoint MWR or Wi-Fi, which may be relevant for small cell backhaul. Satellite backhaul may be applicable in rural areas.

2.7.1.3 Leased Lines

Leased lines are a very viable solution for the backhaul of LTE base stations, as they enable different business models. A leased line provider, offering backhaul to many mobile operators, has another cost position and may be able to justify investments, which a single operator cannot. From a planning and optimization perspective, it is essential to define and monitor the necessary SLAs to assure proper e2e performance.

All the technologies mentioned before can also be offered as leased line. In addition, mixed solution can be realized. For example, the first mile between eNB and a first access point of a leased line provider is owned by the mobile operator followed by a leased line, which is terminated at an IP core site of the mobile operator. The different offers of leased lines are quite complex. Many different SLAs and different QoS implementations are offered. In addition, many different leased line providers exist with different capabilities and they may vary from city to city. In many cases also the first mile dominates the cost of a leased line, as the availability of the needed medium is the same for the mobile operator or a leased line provider.

To standardize the different leased lines in the Ethernet area the Metro Ethernet Forum (MEF) has standardized different connection types. See, for example, Chapter 5 in Metsälä and Salmelin (2012).

2.7.2 *Aggregation and Backbone Network*

In most case the aggregation and the backbone network will already have fiber connections between the active network elements. The respective router or switches will be connected directly via either dark fiber or CWDM or DWDM transport networks. The capacity planning problem area is shifted toward network nodes in this area, as the links have enough capacity.

2.8 Small Cell Backhaul

There is huge variety of different kinds of small cells. In this book no further classification will be given, as this book concentrates on the backhaul aspects. Both for small cells and for the macro cells, two planning problems have to be separated: physical and logical planning. Once the physical connectivity is given, the logical planning can follow the same steps as described in Chapter 5 for the macro base stations. The major difference is the number of base stations, which has an impact on the used address scheme and routing/switching technologies. In addition, the need for further automation is evident and improved SON functionalities are required to set up a network in a cost-efficient way. Small changes in the network architecture are also currently being debated in the industry. The introduction of additional controller network elements may help to optimize network architecture to allow cost-efficient network deployments.

For indoor small cells the physical part is solved easily, as most of the cells will be connected with either copper or fiber. The bigger challenge is the physical connectivity for outdoor small cells. A small cell will be quite cheap compared to a macro base station and the expected revenue per cell will be much smaller than for a macro base station. Thus there is a huge cost pressure for the backhaul. Unfortunately the backhaul has a large fixed cost which does not depend on the bandwidth (digging for cable or fiber, line of sight [LOS] checks, on-site visits, antenna installation, etc.). Using the same backhaul approach as for macro sites does not fit the business case. Thus additional planning/installation approaches, as well as new technologies, are needed. Promising technologies are wireless technologies which are working in nLOS (near line of sight) or NLOS (no line of sight) environment. From a planning perspective, these nLOS and NLOS systems require additional capabilities and tools, as the signal will be transferred via reflection and diffraction. Thus the classic planning approach employed for LOS is not possible any longer.

Automatic antenna steering technologies for LOS systems may also help to meet the required cost position. From a general planning perspective, backhaul will enjoy a higher priority over radio aspects. In many cases a site will not be chosen due to radio requirements but because a cost-efficient backhaul is possible on it.

Depending on the specific usage of a small cell, the backhaul requirements may also be different from the macro base stations. On the one hand, these may be different bandwidth requirement, but on the other delay or synchronization requirements may be different. For example, high-speed handover may be handled entirely by the macro network and thus the synchronization requirements for small cells may be different.

As most of the challenges are due to physical constraints, this book will not consider small cells explicitly. The most challenging part will be the first mile connectivity of the small cells. First choice could be the connection to the next macro base station or another available aggregation point.

2.9 Future Radio Features Affecting Backhaul

This section introduces a few foreseeable future (3GPP Release 12 or beyond) radio features that will possibly affect backhaul transport requirements and dimensioning, i.e. capacity need, latency (S1 or X2) and synchronization.

2.9.1 Inter NodeB CoMP (eCoMP)

Release 11 CoMP practically assumes so-called ideal backhaul, which in 3GPP terminology means a dedicated fiber, to be used between associated eNBs (inter eNB-CoMP) or eNB and remote radio head (intra eNB-CoMP).

3GPP has studied the CoMP performance over non-ideal backhaul within the eCoMP (enhanced CoMP) study item of Release 12. According to the 3GPP (2014) definition, "non-ideal backhaul" refers to those transport technologies widely used by the mobile operators; these technologies are listed in Table 2.5. The latency values given can be interpreted as an SLA type maximum (offered, for example, by the fixed service provider). For comparison the characteristics of ideal backhaul are listed in Table 2.6.

Both distributed and centralized architectures have been proposed for the eCoMP. In the case of distributed architecture, each of the involved eNB has scheduler functionality and

Table 2.5 Categorization of non-ideal backhaul in 3GPP Release 12 work (3GPP, 2014)

Backhaul technology	Latency [ms] (one way)	Throughput [Mbps]	Priority (1 is the highest)
Fiber Access 1	10...30	10...10000	1
Fiber Access 2	5...10	100...1000	2
Fiber Access 3	2...5	50...10000	1
DSL Access	15...60	10...100	1
Cable	25...35	10...100	2
Wireless Backhaul	5...35	10...100 (up to 1000)	1

Table 2.6 Categorization of ideal backhaul in 3GPP Release 12 work (3GPP, 2014)

Backhaul technology	Latency [us] (one way)	Throughput [Mbps]	Priority (1 is the highest)
Fiber Access 4[a]	<2.5[b]	Up to 10000	1

[a] This can be applied between the eNB and the remote radio head.
[b] Propagation delay in the fiber/cable is not included.

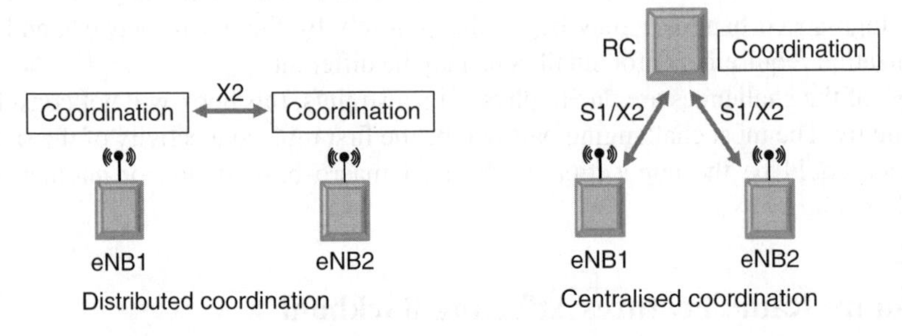

Figure 2.33 Scheduler coordination alternatives (eCoMP)

the coordination information is exchanged between the neighbor eNBs over the X2 interfaces. In centralized eCoMP architecture, the (slave-) eNBs send the required information to a resource coordination (RC) element which makes the scheduling decisions and returns coordination information (muting patterns) to the eNBs. The alternative architectures for inter eNodeB CoMP scheduler coordination are presented in Figure 2.33. As a result of eCoMP studies, it was decided to include in 3GPP Release 12 the solution based on a distributed architecture.

eCoMp Impacts on Backhaul Transport

In the case of distributed coordination architecture, direct X2 connections between eNBs (meshed X2) are preferred, to minimize the latency. Independent of the architecture, the assumption for maximum X2 latency in 3GPP studies has been 60 ms, but reasonable gain may be achievable when the X2 latencies are around 5 ms or less.

2.9.2 Dual Connectivity

In dual connectivity (DC), the UE is connected to two eNBs—the master eNB (MeNB) and the slave eNB (SeNB)—at the same time. This provides possibilities for throughput and system efficiency enhancements with inter-site CA (MeNB + SeNB) and traffic offload to the small cells. In addition, depending on the implemented DC scenario, it potentially improves (simplifies) mobility management because the movement of the UE between the small cells does not appear in the core network (EPC) and handovers are made only when the UE changes the macro area. The principle of DC is illustrated in Figure 2.34.

Out of several architecture alternatives and proposals, the following two were selected for further studies in 3GPP (2013):

- Bearer offload (alternative 1A): own S1s for MeNB and secondary eNB, bearers (of one UE) are separated in EPC, only C-plane is carried over X2.
- Bearer split (alternative 3C): S1-U terminates in MeNB, Small cell/secondary eNB (SeNB) is connected to MeNB via X2 (U-plane and C-plane).

Alternative 3GPP Release 12 DC architectures are described in Figure 2.35. Architectures may coexist, and each eNB should also be able to handle UEs which are not involved (or capable) in DC independently.

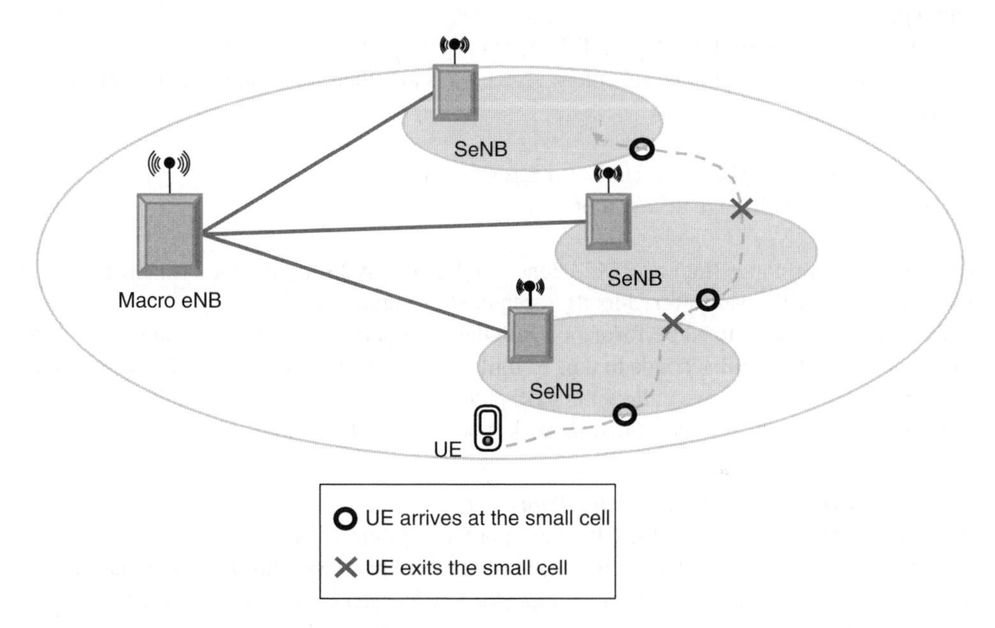

Figure 2.34 Principle of dual connectivity

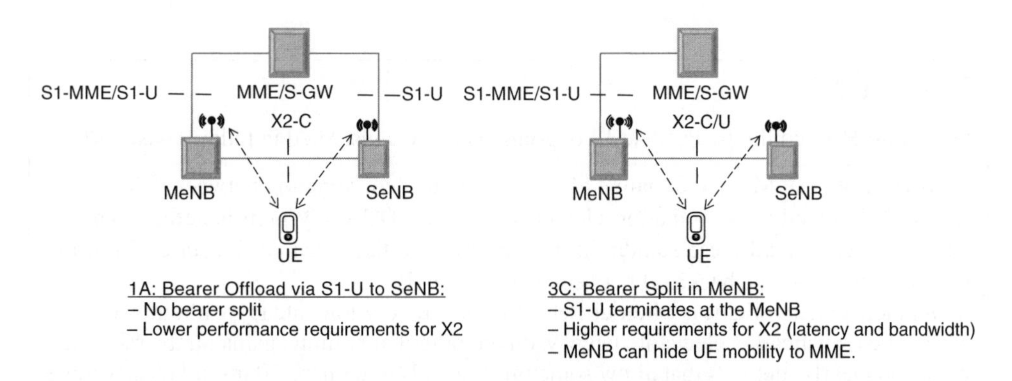

Figure 2.35 Alternative architectures of dual connectivity

Dual Connectivity Impacts on Backhaul Transport

No special backhaul impacts in the case of bearer offload (alternative 1A), the assumption for maximum X2 latency in 3GPP studies has been 60 ms. With bearer split (alternative 3C), the transport capacity demand increases in both S1 and X2, per user throughput gains achievable with up to 30 ms X2 latency (the lower the better).

2.9.3 Dynamic eICIC

In Release 10/11 eICIC (see Section 2.3.2.5), the ABS pattern configuration is semi-static (e.g. on a rate of tens of seconds to minutes) and usually controlled by centralized RRM. In the case of dynamic eICIC, the ABS adaptation can be distributed and based on the local load situation.

Dynamic eICIC algorithm adapts ABS pattern and range extension (handover offset) taking into account macro and small-cell (micro eNB) loads and therefore leading to higher system performance, i.e.:

- better load balancing between HetNet macros and micros
- improvement in cell-edge throughput.

The system performance impact of X2 latency in dynamic ABS muting pattern adjustment as a function of the cell load (offered load) is depicted in Figure 2.36.

To achieve the maximized performance gain the ABS pattern adjustment would need to be done from every few milliseconds to tens of milliseconds. The current (3GPP Release 11) X2 signaling definitions, however, allow the ABS pattern to change every 40 ms at the fastest, and therefore further study and specification work would be needed in 3GPP.

Dynamic eICIC Impacts on Backhaul Transport

Compared to Release 10 and 11 eICIC, the improved system performance means that, in the same proportion, higher bandwidth is needed on the backhaul. Synchronization requirements remain the same. The main benefit comes along the low X2 latency and remarkable performance gain achievable when the X2 latency is 5 ms or less.

Example 2.4

Question: How could the Release 12 ongoing items be considered in today's designs?

Answer: Both eCoMP and dynamic eICIC rely on fast coordination between cells of different eNBs or other coordination elements. Also the DC study item benefits from low latency, so clearly all these standardization items have the potential impact of requiring reduced latency on the backhaul between neighbor eNBs.

In practical networks it is often difficult to achieve very low latency without excessive costs. The realistically achievable latency, on the other hand, limits gains on the radio network and clearly methods that allow some milliseconds to perhaps 10 ms are clearly more applicable than those requiring an "ideal backhaul."

Until the feature standardization has settled, it is very difficult to plan the backhaul in advance accordingly, as also functional split between elements and other architecture decisions have an impact. In general the requirement is toward reduced backhaul latency, and so it is beneficial to study, during the present design, performance in terms of latency, and how it could be further reduced, and alternatives should be considered for designs that necessitate several tens of milliseconds of latency.

2.10 Related Standards and Industry Forums

In the following sections, the main involved organizations and industry forums and their related specifications and standards are introduced.

2.10.1 3GPP

Third Generation Partnership Program (3GPP) defines what transport protocols shall be used for the transport interfaces, but the actual transport specifications and standards are done by other organizations.

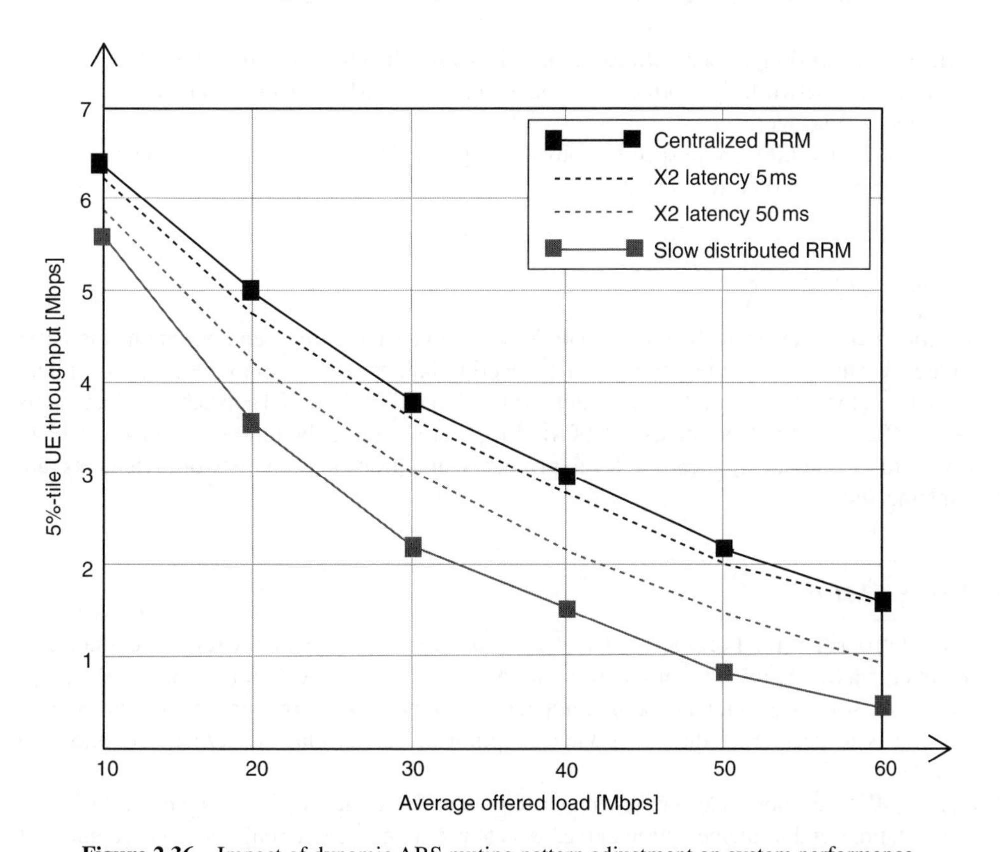

Figure 2.36 Impact of dynamic ABS muting pattern adjustment on system performance

2.10.2 ITU-T SG15

ITU-T Study Group 15 (networks, technologies and infrastructures for transport, access and home) has a subwork group called Question 13/15. Question 13/15 (network synchronization and time distribution performance) studies network synchronization issues related to packet networks with a focus on mobile network needs.

Question 13/15 has specified packet synchronization profiles for IEEE1588 (2008), like:

- G.8265.1 (ITU-T, 2014a): Precision time protocol telecom profile for frequency synchronization
- G.8275.1 (ITU-T, 2014b): Precision time protocol telecom profile for time synchronization with full on-path support
- G.8275.2 (ITU-T, 2014c): Precision time protocol telecom profile for time synchronization with partial on-path support.

2.10.3 IEEE 802

The Institute of Electrical and Electronics Engineers (IEEE) Local Area and Metropolitan Area (LAN/MAN) Standardization Committee (IEEE802) defines the physical and link layer of networks and provides a complete set of standards for carrying IP.

- IEEE802.1 working group defines 'higher' link layer functions for the Ethernet, e.g.:
  - local area network (LAN)/metropolitan area network (MAN) architecture, and
  - network management.
- IEEE802.3 working group specifies physical layer (PHY) and media access control (MAC) for carrying Ethernet over wired links.

2.10.4 IETF

The Internet Engineering Task Force (IETF) is a forum that defines and maintains the basic technical standards for Internet protocols, covering layers from IP up to the general applications like email and Hypertext Transfer Protocol (HTTP). The IETF produces documents called "Request for Comments" (RFCs), which are commonly followed and referred to when, for example, application level protocols and transmission network elements are implemented.

2.10.5 MEF

The Metro Ethernet Forum (MEF) is a global industry alliance with more than 220 members including telecommunications service providers, cable operators, network equipment and software vendors, semiconductor vendors and testing organizations. MEF's mission is to accelerate the worldwide adoption of carrier-class Ethernet networks and services.

The MEF develops carrier Ethernet architectural, service and management technical specifications and implementation agreements to promote interoperability and deployment of carrier Ethernet worldwide.

The key specifications of MEF include:

- MEF 6.2 Ethernet Services Definitions (2014)
- MEF10.3 Ethernet Services Attributes (2013)
- MEF 23.1 Carrier Ethernet Class of Service (2012).

The MEF's specifications for MBH:

- MEF 22.1 MBH Implementation Agreement (2012)
- MEF 22.1.1 MBH IA Amendment (2014).

2.10.6 NGMN

The Next Generation Mobile Network (NGMN) Alliance consists of mobile network operators (members), vendors and manufacturers (sponsors), and universities or non-industrial research institutes (advisers).

The mission of the NGMN Alliance is to expand and evolve the mobile broadband experience, with a particular focus on LTE and LTE-A deployments and their further enhancements.

NGMN publications include:

- LTE Backhauling Deployment Scenarios (2011)
- Integrated QoS Management (2012)
- Security in LTE backhauling (2012)
- Small Cell Backhaul Requirements (2012).

2.10.7 BBF

The Broadband Forum (BBF) was formed by the gradual merger of several different forums that focused on specific technologies. The predecessors of the BBF include forums like: Frame Relay Forum, ATM Forum, DSL Forum and IP/MPLS Forum.

The most recent changes were in 2008, when the DSL Forum changed its name to the Broadband Forum, and in 2009 when the IP/MPLS Forum and the Broadband Forum united and formed the current Broadband Forum.

BBF technical reports include:

- TR-221 MPLS in MBH Networks (2011)
- TR-221 MPLS in MBH Networks Amendment (2013).

2.10.8 SCF

The Small Cell Forum (SCF) aims to support the wide-scale adoption of small cells. Their working groups include: marketing and promotion, radio and PHY, network, regulatory, services, and interoperability as well as several special focus groups.

SCF has released the following backhaul related white paper:

- Backhaul Technologies for Small Cell (2013).

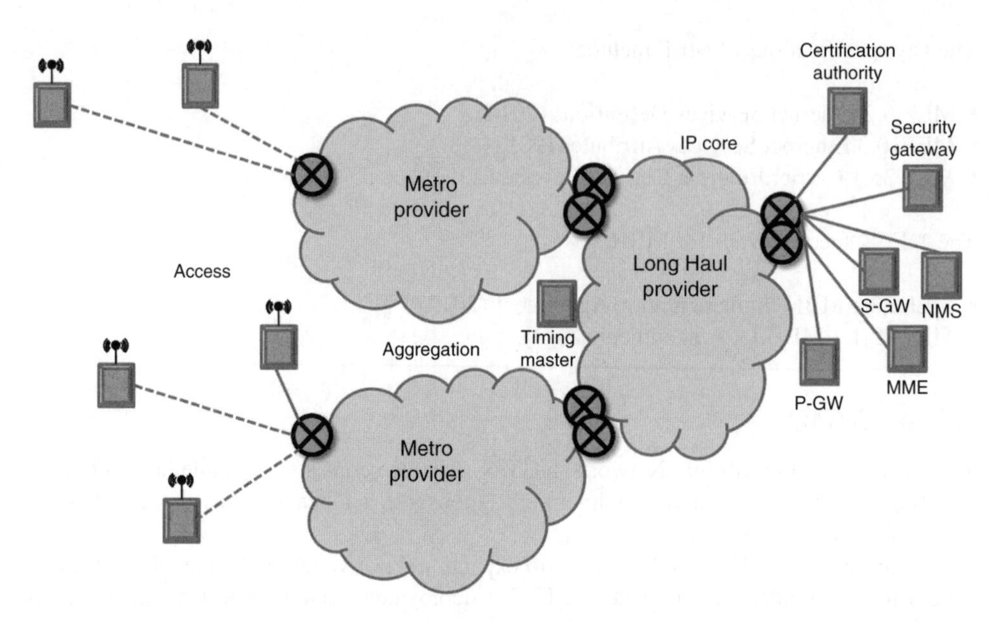

Figure 2.37 Operator example combining different leased line providers and own equipment

2.11 Operator Example

Figure 2.37 gives an example of an operator network. In this case the operator has decided to combine leased lines with own network infrastructure. First of all the operator works with a couple of local leased line providers in the aggregation network to close the gap between the first aggregation point and IP core network. Depending on the geographic regions, different leased line providers are considered. In case the provider cannot offer a cost-efficient first mile connection, an own wireless solution is deployed. The first leased line provider is terminated with own L2/L3 equipment to take part in the statistical aggregation gain and to be able to map different service schemes and implementations of the metro providers to the leased line services of the long haul provider. Another leased line contract is given to a countrywide long haul provider who connects the different aggregation sites to the core sites. Synchronization and security services are provided by the mobile operator. Whether L2 or L3 VPNs are used in the aggregation part depends on the offering of the local provider.

References

3GPP (2015a) *TS29.060: GPRS Tunneling Protocol (GTP) across the Gn and Gp interface*, http://www.3gpp.org/DynaReport/29060.htm, accessed 1st June 2015.

3GPP (2015b) *TS36.413: S1 Application Protocol (S1AP): Evolved Universal Terrestrial Radio Access Network (E-UTRAN)*, http://www.3gpp.org/dynareport/36413.htm, accessed 1st June 2015.

3GPP (2015c) *TS36.423: X2 Application Protocol (X2AP) Evolved Universal Terrestrial Radio Access Network (E-UTRAN); X2 Application Protocol (X2AP)*, http://www.3gpp.org/dynareport/36423.htm, accessed 1st June 2015.

3GPP (2015d) *TS23.203: Policy and charging control architecture*, http://www.3gpp.org/DynaReport/23203.htm, accessed 1st June 2015.

3GPP (2014) *TR36.932: Scenarios and requirements for small cell enhancements for E-UTRA and E-UTRAN,* http://www.etsi.org/deliver/etsi_tr/136900_136999/136932/12.01.00_60/tr_136932v120100p.pdf, accessed 1st June 2015.

3GPP (2013) *TR 36.842: Study on small cell enhancements for E-UTRA and E-UTRAN; Higher layer aspects (release 12), V12.0.0.* 3GPP, Valbonne, France.

Bregni S. and Barbieri L. (2003) Experimental evaluation of the impact of network frequency synchronization on GSM quality of service during handover. *Proceedings of IEEE GLOBECOM*, San Francisco, 1–5th December.

Donnevert J. (2013) *Digitalrichtfunk.* Springer Vieweg, Wiesbaden, 2013.

Holma H. and Toskala A. (2011) *LTE for UMTS: Evolution to LTE-Advanced*, 2nd edn. John Wiley & Sons, Ltd, Chichester, UK.

IEEE1588 (2008) *IEEE standard for a precision clock synchronization protocol for networked measurement and control systems*, July 2008, http://www.nist.gov/el/isd/ieee/ieee1588.cfm, accessed 1st June 2015.

Infonetics Research (2014) Macrocell mobile backhaul equipment and services. 8th April 2014, Exhibit 7.

Integrated QoS Management (2012) *Version 1.3.* NGMN Alliance, December 2012.

ITU-T (2014a) *G.8265.1/Y.1365.1: Precision time protocol telecom profile for frequency synchronization.* July 2014.

ITU-T (2014b) *G.8275.1: Precision time protocol telecom profile for time synchronization with full on-path support.* April 2014.

ITU-T (2014c) *G.8275.2: Precision time protocol telecom profile for time synchronization with partial on-path support.* December 2014.

LTE Backhauling Deployment Scenarios (2011) *Version 1.4.2.* NGMN Alliance, July 2011.

MEF 22.1.1 (2014) Mobile Backhaul Implementation Agreement, Phase 2 Amendment 1: Small Cells, July 2014.

MEF 23.1 (2012) Carrier Ethernet Class of Service Implementation Agreement, Phase 2, January 2012.

MEF 6.2 (2014) EVC Ethernet Services Definitions Phase 3, August 2014.

MEF 10.3 (2013) Ethernet Services Attributes Phase 3, October 2013.

MEF 22.1 (2012) Mobile Backhaul Implementation Agreement, Phase 2, January 2012.

Metsälä E. and Salmelin J. (eds) (2012) *Mobile Backhaul.* John Wiley & Sons, Ltd, Chichester, UK, doi: 10.1002/9781119941019.

Nokia Networks (2014) *LTE-Advanced White Paper,* http://networks.nokia.com/system/files/document/nokia_lte-advanced_white_paper.pdf, accessed September 2014.

Security in LTE backhauling (2012) *Version 1.0.* NGMN Alliance, February 2012.

Small Cell Backhaul Requirements (2012) *Version 1.0.* NGMN Alliance, June 2012.

Small Cell Forum (2013) *Backhaul Technologies for Small Cells,* 049.04.02. White Paper, February 2013.

TR-221 (2011) Technical Specifications for MPLS in Mobile Backhaul Networks (TR-221), October 2011.

TR-221 (2013) Technical Specifications for MPLS in Mobile Backhaul Networks (TR-221, Amendment 1), November 2013.

3

Economic Modeling and Strategic Input for LTE Backhaul

Gabriel Waller and Esa Markus Metsälä
Nokia Networks, Espoo, Finland

3.1 Introduction

The planning and optimization of LTE backhaul has to be based on a company's overall strategy. This sets the scene for high-level and detailed designs. The strategy for LTE backhaul is naturally linked to a LTE radio network strategy since both backhaul and radio network are part of the very same LTE service delivery vehicle. Addressing backhaul and radio network together keeps costs down while maintaining service offering and network performance at their target level.

A long-term intention often includes investments and planning of the usage of capital. Economic analysis is needed for decision making, in order to estimate the value of the project to the company. Financial metrics are calculated to compare two or more technical alternatives from an economic viewpoint. Business case development requires a command of business case modeling as well as an understanding of the subject, such that the model can essentially capture all relevant aspects and present them in monetary terms.

Economic modeling is very useful for all levels of planning, not only for the long-term investment preparation, since now also smaller optimization cycles can be motivated by business benefits and different ideas compared. A model may be prepared by finance professionals with the help of backhaul experts, or as well by backhaul experts with contribution from the finance department.

For this modeling, a cash flow based approach is used in the following sections. What is required for this is that each relevant cash inflow and outflow is identified. Cash inflow could be, for example, LTE service revenue and cash outflows are, for example, investments into network infrastructure or expenses related to the daily operation of the network. The term "relevant cash flow" is important, as one is only interested in those cash flows that are affected by the decision.

LTE Backhaul: Planning and Optimization, First Edition. Edited by Esa Markus Metsälä and Juha T.T. Salmelin.

When discussing costs, the terms "capex" (capital expenditure) and "opex" (operational expenditure) are often used. "Capex" refers to investments in network infrastructure and in general to those expenditures that are expected to contribute to revenue-generating business during a long period. "Opex" refers to the "running costs," that is expenditures which become expenses of the current accounting period. With cash flow based modeling, a distinction between capex and opex is not needed. Still those terms are commonly used and give general insight into the cost structure of the business.

This chapter includes many examples whose purpose is to practice developing business cases for LTE backhaul. The reader needs to replace the values and assumptions for his case as the results of the calculations are not generic but simply reflect the assumptions made.

3.1.1 Role of Backhaul Within LTE

Before developing the cash flow based models, it is useful to review the role of backhaul within the LTE system, in terms of capex and opex. Between capex and opex a trade-off often exists: using capital for the network may enable lower running costs, as leased line cost savings or other ways of reducing operational costs.

In general, incumbent operators often own backhaul infrastructure while new entrants depend more on leasing the required evolved NodeB (eNB) connections. Leasing means there are more running costs (opex) involved, and depending on the market, these costs may be extremely high and even prohibitive for the purpose of a low-cost backhaul. Owning a network, on the other hand, necessarily involves the initial use of capital to build up the network and, of course, will have running costs for operating this self-built backhaul.

In the LTE network, backhaul is a significant source of cost for the LTE network operator. In Figure 3.1, wireless backhaul represents 13% of network capex during 2015.

The estimate in Figure 3.1 shows the backhaul microwave share. In addition, transport functionality is required in other backhaul nodes, routers, switches and security gateways, as well as eNBs. So it is likely that, with all these included, the share of backhaul is larger than the 13% as part of transport costs appear on the radio network side and on the core side.

From operator revenue, network capex is here assumed to account to around 10% of revenue and total opex to e.g. 70% - of course very much operator and case specific.

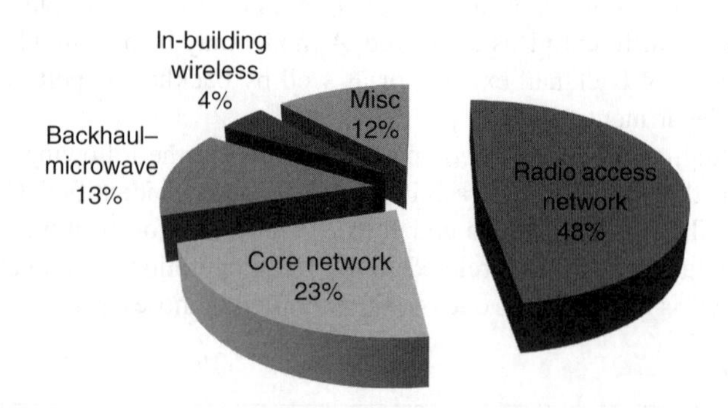

Figure 3.1 2015 Estimate for network capex by network component. *Source*: ABI Research, Mobile Operator Capex, 2015

Figure 3.2 breaks down a telecom operator's opex.

The network opex in Figure 3.2 is 18% of the total opex. Assuming the backhaul share of the network capex is 13% (the microwave backhaul capex presented earlier) and assuming a similar 13% share for backhaul for network opex, one arrives at the figure in Table 3.1 as an estimate of the share of backhaul-related costs in the operator business. This estimate is probably at the lower end and highlights the difficulties in arriving at accurate figures, since costs can be classified in many ways and every network is different.

In Table 3.1, backhaul capex amounts to 1.3% of revenue and opex to 1.63% of revenue. These values certainly vary from network to network. The backhaul opex value could be far greater especially if leased lines in a high cost market are used. For any particular business case model, accurate data from that specific case should be used.

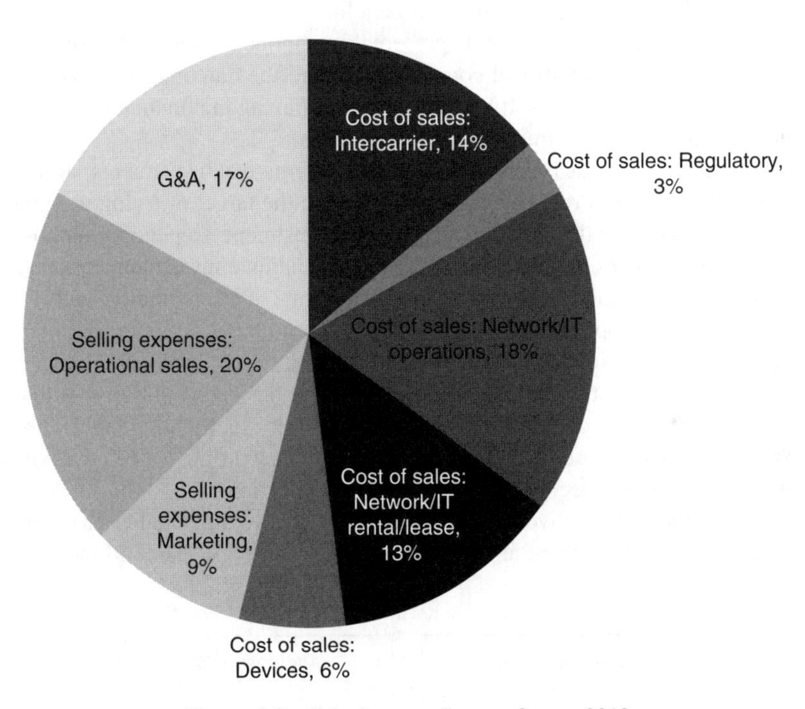

Figure 3.2 Telco's opex. *Source*: Ovum, 2012

Table 3.1 Estimates of backhaul capex and opex as percentage of operator revenue

Capex		Opex	
Network capex out of revenue	10.00%	Network opex out of revenue (0.7 × 0.18)	12.60%
Backhaul capex out of network capex	13.00%	Backhaul opex out of network opex	13.00%
Backhaul capex out of revenue	1.30%	Backhaul opex out of revenue	1.63%

3.1.2 *Why and What to Model*

An investment decision typically concerns purchasing equipment that either increases revenues or reduces costs. For LTE backhaul, new base stations are deployed with related backhaul links to generate more revenue, or the costs of the existing backhaul are reduced by investing in new networks, optimization features or operational processes.

One often compares different approaches in order to select the lowest-cost alternative. During the lifetime of the network, operational costs easily overwhelm the initial equipment cost, so finding the lowest-cost alternative requires analysis of both what is the initial investment needed and then what are the running costs over the years.

A business case or similar type of economic calculation is an essential part of the investment proposal to the management team. Before authorizing consumption of company resources—such as experts' working hours, access to tools and purchase of network equipment—managers want to see that the up-front use of capital and allocation of resources are matched with equal—or preferably much larger—additional benefits.

The purpose of the economic model is to present this in monetary terms. The final investment decision itself is guided by additional considerations and the outcome is not necessarily the same as the business case proposes. Even in these cases, having the financial side visible helps decision makers understand the impacts of their decision.

Figure 3.3 presents the idea of comparing two alternative investments. Relevant cash outflows are subtracted from relevant cash inflows, with the target of trying to maximize the difference and so maximize the value created by the investment. Depending on how the case is formulated, one compares two (or more) alternative candidate investments or one compares a potential investment to the base case of "doing nothing," i.e. continuing with the present solution. For example, an optimization project may be formulated as a comparison against the current way of operating. New network rollouts may have two or more alternatives for backhaul, to be compared against each other. The way the case has been built up should be clearly documented so that the user of the model understands what exactly has been analyzed. From the beginning, the model should, of course, be formulated according to the needs of the users of the business case, i.e. the decision makers in the organization.

In Figure 3.3, option "B" delivers both higher cash inflows and smaller cash outflows, as illustrated by the size of the boxes, and is thus preferable over "A." The inflows between two

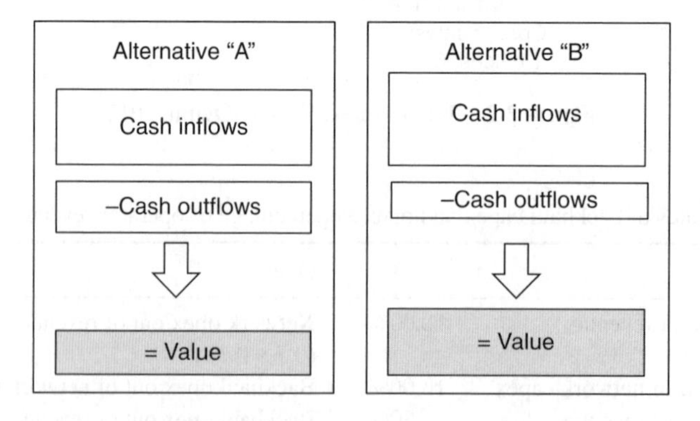

Figure 3.3 Comparison of alternatives

or more alternatives may differ, as well as the outflows. These differences in either inflows or outflows become the decisive factors in the comparison.

Many kinds of proposals for LTE backhaul are important, for example expanding the capacity of network links to support a higher number of subscribers with a good service level, planning for an optimized topology for cost savings or analyzing LTE backhaul management system benefits for faster troubleshooting in case of network failures.

Clearly, some cases can be modeled relatively easily, while collecting information for the modeling of others will be very tedious or costly, so assumptions and simplifications have to be made. Business case developers also need to keep costs down, since with additional effort (and cost) the model can always be subsequently developed by adding further parameters or searching for more accurate data. At some point, however, any additional effort spent in refining the business case will no longer be justified.

No matter how the model is formatted or how much information is acquired, the model should essentially capture relevant additional revenues (or other quantifiable benefits as well as related costs), compare these and present the results to the decision makers, with an insight as well into any key limitations inherent in the model, with the aim of helping the organization to make economically sound decisions.

3.2 Strategic Input for Planning

Chapter 2 divides LTE backhaul planning into strategic planning, high-level design and detailed design phases. While high-level and detailed designs differ from case to case, strategic input certainly is very much company-specific. Because of the longer time horizon, one fact is that strategy tends to involve usage of capital up front, requiring capital budgeting and investment decisions.

Many strategic topics around LTE do not directly affect backhaul planning. One example could be an operator focusing on differentiation by excellent customer service and expanding customer call centers accordingly. Also, many pricing decisions do not directly affect backhaul technical design, other than potentially increasing network traffic load, which of course is a key consideration.

Clearly, LTE backhaul needs to be planned and built according to the operator's business strategy. LTE backhaul strategy then, on one hand, supports the operator's business goals in general and, on other, guides the more detailed backhaul design phases, especially with the bigger design cycles: rollout of new technologies and backhaul networks, significant expansion of existing networks, acquisition of sites, etc.

The operator may emphasize the LTE network's technical performance, for example high-speed service, extensive coverage or reliability of network access. Cost leadership, on the other hand, certainly focuses on finding the very low-cost alternative for the backhaul. Whatever the goals for the network are, they have to be assessed and the backhaul then designed accordingly in any of the planning and optimization cycles.

3.2.1 *Physical Infrastructure*

For the backhaul, long-lasting decisions are those related to the lowest (physical) layer of the network infrastructure: ownership of cable and fiber conduits, rights of way, ownership or lease contracts for equipment sites, radio towers and other premises. There may as well be other rights, contracts or other intangibles that have a strategic nature.

A key topic is that LTE sites already demand 50–150 Mbps capacity, and both peak rates and average traffic is growing rapidly. Meeting these capacity needs requires guidance for what type of physical infrastructure a company has or intends to have in the future, so that there is sufficient bandwidth available in the physical media.

When an operator owns rights to lay cable or owns pipes and conduits for installing cables, further cabling may be fed in for increased bandwidth over the years as and when necessary. Similarly, owning or having long-term leases to wireless backhaul masts and rights to install equipment in these towers and other sites are long-term assets that supply capacity growth, by changing technology and elements without having to acquire new sites.

Infrastructure assets, like the ones mentioned above, clearly form an underlying enabler and at the same time one limiting factor for planning exercises. Whether to own the physical infrastructure thus becomes one of the key strategic considerations for an operator. Acquiring (or divesting of) physical infrastructure takes time. In the long term, an operator's strategic plan may be to develop their own fiber network. In the short term, with regards to planning and optimization, the physical infrastructure will be very much fixed.

3.2.2 Transmission Media

Physical media choices are often restricted to what is available in the region and to what type of media can first of all be made available to the particular eNB site. Fiber is the preferred option for LTE due to the capacity it can provide; however, fiber is not available for every site, at least not immediately. As well, it may be available nearby but not quite right at the eNB connector but, for example, on some street corner. Increasing fiber availability is one strategic choice.

The next best alternative after fiber is wireless backhaul; depending on the bandwidth available for microwave, capacities of several hundred megabits can be supported. Deploying wireless in practice means investing in one's own backhaul network.

Media choices are often not a real consideration for LTE backhaul designs, as there may not be many alternatives at that time, and each individual site will use the best media type available. Even though an operator may strive to acquire fiber infrastructure and sites, and define the strategic intention accordingly, this takes time and the present design cycle will need to be performed without these assets. A clearly expressed goal may simplify high-level design and make that goal more achievable in the subsequent planning rounds. On the other hand, when an operator has suitable infrastructure in place, it is very economical and straightforward to continue using these assets for LTE backhaul, without much need for studying alternatives.

3.2.3 Capacity and Interfaces

A key cost item is the physical interface. Moving up in the Ethernet speeds from 100M to 1G to 10G and 100G is a continuous upgrade process dictated by the capacity needs of the eNB, and the costs associated with the newest and highest data rate options on the other side. At the initial introduction of the highest Ethernet line rate, the costs of the equipment ports are extremely high in the first instance but then become moderate after a while, so that today, for example, 1G is cost efficient with 10G becoming gradually affordable even for the eNBs.

Upgrade of the line rate requires site visits, so considering the capacity raises the question of the time horizon used in the dimensioning phase. Assuming an annual growth rate of, say, 60%, traffic will increase more than fourfold from the initial value after three years, and more than tenfold after five years. This is a huge difference and a balance needs to be struck between excessive up-front investments and costly expansion cycles.

3.2.4 Network Technologies

A central networking technology is the IP protocol. Technology questions are in this sense simple, as all logical interfaces in the LTE network are based on IP, with IPv4 still predominantly used for the backhaul. IPv6 is similarly included with the LTE standards, so it is a possible choice. Many routers are dual-stack, supporting both IPv4 and IPv6, making coexistence of two IP versions easier.

Network technologies—like native IP, MPLS and carrier Ethernet services—are further explored in Chapter 5 (and in Metsälä and Salmelin, 2012). High-level technology guidance may exist which then affects high-level planning.

3.2.5 Network Topology

Physical network topology is formed based on the location of the sites and the way the cables or wireless links connect these sites, making the physical topology rather rigid, as changes occur only by the physical installation of new cabling or from acquiring new sites.

Logical topology, meaning how the traffic is routed, changes more easily based on connectivity arranged by the higher-layer protocols. Connectivity and logical topology may well be dynamic.

Example 3.1

With LTE backhaul, high speeds and low latency requirements prefer "thinner" network topologies, meaning that long chains are avoided and traffic passes from few access links directly to a high-capacity (fiber) network and to the core.

As in Example 3.1, logical topology needs to be designed such that LTE latency requirements are met and that links do not become bottlenecks.

In addition to latency and performance characteristics, several other requirements affect topology: resiliency (with the need for redundant links and nodes) and security (designing the location of security gateways) to name a few. Different high-level requirements may affect the design of both the physical and logical topology.

3.2.6 Make or Buy

One obvious strategic question concerns the operation of the network and the "make or buy" decision. LTE backhaul may be operated in-house or sourced as a service from a backhaul service provider. In managed-service cases, the operation of the backhaul is given to a third party, for example a service company.

Radio network sharing, including backhaul sharing with two or even more operators, is another alternative. The operation of this shared backhaul can be performed as a joint venture or by just one of the parties.

In general, different segments of the backhaul network, as well as certain technologies requiring special competencies, may be operated by dedicated teams and this influences planning as well.

3.2.7 Backhaul Security Aspects

The security landscape of the mobile radio network changed drastically with the introduction of LTE in 2008.

The speed reached new heights due to the enhanced architecture and protocol, causing the traffic to continue its exponential growth. With LTE, management, control and user data traffic were now based on the IP protocol. This meant that much of the malware mechanisms developed for the Internet suddenly were applicable to the mobile network, the traffic there and the management of the mobile network.

In 3G networks the traffic is encrypted all the way from the user equipment (UE) to the radio network controller (RNC). In LTE the traffic is encrypted from the UE just to the base station. The base station itself has become much smaller and many new locations feasible. Often only a minimal physical security exists on these sites, giving an intruder a fairly easy access to the base station.

Often, company-level security guidance exists and the network needs to be designed accordingly.

What are the main threats to the LTE?

Taking control of the eNB means that with smaller size and easier physical access the intruder can access the eNB. In case there are management ports or an un-hardened environment, it might be easy to gain root access. Then, depending on the intruder's business plan, different scenarios are possible. First, adding an additional device that will route the traffic. Then, with a receiver at a convenient location, traffic can be analyzed, identities stolen and either confidential information accessed or some traffic blocked. With a stolen identity fake traffic to premium numbers can be generated, causing financial loss for the legitimate owner.

By establishing *a fake base station* an intruder can control all the LTE traffic in a certain area. The UE will contact the "loudest" base station. In addition to previous threats, attacks against the network can extend to have an impact on the MME and S-GW.

As the encryption endpoint is the eNB, traffic between the eNB and the S-GW is in plaintext and vulnerable to "overhearing." So there are many possibilities to capture the traffic, depending on the communication channel in use. With the captured information. user-sensitive data could be misused, for example.

In case no special authentication mechanism is in place for the base stations, the operator network is also vulnerable to unauthorized traffic that could disturb normal traffic and cause the service to be unavailable for legitimate users (this is known as a "denial of service" [DoS] attack).

These threats are addressed with specific protection mechanisms: having digital identity stored securely in the eNB, securing communications, hardening the eNB and finally implementing security management for the network.

First, digital identity supports authentication of the base station. The risk of fake base stations enrolling into the network is mitigated. The enrollment can be automated, resulting in savings setting up the network as the need for site visits is decreased.

Second, security of the communication is part of the overall LTE design. Secure communication with IPsec can be established between the eNB and the MME/S-GW. All communication (users/control/management) can be secured, mitigating the risk of overhearing traffic as well as the risk of messing up the traffic. IPsec requires that the root of trust be stored securely.

Third, the eNB must be hardened. There should be no active ports waiting to which an intruder could connect his own device, no extra service running in the node, potentially giving the intruder a way in and no known vulnerabilities in the software. The eNB SW should be updated to the latest patch.

Fourth, a reasonable level of security management should be implemented in the network that will alert the operator to any abnormal traffic. Successful security management requires implementations in several different network elements and network layers: eNB, backhaul and radio network and related traffic analysis tools.

3.3 Quantifying Benefits

Some benefits, as well as costs, are difficult to quantify in monetary terms. This is a valid criticism against using economic modeling directly for a decision. Also, the longer the time horizon, the more uncertain are the outcomes of any actions carried out at present. This is why results should always be reviewed and double-checked for validity and consistency.

Starting from the backhaul benefits, three separate types of benefit are discussed. First, service providers receive cash payments for the transmission services sold. In this case (Section 3.3.1) a revenue stream originating from the LTE backhaul exists for the service provider. For the LTE network operator there is no revenue directly originating from the LTE backhaul alone. LTE backhaul is simply one element needed for the LTE service. In the second case (Section 3.3.2), LTE backhaul is modeled as a contributor to the LTE service revenue. In the third case (Section 3.3.3), various costs shall be saved.

3.3.1 Revenue from LTE Backhaul

For transmission service providers, offering leased lines for LTE backhaul is a source of revenue. Service providers sell specific transmission services to the LTE network operator. The entity providing the service can be an affiliated company or another department within the same company.

In these cases it is possible to use the service fee as a revenue item for the LTE backhaul business case, making the benefit side of the business case analysis simple due to the revenue stream that originates from the backhaul. In this case LTE backhaul can be treated more or less as an independent business when looked at from the viewpoint of the service provider.

Backhaul service fees depend on market and region, and factors like capacity provided and other technical service characteristics.

Business case analysis then compares any additional revenues against the expenses required for this revenue increase. Revenues can increase due to the addition of new lines in service, price increases, new customers, new services or enhanced characteristics of existing services.

Example 3.2

A transmission service provider considers improving its service by reducing latency for the service used by the LTE operator for S1 backhaul. This requires an investment in the network by a higher capacity 10G uplink. With a lower latency, a service charge is planned to be increased by 30%. The quantifiable benefit is the 30% increase of revenue, which can, in the analysis, be offset against the expenses of implementing the 10G uplink.

3.3.2 Contribution to Mobile Service Revenue

In the previous section. one had a separate revenue stream originating from the backhaul alone. The mobile network operator's situation is typically different. Backhaul is part of the network infrastructure that is needed for LTE service offering for subscribers, who are then charged for the mobile services used. The LTE network as a whole is needed, as well as other functions—like operator customer service and charging and billing—before any revenue can be collected.

In this situation the benefit is the increase in revenue stream, as the LTE business in its entirety and the business case cost side have to include all types of expenditures that are needed for the increase of LTE service revenue.

Example 3.3

An LTE network operator extends LTE service to a new region, by 200 new eNBs. This is expected to increase LTE service revenues by €X. It is assumed that there is no expansion need to other functions than to the radio network and related backhaul. €X is in this case the quantifiable benefit, i.e. compared with the investment into the network (including both radio and backhaul network).

In Example 3.3, since the expansion is only needed for the radio network, and no other functions require modifications or additional capacity, the business case can be simplified by leaving all those cost items out that do not differ between alternatives. Including all items, however, leads to the same result, and especially if there is uncertainty as to whether any difference exists, it is always correct to include that item into both alternatives. If the item is identical, it will cancel out in the comparison with no impact on the results.

3.3.3 Cost Savings

Many cases are in practice targeting cost savings related to the LTE backhaul as the main benefit. Avoiding a cost that would otherwise occur is a quantifiable benefit. It does not increase revenue, but in the model a cash outflow that can be avoided is a quantifiable benefit. In this way it becomes equivalent to a cash inflow.

Costs can be saved in many ways: making the network more efficient using fewer nodes, using less expensive or fewer lines and aggregating traffic efficiently such that less capacity is needed. Costs can significantly be saved as well by planning the network soundly and making it easy and simple to deploy, operate, maintain and troubleshoot.

Example 3.4

A proposal to retire 100 leased lines and replace them with an own access network is being prepared. The quantifiable benefit in this case is the cash saving related to the fees of the 100 leased lines. This saved cost is offset against all expenditures needed for the own access network.

3.4 Quantifying Costs

In cash flow based models, the benefits discussed in Section 3.3 become the cash inflows while all types of cash outflow needed to realize these cash inflows constitute the cost side. The terminology relating to the cost side of business cases is often confusing, as terms like "expenditure," "expenses" and "costs" are sometimes used interchangeably. For cash flow based models there is no need for more than to identify all cash outflows that occur and are relevant, whether they then in financial accounting terms become expenses (on the profit/loss statement) or assets (on the balance sheet).

3.4.1 Equipment Purchases

With LTE rollouts, backhaul equipment are purchased along with base stations and core elements. The purchase price of this equipment represents a cash outflow. This item is one of the most straightforward to quantify, as the purchase price is known.

List prices may differ from the actual prices and so it is the price less any discounts that is relevant. The equipment itself could, for example, be a wireless or optical transmission device, a switch or router, a security gateway or some other backhaul device.

When networking equipment is purchased, spare parts, related element or network management devices and new interfaces to other elements may be needed. If these items and upgrades in other elements are prerequisites to having the newly purchased backhaul equipment operating within the network, the costs of this equipment are included in the cost side of the business case.

3.4.2 Economic Lifetime

The time horizon used in the business case was already mentioned earlier. For equipment purchases, a related topic is the economic lifetime of the device.

Eventually all equipment physically wear out and break down. Even before that happens, typically it becomes uneconomical in operation, due to limitations of its features or functionality, or due to technological evolution or new, more-efficient, ways of providing the same function. In the LTE backhaul, an example could be a device that had a hard limit on bandwidth supported so that it needed to be replaced with a higher bandwidth solution.

This is why the economic lifetime of the equipment is used in the economic model. Economic lifetime is the amount of time the equipment or machine is expected to be economically viable in use. Uncertainty in the expected economic lifetime leads to being conservative and only expecting a time period that is very likely to be reached. If the business case is based

Table 3.2 Opex types relevant to the LTE backhaul

Item	Description
Site costs	Site rents, electricity, leased line fees, etc.
Proactive backhaul O&M	Monitoring, preventive maintenance, network assessments
Reactive backhaul O&M	Troubleshooting and correcting faults when they occur
Planning a new network	First-time planning of a new network
Operational planning and optimization	Ongoing smaller planning cycles
Network implementation	Implementing network plans and integration of new features into production network
Service provisioning	Setting up backhaul service for external or internal customers

on cost savings the acquired equipment shall bring, then these cost savings can at maximum be calculated for the economic lifetime of the equipment.

In some cases the device may still have residual value at the end of its lifecycle, for example if it can be sold instead of being scrapped. If this is assumed, the residual value is included as a positive cash flow at the end of the equipment's lifetime. If there is a cost for scrapping, then the residual value is negative. Often, the residual value is assumed to be negligible.

3.4.3 *Operational Costs*

Perhaps the most difficult estimate to make is that of the cost to operate the network. The equipment cost, as stated, has a precise, known value, and thus equipment prices can be compared directly. There are many cost items during the operational phase that are relevant but may be difficult to quantify and depend on various assumptions being made.

Operational expenditures have been studied e.g. in Verbrugge et al. (2005). Loosely based on this, cost types specifically relevant to the LTE backhaul are defined in Table 3.2 in order to have an example of what cost items one needs to study and include in business cases. For cash flow based business cases, the category of the cash outflow is not of interest. It is enough that there is a relevant cash outflow.

A very basic first cost class in Table 3.2 is that related to sites: site rents, electricity and leased line fees and other similar items. Backhaul sites are often shared with base stations so both base stations and transmission equipment contribute to these costs.

"Proactive backhaul (O&M)" refers to maintenance and monitoring the network in the fault-free state of its normal operation. When a failure in the backhaul occurs, it needs to be found and corrected. This is separately indicated as "reactive backhaul O&M."

Planning also consists of two items, the first one related to the planning of a new network and the second one to the ongoing operational planning.

Implementation of the network consists of implementation of the operational plans as well as the implementation of new network plans.[1]

Finally, backhaul service provisioning means setting up connectivity for the LTE backhaul service consumed by the external or internal customer.

[1] Implementation expenditures (impex) may as well be separated from opex, as implementation of the network is closely related to the initial purchase of the new network devices (capex). Implementation is a one-time item like equipment purchase.

Many business cases related to backhaul are structured with operational cost savings being viewed as the benefit and so being able to quantify these costs is very important. Difficulties arise when not all items are readily quantifiable. Items like rents, leased line payments and the cost of electricity are usually simple to calculate accurately. The amount of working hours needed to perform certain tasks is more difficult to estimate.

3.4.4 Other Costs

For LTE backhaul there may be yet more items that affect cash inflows or outflows. Poor customer experience due to backhaul congestion causes customer churn and lost revenue. This particular item can be alternatively called "cost of poor quality," as there now is a defect in the LTE backhaul. If backhaul would not be congested, there would be no lost revenue.

Example 3.5

With LTE, data service has become much more important than what it used to be in previous voice-oriented networks. LTE customers expect a good service and access to favorite applications also while using the "background data." If this expectation is not met, there is a risk of customers switching to another network operator.

In Example 3.5, poor service may be due to backhaul congestion. Estimating how much revenue (due to customer churn) is lost because of the backhaul is, however, no easy task.

An important concept related to business cases is a sunk cost, which refers to a past cost (a cash outflow that has already occurred) and thus is irrelevant to the present decision. For business cases one is interested only in the quantifiable benefits and costs that are affected by the decision. Since past costs can usually not be affected by any decision in the present moment, any amount that has already been paid and cannot be recovered by the present decision is irrelevant.

Example 3.6

A network upgrade proposal is prepared with two alternatives, A and B. Costs related to both alternatives are presented. Studies related to project A were started already two years ago, which is why the proposal includes costs from the previous two years, as well as the remaining costs. Project B has only been studied for six months, and the preparation costs from these six months are included, as well as the remaining costs.

Question: Is this correct?

Answer: No. Costs that have incurred already cannot be recovered. The business case for project A should be modified by removing the past (sunk) costs from the previous two years, and similarly the costs of project B should not include costs from the previous six months.

In retrospect it sometimes turns out that project lifetime costs are so high that the project should not have been started at all. However, for the present moment the relevant question is whether the company is better off carrying out the remaining work (and remaining costs) and then collecting the additional revenues or other benefits, or whether the project should be discontinued. How much effort has already been spent on the project is irrelevant for this decision: the question concerns the *remaining* effort. If the remaining effort can be justified by the expected benefits, the company should continue and carry out the remaining tasks and costs. Otherwise, the project should be terminated.

3.5 Case Router

In this section, the concepts of net present value (NPV) and payback period are introduced with a simple example. The calculations are based on cash inflows and outflows that do not depend on financial accounting.

> **Example 3.7**
>
> *Router Investment*
>
> In this example it is proposed that a new site router be purchased to consolidate traffic from multiple LTE base stations, replacing an old device. The main benefit would be that traffic could then be combined to a single link instead of three separate links. See Figure 3.4, where the right-hand side represents the target state, where the new router collects traffic from base stations into a single common uplink.

The router costs 80, and allows saving LTE backhaul operational costs by 50 annually from there on for five years, due to the reduction of the number of links.

The case is constructed to compare the proposed investment to a new device (the right hand side of Figure 3.4) with continuing as in the present moment (the left-hand side of Figure 3.4).

3.5.1 Cash Flow

Cash inflows and outflows are documented into Table 3.3, with cash cost savings treated as positive flows (cash inflows).

When company pays for a product or service, it is a cash outflow that is marked with a minus (–) sign. When a company receives a payment as sales revenue, it is a cash inflow, a positive value. In this case, cash cost savings constitute the equivalent of positive cash inflows. A corresponding graph of Table 3.3 is given in Figure 3.5.

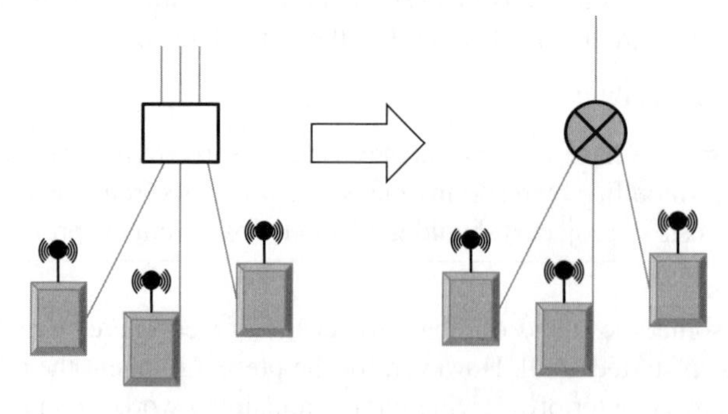

Figure 3.4 Router investment

Table 3.3 Annual cash flows of Example 3.7

	Cash cost savings	Equipment	Net cash flow
Year 0	0	–80	–80
Year 1	50	0	50
Year 2	50	0	50
Year 3	50	0	50
Year 4	50	0	50
Year 5	50	0	50

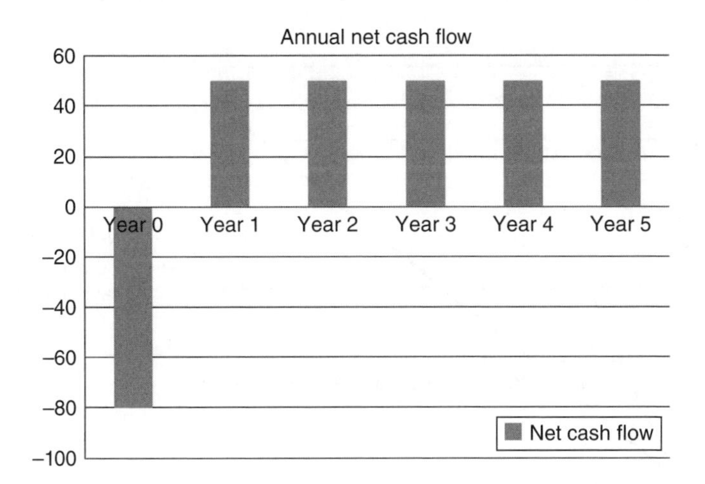

Figure 3.5 Annual net cash flows of Example 3.7

As seen in Figure 3.5, in the initial year (marked "Year 0"), a cost of 80 has to be sacrificed to purchase the router, without any balancing positive cash inflow yet occurring. After this, cash costs are saved for five consecutive years, and for a positive investment decision these future cash inflows should exceed the cash outflow of the initial year. The cash flows are assumed to occur at the end of each year, initial investment (–80) at the end of Year 0 and the first operational year being the one marked "Year 1," so that at the end of Year 1 the cash flow is +50.

3.5.2 Payback Period

The payback period is a simple metric that gives the duration of time needed to recover the amount used for the initial investment. The shorter the value, the better the investment is.

For Example 3.7, annual net cash flows are calculated by subtracting the annual cash outflow from the annual cash cost savings. Cumulative net cash flow is shown as the last column of Table 3.4 and a corresponding graph is given in Figure 3.6.

From Table 3.4 and Figure 3.6, one notices that at the end of the first year of operation cumulative net cash flow is still negative (–30). At the end of the second year, the value is positive (20). Thus, the payback period in this case is somewhere between one and two years, as can be observed from the graph as well. Between the first and second years, the balance is improving by the speed of 50. Exactly 30/50th of the second year is needed to recover the

Table 3.4 Cumulative net cash flows of Example 3.7

	Cash cost savings	Cash outflow (Equipment)	Annual net cash flow	Cumulative net cash flow
Year 0	—	–80	0	–80
Year 1	50	—	50	–30
Year 2	50	—	50	20
Year 3	50	—	50	70
Year 4	50	—	50	120
Year 5	50	—	50	170

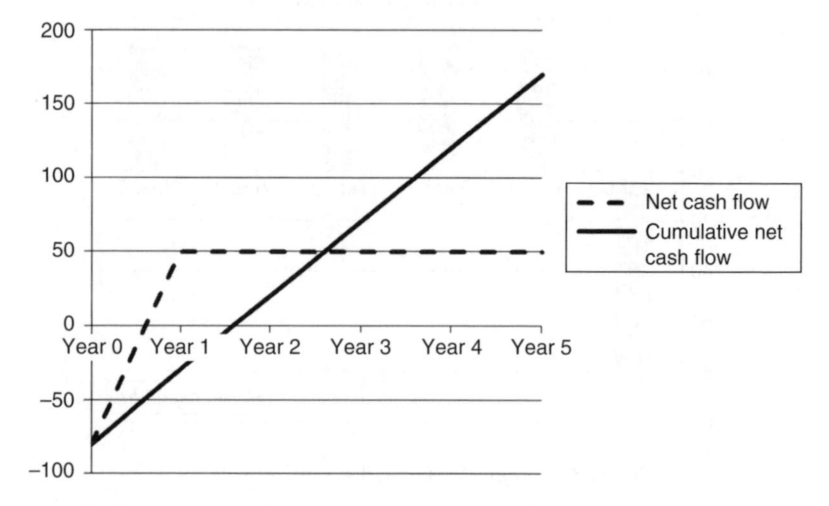

Figure 3.6 Example 3.7 net cash flow and cumulative net cash flow

remaining balance of –30 at the end of first year of operation. Thus, the payback period is 1+30/50 years, or 1.6 years, at which moment the cumulative net cash flow is zero (the break-even point).

Conceptually, the payback period is easy to understand. After 1.6 years, the company has generated cash savings in operations equal to the purchase price of the router.

Some drawbacks related to the use of payback period exist. First, it does not take the time value of money into account. Also, the payback period does not tell how profitable the investment is. One only knows the duration of time needed to recover the initial capital outlay.

Example 3.8

Modified Example 3.7

Consider that in the router investment introduced in Example 3.7 it is projected, during the business case analysis phase, that during the third year a further purchase of redundant chassis and further line cards will be needed, due to traffic growth, in order to maintain the annual cash cost savings of 50 for the subsequent years (4 and 5). This is estimated to cost 80. What is the new payback period?

From Table 3.4, cumulative net cash flow at the end of the third year is 70. When the additional purchase price, 80, is subtracted from this, the result becomes –10. In this situation, the payback period cannot be meaningfully calculated. The payback period is suitable for typical cases where cumulative cash flow stays positive after having reached break-even the first time.

Despite these limitations, the payback period it is often used, especially when complemented with NPV.

3.5.3 Net Present Value (NPV)

When making a deposit in a bank, one earns interest for the principal. Assuming, for example, an annual interest rate of 3%, €1000 grows into €1030 within one year. Similarly, a promise to receive €1000 a year from today is worth less than €1000 in hand today (with any positive interest rate). With the 3% interest rate assumed, €1000 in a year from today would be worth €1000 × 1/(1 + 0.03) or €970.87, since this amount grows to €1000 within a year, when the annual interest of 3% is added to the principal at the year-end.

With the interest rate is known, one can shift cash amounts from one year to another. Future cash amounts are brought to the present time (or other earlier time instant) using "discounting." Similarly, cash amounts can be moved to future values by "prolonging."

The equation for discounting to present value (P) is:

$$P = F_0 + \frac{F_1}{(1+r)} + \frac{F_2}{(1+r)^2} + \frac{F_3}{(1+r)^3} \ldots = \sum_{i=0}^{N} \frac{F_i}{(1+r)^i} \tag{3.1}$$

where

$$P = Present\ value\left(sum\ of\ discounted\ cash\ flows\right)$$

$$F_i = cash\ flow\ at\ year\ i$$

$$N = number\ of\ years$$

$$r = interest\left(discount\right)rate$$

Equation (3.1) assumes that interest is added to the principal (not withdrawn) and so it is compounding.

In general, the period could be any period when the interest is calculated; r then denotes the interest during that period and N denotes the number of these periods. Often a period of one year is used, in which case the annual interest rate is used.

With this information the present value of cash flows can be calculated. Returning to Example 3.7, the cash flows from Table 3.4 are used. The only parameter missing is the interest rate. This topic is revisited later, and for now present value is calculated with an annual rate of 12%.

For the first year the discount rate is $1/(1 + 0.12)^1$, for the second year $1/(1 + 0.12)^2$, etc.

In Table 3.5, NPV is the sum of the present value of annual cash flows for all the years considered. The present value of the annual cash flow is obtained by multiplying the annual net cash flow by the discount factor separately for each year. All cash flows are expected to occur at the end of the year.

Table 3.5 Net present value calculation

	Equipment	Cash cost savings	Net cash flow	Interest rate	Discount factor	Present value
Year 0	–80	0	–80	0.12	1	–80
Year 1	0	50	50	0.12	0.89285714	44.64
Year 2	0	50	50	0.12	0.79719388	39.86
Year 3	0	50	50	0.12	0.71178025	35.59
Year 4	0	50	50	0.12	0.63551808	31.78
Year 5	0	50	50	0.12	0.56742686	28.37
NPV						100.24

The sum of these discounted cash flows (DCF) is the NPV. NPV gives the value of the investment at the present moment (end of Year 0). Table 3.5 shows that the investment with the assumptions given is economically desirable, since it adds value to the business of 100.24. Compared with the cumulative cash flow calculation in Table 3.4, there is a difference in the sum of cash flows due to the discounting in Table 3.5. Without discounting (as was used for the payback period), the sum of cash flows equaled 170, while now for NPV one notices the result of 100.24. Future cash flows diminish in value when they are brought back to the present moment, when the interest rate is positive as usual.

NPV overcomes the drawback of the payback period, where the time value of money is not considered. Also, now with NPV, multiple investment proposals can be compared, as in Example 3.9.

Example 3.9

The transmission department may expand the backhaul for LTE either in Region A or in Region B. The department head is presented the proposals: for Region A expansion, NPV is 120. For Region B expansion, NPV is 85. The expansion of the LTE backhaul in Region A should be carried out, because of the higher NPV of that project.

All things being equal, the company should select the proposal with the higher NPV.

NPV can also become negative, which means the proposed investment destroys value. Proposals with negative NPV should be rejected.

Example 3.10

While discussing the case of Example 3.9 further, a comment is made that for Region A the calculated payback period is 2.4 years, while for Region B it is only 1.9 years.

Question: How should the comment be answered? Should the decision be changed?

Answer: The decision should not be changed. NPV of expansion of Region A is higher and thus it is preferable.

As brought up in Example 3.10, the payback period and NPV may show different preferences. NPV is considered theoretically superior to payback period due to the limitations of the payback period. Of course, seldom can management decide on investments simply based on a single metric—NPV or otherwise. Still, NPV is very useful tool for estimating the economic feasibility of the project.

Example 3.11

For the router investment case (Example 3.7), the new router replaced an existing device, which was to be scrapped. But the old device still has a value of 20 in the balance sheet, and someone points out that this write-down charge has not been included in the NPV calculation.

Question: Is the write-down value relevant to the NPV calculation?

Answer: There is no cash outflow related to the write-down of the old device. Whatever value the old equipment may still have in the balance sheet is irrelevant to the NPV calculation. If the old device could be sold for a price, the received price should be taken into account as a positive cash inflow.

Example 3.11 highlights that with cash flow based business cases charges in financial accounting are not relevant—unless they cause cash outflows. Dependencies to financial accounting shall be visited briefly later with "return on investment" (ROI).

3.5.4 Selection of the Interest Rate

In the above calculation of NPV, 12% was used as the interest rate. The higher the interest rate is, the less valuable future cash flows become when they are discounted to the present moment. In general, the more risky the investment is, the higher the rate should be. NPV can be interpreted to earn the stated amount as present value when the investment is financed with capital earning a 12% return rate.

The rate of return requirement (interest rate to be used) is modeled as comprising a risk-free interest rate r_f and a risk premium factor (r_p). The risk-free interest rate is the interest rate of a perceived "risk-free asset" (e.g. a short-term government bond). The r_p is then the component that takes into account risks specific to the project in question. The r_p reflects the rate of return requirement of an investor who is willing to finance the investment, provided he receives a rate according to the sum of the risk-free rate (r_f) and the r_p, to compensate for the risks specific to the project.

Another way to view the required rate of return is to estimate the cost of capital for the company carrying out the investment (for example the LTE backhaul network operator). Weighted average cost of capital (WACC) is calculated with the help of the following components:

Cost of own capital (return requirements for shareholders' equity) = c1
Cost of loaned capital (interest rate of loans) = c2
Share of own capital = s

The equation for WACC is

$$WACC = s \times c1 + (1 - s) \times c2 \tag{3.2}$$

Assume, for example, that c1 (cost of own capital) = 0.15, c2 (cost of loaned capital) = 0.05 and that 50% of capital employed is own. Using Equation 3.2, one calculates WACC as:

$$\text{WACC} = (0.5 \times 0.15) + (0.5 \times 0.05) = 0.095 = 9.5\%.$$

With this value as the discount rate (interest rate), the company can service its loans (with a 5% interest rate) and meet the return requirements for its own capital.

When discussing the rate of return required, inflation becomes an issue. Real return on any investment is also affected by inflation. Rate of return requirements will be higher in case of inflation. However, if future cash flows are also affected by the same inflation rate, the net effect to NPV calculation is zero.[2]

For further reading on investments, see Bodie et al. (2011).

3.5.5 *Internal Rate of Return*

The technique of DCF used in NPV calculation can as well be used to find out the internal rate of return (IRR) of the investment. In the NPV calculation for the router investment case (Example 3.7), with a discount rate of 12% NPV was still largely positive, so obviously the discount rate could have been even higher before NPV becoming negative. IRR is exactly the rate with which NPV becomes zero. It tells then the amount of return the investment earns. As such, the higher the IRR, the more beneficial the investment is (all else staying equal).

By trial and error, one finds that IRR in Example 3.7 is 40.2%. The equation is that of the NPV, where the sum of DCF (present values) is set to 0 and one solves for the interest rate, r.

$$F_0 + \frac{F_1}{(1+r)} + \frac{F_2}{(1+r)^2} + \frac{F_3}{(1+r)^3} \ldots = 0, \textit{where one solves for } r \qquad (3.3)$$

The discount rate, r, where the above sum of DCF is 0, is IRR. IRR, or r, as marked in Equation 3.3, is in practice found with a calculator application.

With IRR calculation, it is assumed that the cash flows received are re-invested at the IRR rate (i.e. 40.2% in Example 3.7). Since the interest rate used in the NPV calculation is the hurdle rate for investments and projects may in general be accepted at the hurdle rate, it is not guaranteed that proceeds from projects can be re-invested at a rate higher than this hurdle rate.

IRR also has a limitation with cash flows where the cash flows change signs multiple times. Typically, however, when the initial cash flow is negative and after some years the net cash flow turns and stays positive, there will not be this issue.

3.5.6 *Return on Investment and Further Metrics*

Often, heads of business units have incentives that are tied to the profit the business shows in financial accounting. NPV does not measure profit or profitability for any accounting period. In Example 3.7, NPV was calculated over a five-year period, but there was no statement on how the investment affected the reported profit of the company.

Financial statements, like profit/loss statements and balance sheets, are prepared for the public according to accounting standards. Luckily, one does not need to master these when calculating NPV and payback period with cash flow based models, since business cases are prepared for company internal decision making and so need not adhere to accounting standards.

[2] If this is not the case, inflation needs to be taken into account. For further information, see the literature, for example the references listed at the end of this chapter.

However, other metrics, like ROI are, in addition to NPV and payback period, used for internal decision making. ROI uses profit (net income) from the profit/loss statement as its input. This creates a dependency to stated financial reports. With ROI, suddenly one needs to know the net income reported, and the calculation of net income is in turn dependent on several financial accounting rules.

In the router investment case (Example 3.7), the router product costs would be capitalized in financial accounting, meaning that since the router is expected to contribute to revenue generation for five years the initial cost would not be all expensed immediately, but rather depreciated during its economic life. If depreciation is aggressive, the first years will show a reported profit that is "too low" due to high depreciation charges, but after the asset is completely depreciated the profit as reported will be "too high" because suddenly there will be no further charges. Note that the depreciation charge does not cause any cash outflow, and instead in the NPV and payback period calculations the initial purchase price was taken into account when the cash outflow initially occurred (when the purchase price of the router was paid).

With big up-front cash outflows—such as sizeable investments that are expected to be operating over long periods—net profit reported according to financial accounting will be modest in the first years, if not negative, as the gains will be collected gradually over a number of years and accounting periods. If managers are promoted frequently based on profit shown, this creates an incentive to maximize performance of the business as reported in the short term.

When new investment renders old equipment useless and the old equipment is still not completely depreciated, the remaining balance of that asset needs to be written off to the profit/loss account in the financial statements, creating a charge without a cash outflow, affecting calculation of ROI but not NPV or the payback period.

Example 3.12

Question: Consider Example 3.7 and the comment of the write-down of the old device. What type of impact (if any) there would be to the ROI?

Answer: ROI is affected. Any remaining balance sheet value of the old device is expensed in the profit/loss account (according to specific financial accounting rules) in case the device is removed from operation. This affects the profit reported, and consequently it affects the ROI calculated. Profit for the accounting period where the write-down takes place will be reduced and ROI will have a smaller value for that specific accounting period. (Note that the write-down does not cause a cash outflow, and so NPV is not affected, as discussed in Example 3.11.)

As shown in Example 3.12, ROI is influenced by financial accounting rules. The amount of write-down depends on how the asset has been valued and depreciated during earlier accounting periods, and on financial accounting standards. Net income as reported in the profit/loss statement also depends on how much tax the company pays, which then in turn depends on how the taxable income is calculated, etc.

For investment calculations NPV is in general a better tool because accounting rules and practices do not affect the calculation result. NPV gives the decision maker a clear view of whether the investment is value-adding or not, and by how much.

3.6 Wireless Backhaul Case Study

3.6.1 Case Definition

Fifteen new LTE eNBs are added, and the backhauling option is to be selected. There are two options to consider:

Alternative A: Leasing the access network lines from a third-party service provider, for each of the eNBs.
Alternative B: Building and operating an own wireless access network for these eNBs.

In Figure 3.7 with leased lines, each hub site (Hub 1, Hub 2 and Hub 3) connects five base stations (eNBs) to the network using leased lines (abbreviated LL on the figure). Each leased line costs 1000 units each month based on current prices for a three-year contract period. A fixed one-time fee of 3000 per commissioned leased line is incurred initially. Due to price erosion, when a new contract is negotiated after three years, the price is estimated to be 20% lower than the initial price. The SLA (service level agreement) given is monitored in order to manage quality and availability. An annual additional cost of 1000 is assumed per leased line by the operational team for the SLA monitoring.

No new equipment is required, since hubs exist and can be directly connected to the leased line service. As well, the eNBs have suitable interfaces that connect directly to the leased service.

A business case for adding the 15 eNBs to the radio network, concerning how much additional revenue is generated with that investment and what costs are involved (eNBs, their commissioning, transport, etc.), is drawn up. The result is positive. In that calculation, it is assumed leased lines will be used, at least initially.

The task is to evaluate alternative access for the first mile of these eNBs, using similar assumptions as in the radio network business case. If one manages to reduce the cost from the

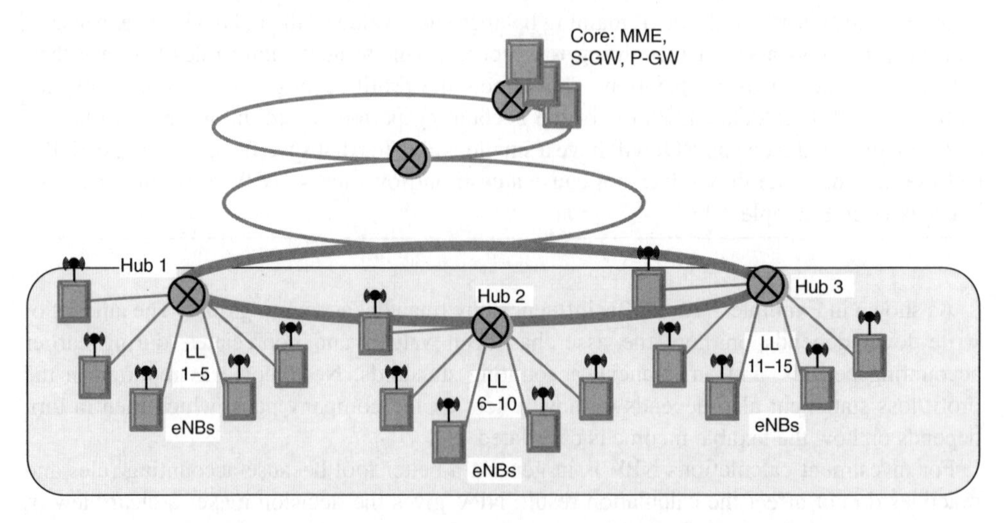

Figure 3.7 Alternative A: Leased line access (shaded area considered)

base case of leased line access, the overall business case can be re-calculated with this reduced backhaul cost, thus improving it.

The cash cost savings of not using leased lines are modeled as benefits. Ultimately, both scenarios concern only costs (cash outflows), either as paying leased line fees or as paying for wireless backhaul products and their operation. With a transport-focused comparison case leased line cash cost savings are modeled as benefits for the wireless access case. NPV will be positive if costs can be saved compared to the leased lines and, if not, leased lines will remain the preferred solution.

Figure 3.8 shows a wireless access network alternative.

In Figure 3.8, the network topology is similar to that of Figure 3.7. The only difference is that each eNB connects to the hub site using wireless link (abbreviated WL on the figure), with the wireless link consisting of indoor and outdoor units (IUs and OUs) at both ends of the link.

For wireless access, 30 new wireless transmission equipment (one IU and one OU at each end of the 15 links) with a total purchase price of 130,000 is required. Initial implementation cost for 15 hops is estimated to be 30,000 (including planning, site implementation, training, initial commissioning, etc.), based on rollouts of similar equipment in other regions. Additionally, an existing wireless network management station requires a license upgrade from the supplier to support the additional equipment being managed. This costs 5000 annually. For the wireless spectrum used, there is an additional annual license fee of 1000 per link.

The operation of the wireless backhaul requires additional capacity from the operations team. This is estimated to increase the annual operational cost by 90,000.

The new equipment will likely have a physical operating lifetime of 20–30 years; however, the radio network expansion was calculated using a time horizon of five years. For a five-year period, the capacity provided by either leased lines or wireless transmission is estimated to be adequate, with a projected significant growth of traffic. Within the analysis, a five-year time horizon is thus assumed, to reflect the radio network case assumption. After five years, the equipment is expected to be removed from the operation and have no terminal value. In practice, the equipment could perhaps be re-installed further to serve lower-capacity sites but it is decided not to account for any residual value.

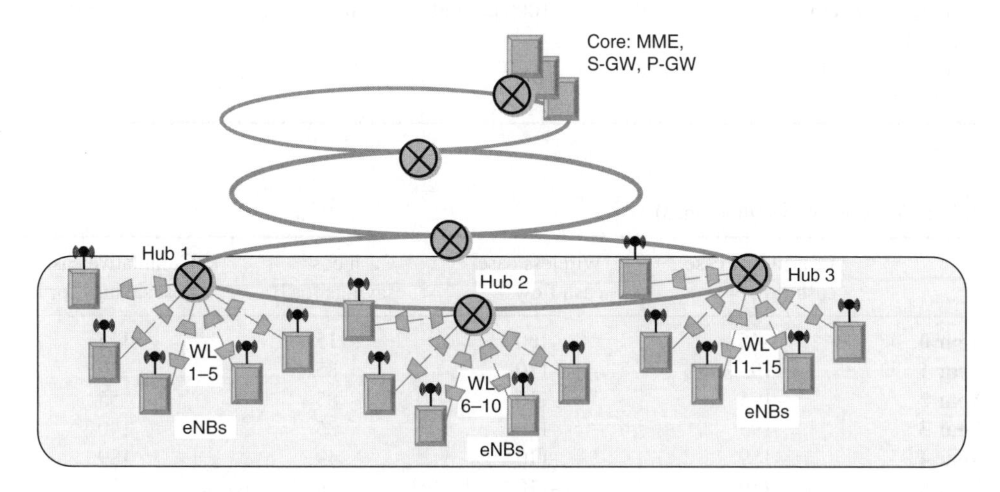

Figure 3.8 Alternative B: Wireless access (shaded area considered)

All else in the network stays equal except the access lines in the shaded area, so there is no need to consider other segments of the transport network, any radio network topics or any other revenue or cost items in the operator's business. Only items differing between the approaches are included.

Relevant costs for both options are shown in Table 3.6.

3.6.2 *Payback Period*

After having established the facts and key assumptions as above, cash flows can be included, as in Table 3.7.

Table 3.6 Related cost items for Alternative A (leased lines) and Alternative B (wireless)

Cost item	One-time or recurring?	Amount	A: Leased line case (Sum for 15 links)	B: Wireless case (Sum for 15 links)
A: Leased line initial installation	One-time	3000 per line	45,000	0
A: Leased line monitoring	Annual	1000 per line	15,000	0
A: Leased line fee	Monthly	Years 1–3: 1000 per line Years 4–5: 800 per line	Years 1–3: 180,000 annually Years 4–5: 144,000 annually	0
B: Wireless products purchase	One-time	130,000	0	130,000
B: Wireless products implementation	One-time	30,000	0	30,000
B: Wireless products operation	Annual	90,000	0	90,000
B: Wireless spectrum license fee	Annual	1000 per link	0	15,000
B: Wireless products mgmt station	Annual	5000	0	5000

Table 3.7 Cash flows (thousands)

	Leased line case cash flow	Wireless case cash flow	Cash cost savings for B	Cumulative cash cost savings for B
Year 0	–45	–160	–115	–115
Year 1	–195	–110	85	–30
Year 2	–195	–110	85	55
Year 3	–195	–110	85	140
Year 4	–159	–110	49	189
Year 5	–159	–110	49	238

Note that, for both leased lines and wireless, cash flows are negative, as all items considered are cash outflows, and the revenue side (LTE service revenues) was left out of the model as it is assumed to be independent of what type of backhaul is used. In the "Cash cost savings" column, the cash costs of wireless backhaul (Alternative B) are compared to the cash costs of leased lines (Alternative A), and the column tells how much cash costs are saved with wireless in this case. As can be seen in the table, the net cash cost savings are positive for every year for Alternative B, except for the initial investment year.

Savings of leased line cash costs occur for the duration of the five years. For Years 1 to 3, the cash cost savings amount to 85,000, consisting of the difference between the leased line fees (and additional monitoring) for 15 leased lines and the cost of operating the wireless backhaul. For Years 4 and 5, the cash cost savings are smaller, due to the 20% price erosion of the leased lines, and the amount is 49,000.

From the column "Cumulative cash cost savings" one obtains the payback period as 1 + 30/85 years, or 1.35 years, for the investment in the wireless access (Alternative B).

Also the cumulative net cash flow (Figure 3.9) at the end of Year 5 is 238,000, which leads one to assume that the investment in Alternative B is indeed value-generating. However, one should confirm this with NPV.

3.6.3 NPV

Discounted annual cash flows are obtained by multiplying the annual net cash flow by the discount factor, as in Table 3.8.

NPV in Table 3.8 gives the value of the investment at the present time (end of Year 0), the investment here again meaning the purchase and installation of wireless transmission (Alternative B) as compared to using leased lines (Alternative A). The NPV calculation shows that Alternative B is economically desirable, since it adds value to the business by 148,100. The decision maker is in this case likely to approve the investment in wireless backhaul instead of leasing access lines, and the overall business case for the radio network expansion should be recalculated with the improved backhaul solution, since now the radio network business case improves by the very same 148,100 (assuming everything else stays equal).

Example 3.13

Question: In this wireless backhaul comparison case, what would the NPV be if the time horizon were only three years? Why would one want to consider a shorter timeframe?

Answer: Assuming everything else stays the same, one can readily use DCF as calculated in Table 3.8, taking into account the initial investment (–115,000) and then DCF from three consecutive years (1, 2 and 3). The result is 89,150.

Recalculating NPV over a three-year period instead of five reduces risks related to future cash flows, as now it is expected that no cash cost savings occur after Year 3.

In general, calculating NPV with different input data (e.g. differences in expected cash cost savings or expected revenues) encourages the investigation of alternative scenarios, such as an optimistic scenario, a realistic scenario and a pessimistic scenario.

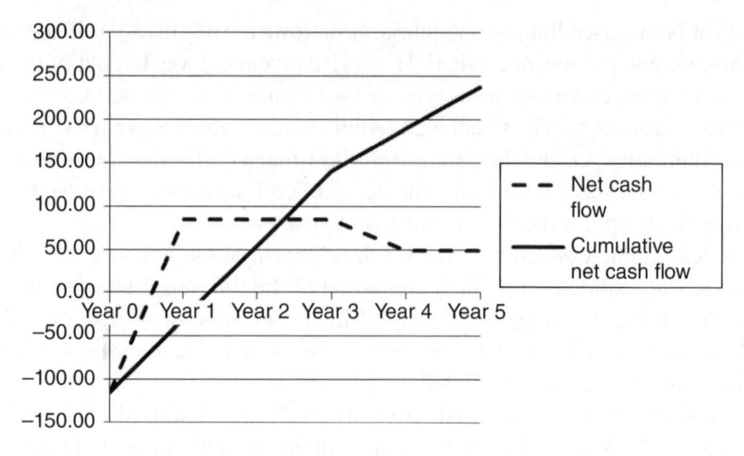

Figure 3.9 Cumulative net cash flow (thousands)

Table 3.8 NPV of Example 3.8 (thousands) for Alternative B (compared against Alternative A)

	Net cash flow	Discount factor (r = 12%)	Discounted cash flow
Year 0	−115	1	−115
Year 1	85	0.892857	75.89
Year 2	85	0.797194	67.76
Year 3	85	0.71178	60.50
Year 4	49	0.635518	31.14
Year 5	49	0.567427	27.80
SUM (NPV) =			148,10

The purpose of this comparison case between Alternatives A and B was to walk through the calculation of the payback period and NPV with given assumptions and observe how input data are fed into equations and how one arrives at the results. As such, there is no intention or even possibility to generally highlight a preference for an alternative, since every network, operator situation and market environment is different. What is essential is that the presented methodology for calculating NPV and the payback period can be generally applied to any LTE backhaul planning and optimization project, when the assumptions are developed according to that specific case.

References

ABI research (2015) Mobile Operator Capex.

Bodie Z., Kane A., Marcus A. (2011) *Investments*. McGraw-Hill/Irwin, New York.

Metsälä E. and Salmelin J. (eds) (2012) *Mobile Backhaul*. John Wiley & Sons, Ltd, Chichester, UK, doi: 10.1002/9781119941019.

Ovum (2012) *A Growth Opportunity for Vendors: Telco Opex* (TE008-001254).
Verbrugge S., Pasqualini S., Westphal F.-J. et al. (2005) Modeling operational expenditures for telecom operators. *Proc. Optical Network Design and Mode*, 455–66.

Further Reading

Drury, C. (1996) *Management and Cost Accounting*. International Thomson Business Press, Stamford, CT.
Horngren, F. (2000) *Cost Accounting*, 10th edn. Prentice Hall, Upper Saddle River, NJ.

4

Dimensioning Aspects and Analytical Models of LTE MBH Networks

Csaba Vulkán[1] and Juha T.T. Salmelin[2]
[1] *Nokia Networks, Budapest, Hungary*
[2] *Nokia Networks, Espoo, Finland*

4.1 Introduction

The role of the mobile backhaul (MBH) is to provide transport connectivity between the eNBs and the core network elements. During the dimensioning process, the optimal MBH architecture, physical and logical topology and transport services are created with the scope to support the services provided by the LTE network and to enable the seamless operation and management of the LTE system. The large variety of services and applications an LTE user should be able to reach—ranging from VoLTE to over-the-top (OTT) applications such as YouTube©, Flickr© and Facebook©—mandates a careful and complex planning process where network costs, optimal configuration and customer experience must be considered at the same time. Seamless network operation means not only the U-plane but also the C-plane can operate efficiently and securely, LTE services can be established with low latency and end-user applications are supported according to their own quality requirements, whereas the operator can continuously monitor and manage the network. Accordingly, the underlying transport network is a multi-purpose system that should be dimensioned to serve all the U-plane, C-plane, M-plane and synchronization plane (S-plane) traffic at the same time (Figure 4.1). While the U-plane traffic represents the dominant share of the load the transport network has to serve, during the network dimensioning the special requirements of the control, management and S-plane traffic should be considered as well. Additionally, the transport network itself has its own control and M-plane traffic that has to be considered when the required transport resources are calculated. Throughout this chapter, the modeling and dimensioning aspects of an MBH serving LTE U-plane traffic are discussed.

LTE Backhaul: Planning and Optimization, First Edition. Edited by Esa Markus Metsälä and Juha T.T. Salmelin.

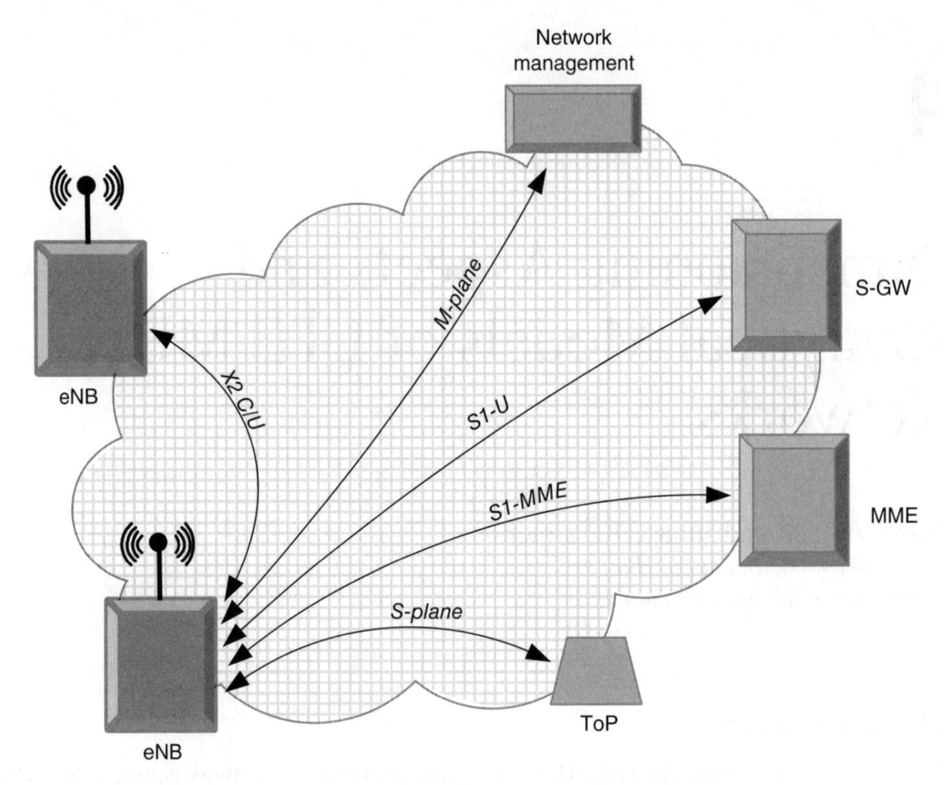

Figure 4.1 The connectivity and interfaces of an LTE eNB

The dimensioning of the MBH is an iterative and complex process, requiring the use of dedicated models capable of capturing the relevant behavioral aspects of the user (mobility, traffic demand, etc.), the traffic (including the applications) and the network. The architectural and technological choices and selections made before and during dimensioning influence the dimensioning model selection as well.

As a starting point of the transport network dimensioning, the traffic endpoints from the MBH perspective—that is the LTE network elements, their number, location and logical relation (i.e. the LTE interfaces)—must be defined. For each eNB, its maximum air interface capacity, its S1 interfaces—that is its S-GW and MME connections—and the X2 interfaces to its neighbors should be specified. These are the result of the radio planning process and represent the e2e connectivity requirement the transport network has to support. Accordingly, this information is the main input of the topology planning. If the automatic neighbor relation (ANR) functionality is part of the LTE deployment, further X2 interfaces might be activated/removed automatically through it. The MBH should be prepared to handle the additional demand for transport resources generated by the ANR. Additionally, for accurate planning, the traffic mix—that is the LTE services and applications—the expected volume and the composition of the OTT traffic has to be given per eNB level as well. This defines the demand the network should be able to serve. Based on these and the LTE QoS architecture that includes the mapping of the services and applications to the LTE QoS classes and the QoS requirements, a rough dimensioning (with simple Erlang formulas and processor sharing models) of the bandwidth requirement can be performed to be used as a starting point for detailed dimensioning. The accuracy of dimensioning largely depends on the accuracy of the predicted/forecasted traffic mix. The unavailability or inaccuracy of the above information reduces the efficiency of dimensioning; however, it is only a necessary but not sufficient enabler of accurate planning. The selection of a modeling/dimensioning approach able to capture the system behavior at the right level of detail is also important.

Nowadays, the OTT services and applications dominantly having Transmission Control Protocol (TCP) as the transport layer protocol represent the overwhelming share of wireless traffic. The Internet technology evolution and the wide use of cloud-based services, social media and content delivery networks enabled users to act not only as content consumers (downloading Web pages, software packages, pictures, audio and video) but also as content producers (sharing self-produced multimedia content, tweets, pictures, personal content, uploading files to cloud storage systems, etc.). This and the nature of the underlying TCP protocol that is operating via a self-clocking mechanism, where the data transmission rate of the source depends on the feedback arriving from the receiver in the form of acknowledgments, imply that the UL and DL bandwidth cannot be dimensioned by completely decoupling the UL and DL traffic. It is important to note that, due to the self-clocking nature of the TCP, there is a direct relation between the network capacity, the QoS architecture and the data rate of the U-plane traffic, that is if one estimates the average rate of the OTT traffic and defines that as the transport capacity, in reality the system will not be able to serve the estimated traffic mix. Accordingly, the traffic models used in dimensioning, discussed in detail later on, indeed influence the outcome.

If the per eNB bandwidth estimation is available, for example as a result of the rough dimensioning method described above, dimensioning can continue by defining the physical and logical topology. In this phase, constraints such as the available sites, infrastructure elements, coexisting traffic (such as 2G, EDGE, 3G, HSPA) or the earlier investments act as influencing factors and can restrict the amount of valid alternatives. During the topology design phase, three additional aspects should be considered: transport security solution (e.g. IPsec), transport reliability and availability requirements and the potential source of multiplexing gain. Transport reliability and the potential resource savings enabled by the multiplexing gain are discussed below in detail in Sections 4.7.2 and 4.5.

Once the initial physical and logical topologies are available, a more detailed bandwidth dimensioning can take place. The scope of this phase is to calculate the required resources (logical and physical) by considering the traffic demand, the transport topology, the traffic routes and the QoS/QoE requirements. During this phase, detailed network and traffic models can be applied. This is the phase when the transport QoS architecture and the mapping of the LTE QoS classes to the transport QoS classes can be considered. As a result of this phase, an accurate bandwidth calculation (including the transport and IPsec protocol overheads) for each transport service will be available.

With this result, the logical and physical topology can be revisited, and an optimization of the overall topology can be executed. Finally, after several iteration steps, the calculated capacity can be converted to the real capacity with the right level of granularity.

This chapter provides an overview of the dimensioning paradigm, QoS and reliability driven approaches, the QoE considerations, multiplexing gain, traffic and network modeling. However, it does not offer an exhaustive survey of all the valid user, traffic and network modeling methods that can be used for LTE MBH dimensioning but instead covers the problems and aspects considered relevant.

4.2 Dimensioning Paradigm

In simple terms, MBH networks should be dimensioned to provide reliable service to the users. Users should be able to connect to the network and use the services for which they have subscribed anytime and everywhere under the coverage area, whereas service outages or connectivity problems should be rare. The services should be available with proper quality to allow good user experience. Accordingly, during the planning process, two distinct sets of requirements must be considered, that is, the dimensioned network should meet both the reliability and the QoS/QoE expectations. The former is described by service availability, resilience, strict maximum downtime and recovery time parameters. The QoS requirements are often defined separately per each QoS class indicator (QCI) with parameters such as delay/latency, delay variation, loss, guaranteed bit rate and minimum throughput. However, good QoS does not necessarily mean good QoE; therefore, during the dimensioning process, the QoE targets should also be considered in addition to the QoS targets.

LTE service availability depends on the MBH itself but also on factors external to it, such as: the radio coverage; the capability of the control layer to support low-latency connection establishment, like packet data protocol (PDP) context and radio access (RA) bearer; the efficiency of the provisioning of the QoS parameters (according to the subscription, subscriber profile); and the capability of the radio network to execute handovers, etc. Accordingly, the corresponding system level requirements should be divided between the radio and the MBH. As a rule of thumb, the transport network should be able to serve the amount of connections/users that have been granted access by the system, and the blocking rate requirements defined for the system should be met. The MBH-specific requirements are related to the transport services. The reliability considerations are on one hand tightly coupled to the backhaul topology (its physical and logical architecture) and on the other to the selected networking technology. The logical transport architecture/topology and partly the physical topology depend on the traffic demand and the services the network has to serve, in other words on the resource and bandwidth need of the traffic served by each eNB. The amount of MBH resources required by an eNB depends on the traffic mix, and the QoS requirements can vary widely between the average traffic and the peak data rate an eNB has to serve, and is different for DL and UL. In the majority of cases, the bandwidth required to serve the UL traffic is less than the bandwidth requirement of the DL traffic. On the other hand, asymmetrical transport connectivity leads to the DL/UL asymmetry problem, where the DL throughput downgrades whenever the capacity limited UL is overloaded (Szilágyi and Vulkán, 2014).

Based on these, dimensioning is not a linear process but rather an iteration, where several steps are required in order to create the right topology and to calculate the total amount of resources. A detailed overview of the process is given in Figure 4.2.

From a planning point of view, the reliability criteria are addressed by selecting the right transport technology and by designing the network topology that has enough redundancy so that the link and node failures have only a minimal impact on the network services and in case of failures the service can be recovered quickly. This, however, increases the costs, as resilience to multiple sequential link failures can be guaranteed only by creating dense network topologies. In order to reduce the costs, a common approach is to provide "1+1 protection" with bandwidth guarantee for the high-priority services that can handle one single failure.

The recovery times depend on the selected transport technology—for example Internet protocol (IP), multiprotocol label switching (MPLS) and Ethernet—and has two major

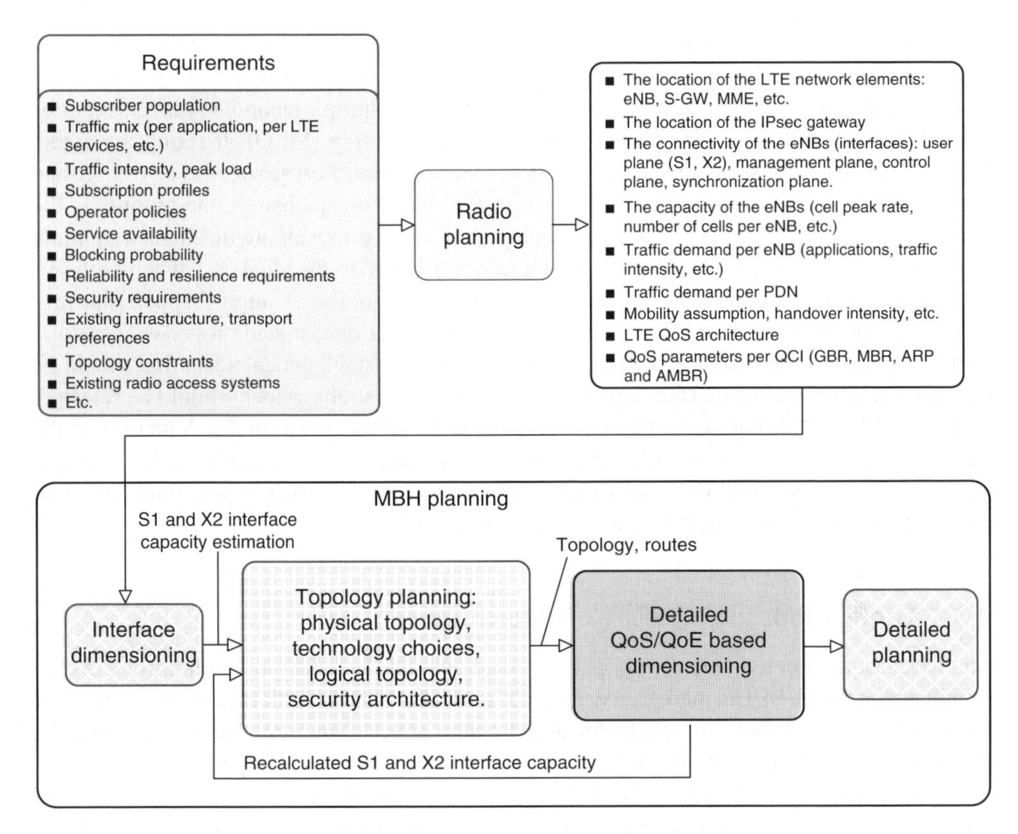

Figure 4.2 The dimensioning process

components: the failure detection time (or latency) and the reaction time. Efficient link failure detection mechanisms such as bidirectional forwarding detection (BFD) can reduce the latency to milliseconds with the penalty of a slightly increased link utilization. In a routed IP network the recovery takes considerable time as it requires that the routing tables in each device converge (provided there is enough redundancy in the first place). This can be improved by proper routing domain design (designing the border of autonomous routing domains) and by applying fast reroute techniques such as IP fast reroute (e.g. not-via) that sustains the IP layer connectivity until the convergence is completed. If the transport connectivity is provided through MPLS-TP (multiprotocol label switching traffic profile), carrier Ethernet, etc., tunnels or pseudowires, the efficiency of the path switch mechanism (from the default path to the protecting path) provides a lower bound to the recovery time. The transport network can be designed with permanent path protection (e.g. by preconfiguring the secondary/protecting tunnels) or by applying dynamic path reroute mechanisms. The protection can be limited only to the priority services or extended to each service class. These considerations have a direct impact on the physical and logical topology. The latter depends also on the type of transport service, that is point-to-point (E-line), point-to-multipoint (E-tree) or multipoint-to-multipoint (E-LAN). Additionally, the favored resource provisioning approach with bandwidth guarantee versus overbooking gives another dimensioning parameter. If the scope is to harvest geographic, time and packet level diversity through statistical multiplexing, the choice of technology layer where the grooming is executed (optical, Ethernet, IP/MPLS) has a direct impact on the resulting network. The optimization of systems with multiple technology layers requires special models and algorithms such as multilayer optimization (MLO). It should be noted, however, that in contrast to the planning process, when the transport services to be established are known, in a real system, the transport services will be established one-by-one as the network is extended when new eNBs are deployed, thus the provisioning mechanism running in the network can only provide suboptimal results compared to the ideal case when each service is provisioned in one step. Due to suboptimal provisioning, it might happen that new services cannot be established even though the network was dimensioned properly. Therefore, during topology dimensioning, different transport service establishment scenarios should be evaluated and the worst case (that requires the highest amount of services) should be selected.

It should be noted that the resource allocations and traffic separation are done only at the logical level, that is they mean only administrative separation, whereas traffic carried in separate tunnels will be eventually multiplexed together on the physical layer, thus QoS will depend on the transport schedulers and on the traffic itself.

4.3 Applications and QoE: Considerations

The LTE is a distributed, completely packet based technology, where the radio protocols are terminated at the eNB. The increased data rate it provides together with the device technology makes it a valid alternative of the fixed-line Internet service. Nowadays, users are expecting that they can access their favorite Internet applications and services at anywhere and anytime, taking advantage of the benefits of mobility. Additionally, they are acting as not only content consumers but also producers as well. This tendency is reflected by the popularity of cloud-based services as well (Dropbox©, etc.), where the uploaded traffic can be significant. Accordingly, the LTE traffic is dominated by OTT traffic not by legacy telecom services such as voice and short message service (SMS). Although important from a revenue point of view, these do not represent significant volume. As the radio protocols are terminated at the eNBs, there are no special requirements or traffic patterns to be considered during the MBH dimensioning process rooted in the operation of the radio protocols such as delay constraints due to the radio link control (RLC) acknowledged mode (AM) timeout timers in case of HSPA or the deterministic transmission time interval (TTI) and fixed maximum rate of the dedicated channels (DCHs).

Through the packet scheduler, the radio interface operation still has impact on the way the U-plane traffic behaves, but this impact is far more limited than was the case for earlier 3GPP technologies. Dimensioning should follow this tendency, in addition to the already discussed reliability and resilience considerations, and aim to calculate the required resources to serve the OTT traffic with an acceptable level of quality. Therefore, proper insight into the anatomy of the traffic generated by the various OTT applications is mandatory. It should be noted that LTE users will find the services provided by the operators satisfactory not if the data rate according to their subscription is available but if their favorite applications and services are accessible in the way (at the location and time) that is most convenient to them. This is a significant challenge as their favorite set of applications might vary significantly, depending on their age, social status, etc. Defining the relevant subscriber profiles (from an application preference point of view) is a key element of successful network design; however, it is beyond the scope of the MBH dimensioning as it should be regarded as an input to it.

There are a wide variety of OTT applications whose popularity ebbs and flows, depending upon the current trend; however, the data volume they represent is overwhelming. According to DOMO (2015), the total Internet population now represents 2.4 billion people. Every minute of every day, YouTube users upload 48 hours of new video; Google receives over 4,000,000 new queries; Facebook users share 2,460,000 pieces of content; Apple users download 48,000 apps; Twitter users post 277,000 times, etc. Obviously, a given LTE network must face a small fraction of this demand; however, even if this is scaled down to its targeted subscriber base, the volume remains significant.

Special QoE requirements can be defined/formulated for each application. Good YouTube or any video experience requires continuous playback without stalling, that is the network should be able to guarantee at least a bandwidth that equals or exceeds the media rate of the video, including the protocol overheads. Higher bandwidth does not improve the user experience; lower bandwidth for a time longer than the depletion time of the playout buffer would cause stalling and experience degradation. A good Web-browsing experience requires that the requested Web page (consisting of dozens of objects, many of them are transported via parallel TCP flows) is downloaded in a timely manner. User satisfaction significantly decreases as the download time increases, thus the download time has direct business impacts as well (Website Optimization, 2015). A third example is the interactive applications like online gaming or chat where not the throughput or the download time but the latency is the key parameter. Good user experience in this case requires low latency. The common characteristic of the OTT applications is, however, that they primarily use HTTP for content delivery, that is the HTTP can be regarded as an integration layer, whereas the dominant transport network layer protocol is TCP. Capturing the relevant behavior of these protocols is an important ingredient of a proper dimensioning process. The next section provides an overview of the relevant aspects of the TCP and a discussion of the anatomy of an HTTP session and YouTube download.

4.3.1 Transmission Control Protocol

Transmission Control Protocol (TCP) is the dominant transport layer protocol within the Internet and, as such, it is the primary protocol used by the OTT applications for data transfer. It is fair to assume that the majority of data traffic an LTE network has to handle is TCP based. Accordingly, the operation of the TCP and its interaction with the system (both with the air interface and the MBH) and with the concurrent flows play an important role in the overall system performance; therefore, it should be considered during the dimensioning process by applying proper models. TCP is an adaptive and dynamic protocol where the data source probes the system for the available resources and adapts its transmission rate based on the feedback from the receiver. This behavior is rooted in the scope of the TCP that is to offer a connection oriented, point-to-point, full duplex and reliable service for data transfer over the Internet or over any IP-based network that is by nature an unreliable media. Data transfer requires connection establishment (three-way handshake) during which the two communicating entities announce the relevant information to each other, including the maximum window size that imposes an upper bound on the data rate and the burst size of the connection. As the scope of the TCP is to provide reliable and efficient data transfer but the source has no information on the available bandwidth (that is dynamically changing in a packet environment anyway), it uses adaptive mechanisms (congestion control) to find the suitable transfer rate. Accordingly, the source maintains a congestion window that is increased or decreased depending on the feedback originating from the receiver. This and the receive window announced by the receiver define the maximum amount of data in transit (without being acknowledged by the receiver), which is the burst size between the communicating entities. The data rate is the ratio of this burst size (referred to as the "window size") and the RTT. After the connection setup, the source starts transferring data with the so-called slow-start mechanism which increases the congestion window usually from a single segment exponentially at each RTT. Eventually, after reaching a considerable rate (defined by the slow-start threshold), the source enters the congestion-avoidance mode where the data transfer is done according to a less aggressive increase mechanism, referred to as "congestion avoidance."

Depending on the TCP version (e.g. New Reno or CUBIC), the congestion window of the TCP is increased by one segment after each RTT or according to a cubic function. This increase of the amount of in-transit packets is continued until the maximum window size (announced during the connection setup) is reached or a loss is detected. Upon data loss, depending on its severity, the TCP source enters the "timeout" or "fast recovery" state. In both cases, besides retransmitting the data considered lost, the window size and thus the transmission rate of the source is reduced, either to zero (timeout) or to a given percentage (70% in case of CUBIC) of the original rate (fast recovery). It is important to note that, in order to preserve the fairness among the concurrent TCP flows, transport devices implement buffer management algorithms such as random early detection (RED) or controlled delay (CoDel) that operate by discarding segments from randomly selected flows prior to full buffer load. This triggers fast recovery at the source affected by the discards and de-synchronization of the TCP sources. The latter increases the multiplexing gain as well. Accordingly, the target of dimensioning is not to prevent such random discards but to calculate the resources in such a way that the likelihood of cases when the sources are not able to regain their throughput after discard but enter in exponential back-off, is minimised.

The available resources per flow can be decreased to the congestion collapse point, where an increased number of connections enter into exponential back-off simultaneously. As shown in Figure 4.3, as the load is increased (the available nominal bandwidth/fair share per connection is decreased) by starting more and more connections over the same bottleneck link, the per connection throughput (measured for those connections that are not in exponential back-off) greatly exceeds its fair share (available nominal bandwidth per connection). It is also visible that an increased buffer size can improve the average throughput.

After the congestion collapse point, the ratio of the throughput experienced by the connections that can avoid entering into timeout and the fair share shows exponential increase. This is due to the fact that these connections are able to take advantage of the bandwidth left unused by the connections being in timeout and exponential back-off (Figure 4.4).

The ratio of connections in exponential back-off (timeout) is shown in Figure 4.5. The congestion collapse appears when the ratio starts to increase exponentially. The scope of the dimensioning should be to find the bandwidth that is enough to serve the maximum traffic load (input for dimensioning) with the target level of service (throughput in this case).

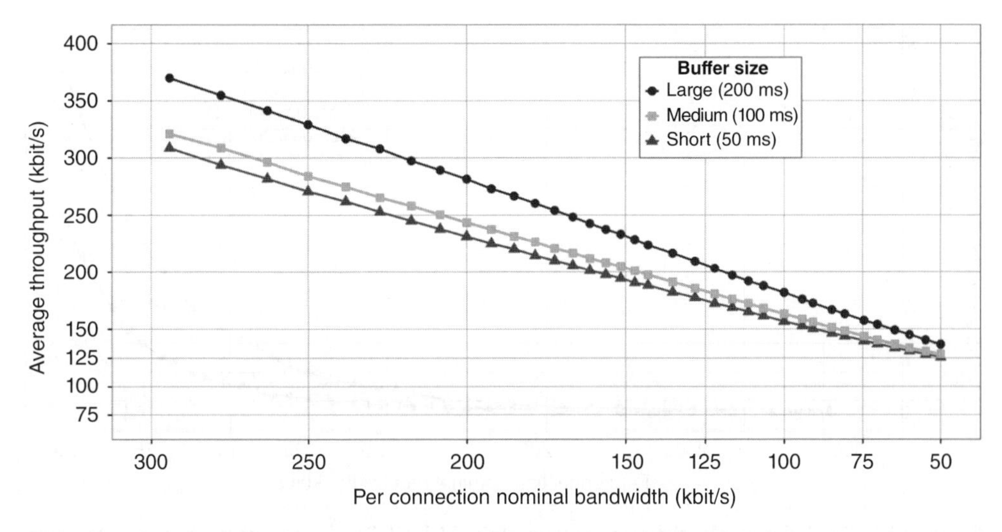

Figure 4.3 Simulation results showing the measured throughput of TCP CUBIC connections not in timeout over a 10 Mbps transport link serving a first in, first out (FIFO) buffer with maximum delay thresholds set to 50, 100 and 200 ms. The buffer was managed by RED with minimum threshold set to 50% of the maximum delay and maximum drop probability set to 10%. The number of simultaneous TCP connections was increased from 30 to 200. The available per connection nominal bandwidth is the link capacity divided by the number of TCP connections

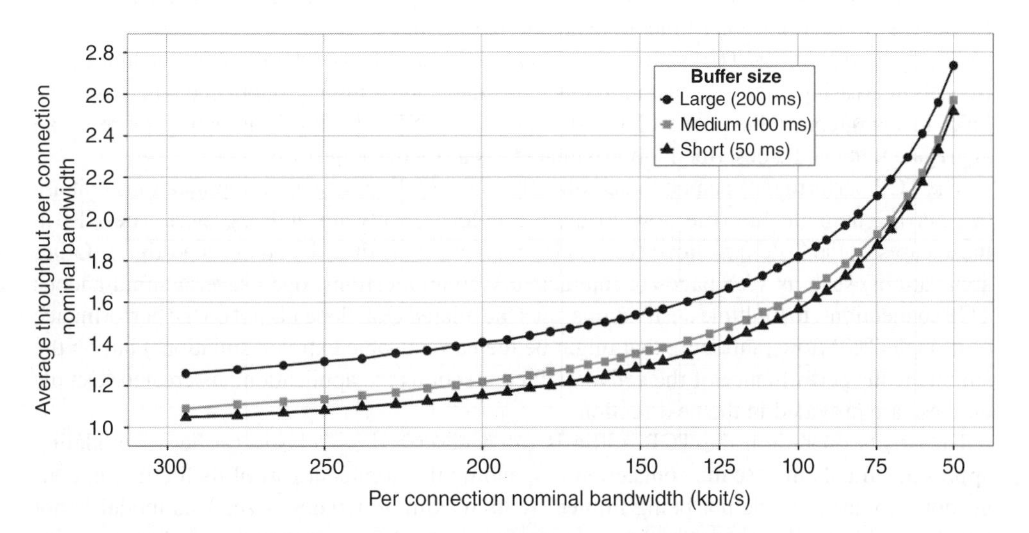

Figure 4.4 Simulation results showing the ratio of the measured throughput and the available (nominal) bandwidth of TCP CUBIC connections not in timeout. The simulation setup was the same as the one shown in Figure 4.3. As the load is increased by starting more and more connections, an increasing number of connections enter in timeout, thus the available bandwidth is taken by those connections that are not in timeout

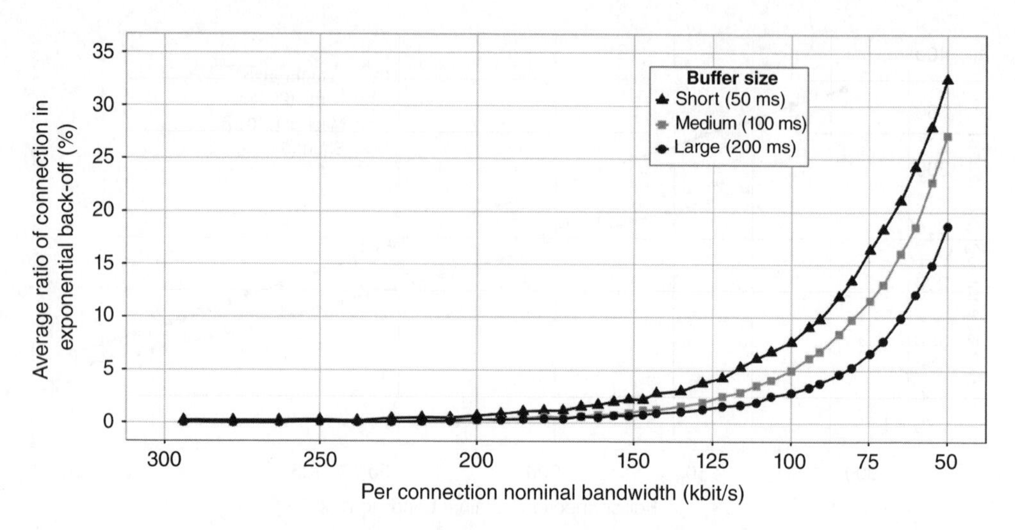

Figure 4.5 Simulation results showing the ratio of the TCP CUBIC in exponential back-off. The simulation setup is the same as the one in Figure 4.3. As the load is increased by starting more and more connections, increased numbers of connections enter timeout

Additionally, the bandwidth should be enough to serve the maximum amount of simultaneous TCP connections without congestion collapse.

It should be noted, however, that due to buffer management mechanisms such as RED or CoDel the connections enter timeout in an alternating manner.

The dynamism of the TCP sources is becoming more and more dominant as the amount of flows is decreasing, that is, at last mile links, this behavior is more pronounced whereas at the links that aggregate the traffic of tens-hundreds of eNBs the traffic is acting more as an aggregate that can be described (approximated) as a stochastic process.

The TCP performance can be linked directly to the QoE in the case of applications that are transferring a single file (file downloads, file uploads, software updates, etc.). The KPI in these cases is the "file download time," also referred to as the "document load time." Other applications, such as Web pages or interactive Web applications, open several simultaneous TCP connections, to multiple destinations, thus the related QoE depends not on the performance of a single TCP flow (although, that might be the case in some extreme situations) but on the simultaneous performance of the active TCP connections the application has created. Details on these are provided in the next section.

Due to its operation, the TCP traffic is often described as "elastic traffic," a modeling approach suitable in case the connection is spending the dominant part of its lifetime in congestion avoidance mode not being limited by its maximum window size. This model is not suitable when small amounts of data are transferred or when the connection is limited by its maximum window size. In these cases the TCP connections are not acting as elastic traffic. The ideal elastic traffic source resembles a TCP connection with infinite window size being able to adapt its rate without any delay to the available resources. Further details on the corresponding models are given in the subsequent sections.

4.3.2 Web Browsing

For simplicity reasons, throughout this section the term "Web browsing" refers to the activity of the user to download a given Web page excluding any complex activity of engaging in interactive dialogues with the server side of the applications. Such scenarios would require that users interact with the content, which itself is an important factor of the end user's expectations and experience. This can vary greatly depending on the application itself and require special models that should incorporate both the user's behavior and the architecture of the application as well. Therefore, these are beyond general planning and dimensioning methods, and are not discussed here. One exception is made, namely YouTube, which is a dominant Web application, representing a significant share of the total amount of content transferred through the Internet, and due to its simpler architecture (when compared with complex multimedia and social networking Web applications) it is easier to model.

The KPI of the Web browsing is the (Web) page (down)load time, that is the time spent from sending the initial request to until each resource element (e.g. text, image, scripts, style sheets) of the requested Web page is loaded. Obviously, the page load time is influenced by other factors than the available network resources, such as the operating system, the browser version of the receiving device, the Web server responsiveness and Internet latency. In this section, "Web page download time" refers to the time required to transfer the requested content through the MBH, and all the factors external to the network itself are excluded.

Web pages are downloaded over the HTTP, which is an application protocol used by almost every OTT application. According to the HTTP Archive (2014), as of mid-June 2014, the average Web page size is 1800 Kb, composed of 96 objects fetched through 37 TCP connections. The anatomy of an HTTP object download (e.g. the main page of a Web server) is shown in Figure 4.6.

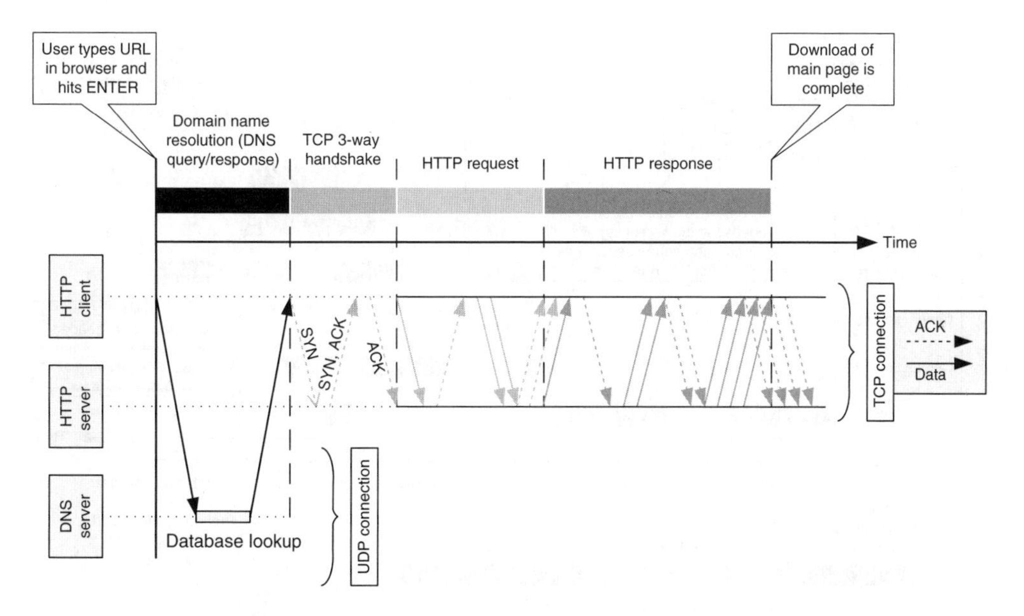

Figure 4.6 The anatomy of an HTTP object download

As the figure shows, the download consists of several stages: domain name resolution, TCP connection setup, HTTP request and finally HTTP response. As discussed, a Web page consists of several objects downloaded through simultaneous TCP connections (Figure 4.7).

The high number of TCP connections opened to download a single Web page that itself is relatively limited in size (1800 Kb) causes that the TCP connections are spending the majority of their lifetime in slow-start and some are even terminated before reaching high data rates.

According to KISSmetrics (2015), mobile Internet users expect that there is no significant difference between the Web-browsing experience over mobile phones and desktops. User patience regarding the Web page download time is limited, according to the same source; some of the users tend to abandon the page if it is still not completely loaded within 10 seconds.

A plausible approach to providing a rough estimate of the required link capacity to download a Web page is to use the download time as a dimensioning requirement (target download time) and calculate the capacity as a function of the estimated average Web page size and the target download time. Figure 4.8 gives an example of the calculated capacity in case of a Web page (amazon.com) with size of 1450 Kb, consisting of 97 objects in function of the target download time. The result of the calculation was evaluated with simulations: the download time of the same Web page was measured over an LTE network where the transport capacity was set to the calculated value. The simulated Web browser (Opera Mobile 11) was using six TCP connections (with the selective acknowledgment [SACK] option enabled) to download the Web page. The HTTP 1.1 pipelining option was enabled that allows the source to transfer distinct objects one by one through the same TCP connection. The content server was Linux, with a default TCP CUBIC implementation. The topology consisted of a single eNB connected via a direct transport link to the S-GW. The link capacity was set to the calculated values using the target Web page download times (5–60s). As shown in Figure 4.8, this simple calculation gives acceptable results in case the target download time is set above 10 seconds.

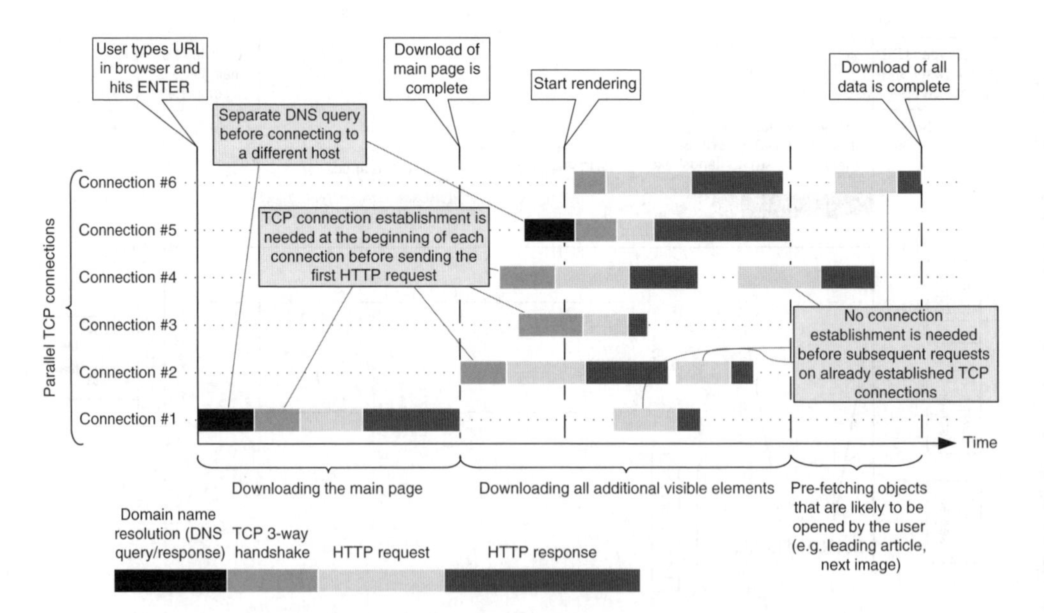

Figure 4.7 The anatomy of a Web page download

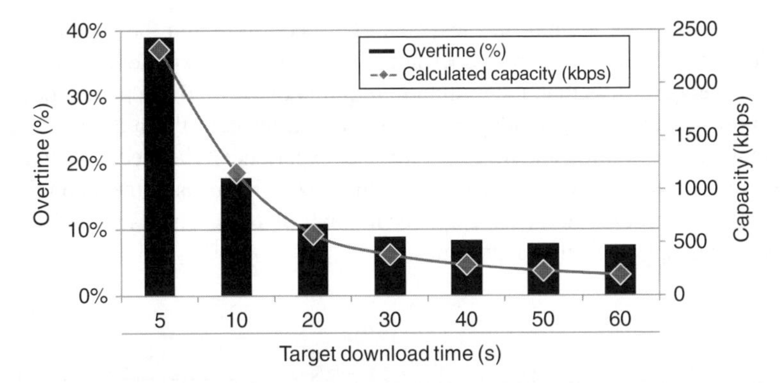

Figure 4.8 Simulation evaluation of the simple capacity calculation. The calculated capacity is shown in the function of the target download time (the quality requirement). The results show that at reasonable target download time values the overtime of the Web page download versus the target download time is less than 20%

In these cases the real download times are less than 20% above the target download times (quantified as "overtime" on the figure). The overtime is due to fact that the calculation is not considering the slow-start phase (when the sources are not able to send with full speed). At high link capacities (calculated under strict download time requirements) the domain name system (DNS) lookup and the slow-start phase take significant time, whereas due to the low amount of data that is transferred over six TCP connections the download rate is not reaching the available link capacity, that is the download rate is limited by the RTT (Figure 4.9a). This illustrates that the system latency is a relevant component of the overall Web page download time especially if the expectation for the latter is relatively strict. When the calculated link capacity is lower as a result of a more relaxed download time requirement, the TCP connections spend the major part of their lifetime in congestion avoidance mode, being able to utilize the available capacity more efficiently; therefore, the overtime compared to the total download time is not significant (Figure 4.9b).

The results indicate that elastic traffic models (discussed below) are not suitable for TCP modeling when downloading only small amounts of data per connection as, owing to the assumption of ideal elasticity, they cannot capture the slow-start phase of the TCP connections.

4.3.3 Video Download

YouTube is a prominent OTT application for video content delivery. According to DOMO (2015), users upload 48 hours of new video every minute of every day. This content is geographically distributed among many content delivery nodes around the globe in order to enable quick access to it regardless of the location of the users. Users are using either Web browsers or dedicated applications to access the content that is delivered via TCP connections in several formats such as MPEG4 (Moving Picture Experts Group), Flash, HTML (Hypertext Markup Language), etc. either in the form of equal-sized data chunks or as a monolithic object. A typical data chunk size is 1.7 Mb. Proper user experience requires that the content is available within a short time and the playback never stalls. As each video is encoded with a given media rate, the minimum quality requirement is that throughout the whole session the system is able to guarantee a throughput not less than the media rate and the protocol overheads. Providing higher amount of resources does not increase the user experience, whereas a bandwidth that on average is less than the minimum required one will give a poor user experience. Transient shortages in the bandwidth, provided that on average the required resources are available, might cause stalling and thus degrade user experience. The media rate can vary over a large interval, as shown in Figure 4.10, that summarizes the measurements on the YouTube media rate distribution executed during the second half of 2012. The itag classifies the video format, resolution and other media attributes.

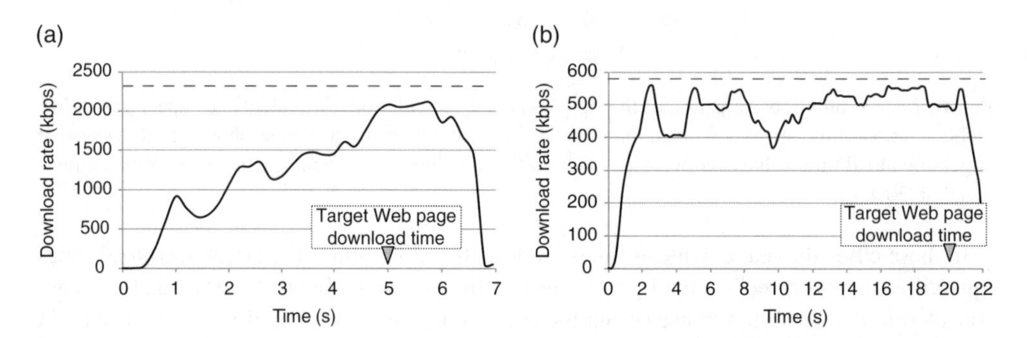

Figure 4.9 Web page download rate as a function of time. The same Web page was downloaded over an LTE system with a point-to-point transport link connecting the S-GW and the eNB. The link capacity was calculated with the simple capacity calculation based on the target Web page download time set to 5 s (a) and 20 s (b)

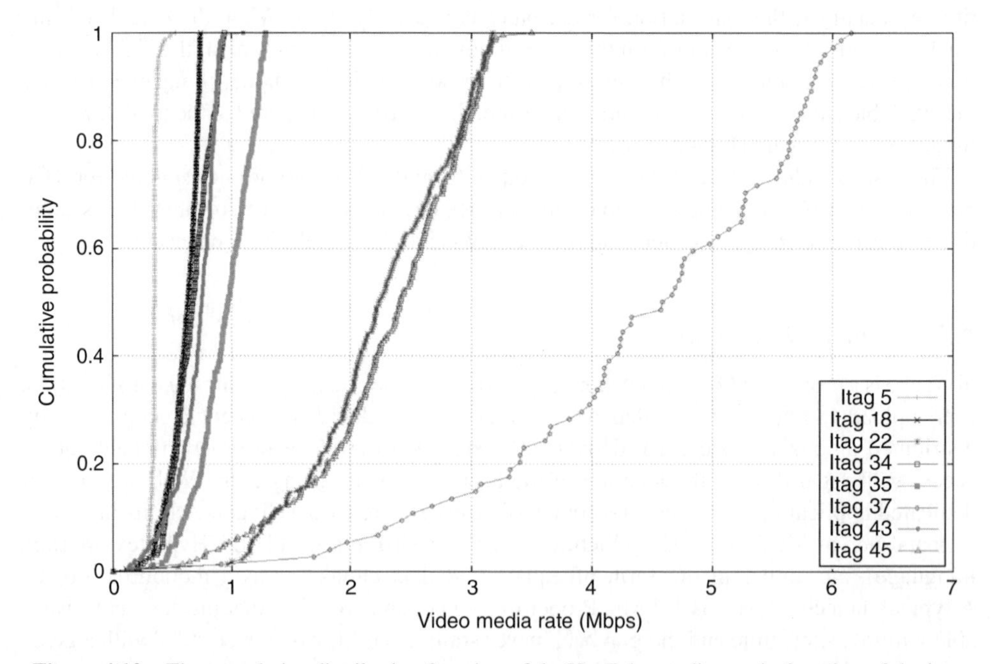

Figure 4.10 The cumulative distribution function of the YouTube media rate in function of the itag

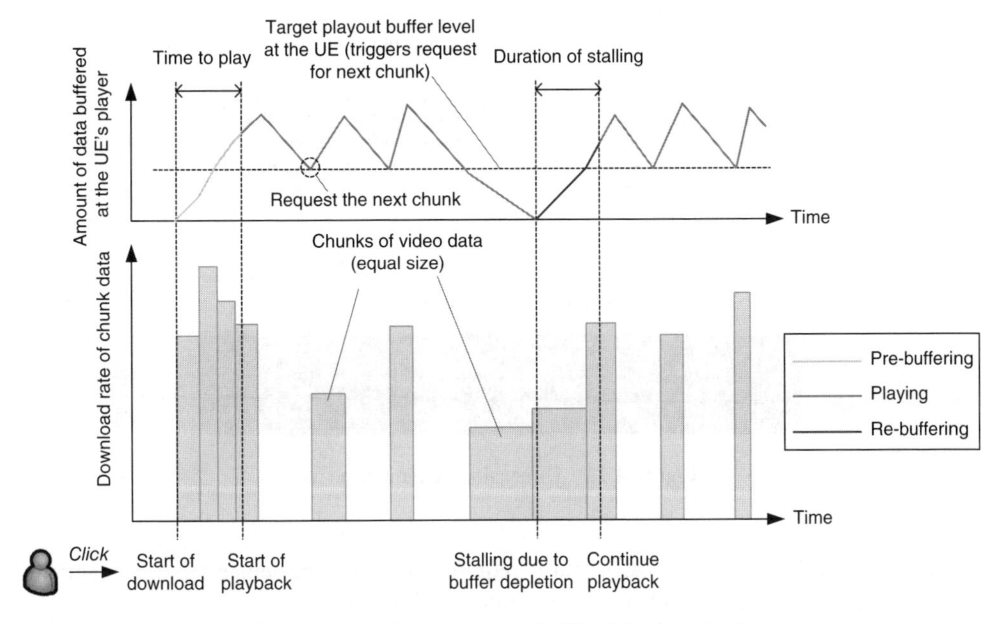

Figure 4.11 The anatomy of a YouTube download

The anatomy of a YouTube session is shown in Figure 4.11. It is visible that there is an idle period between the download of two consecutive data chunks whenever the download rate exceeds the media rate. If the download rate falls below the media rate, the chunks are overlapping in time and the UE's playout buffer may eventually deplete. Media delivery will resume after a re-buffering time during which enough data are downloaded.

As the YouTube traffic represents a significant share, volume wise, in the overall traffic any LTE network has to serve, its specific requirements should be considered during dimensioning. A simple but efficient model of these sessions consists of a user level model that captures the overall behavior of a YouTube consumer (number of downloaded videos per session, idle period between two consecutive downloads) and parameters that describe the content itself (the distribution of the media rate, the distribution of the duration and the chunk size). The chunk size is needed only in cases of detailed modeling (as described later on). As auxiliary parameter the user interest in downloading the whole content can be given with a probability. This is important, as a significant number of the videos that are started are never actually finished. Including this parameter would prevent over-dimensioning. The quality requirement for dimensioning is to provide at least a bandwidth that equals the media rate plus the overheads.

4.4 Dimensioning Requirements

The scope of dimensioning is to achieve the highest level of reliability over the optimal network topology by using the lowest possible amount of resources. The amount of resources depends on the level of protection and on the amount of traffic to be served with the predefined level of QoS/QoE. Thus the traffic mix (what is the demand the network has to serve) and the

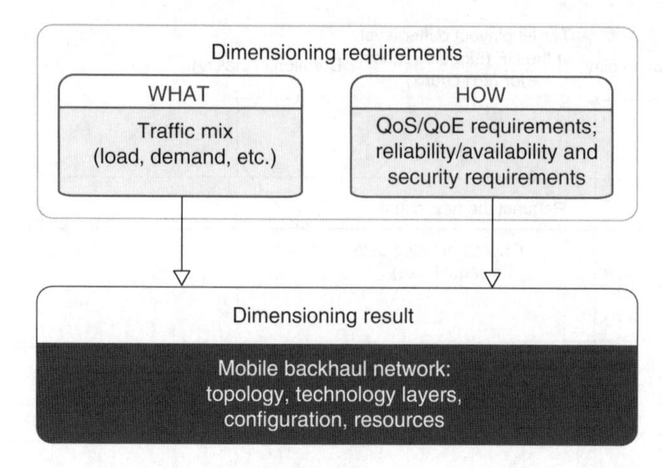

Figure 4.12 Dimensioning requirements

QoS/QoE and resilience/availability/security requirements (how the demand should be served) are the primary inputs of the network dimensioning (Figure 4.12).

The level of detail and the accuracy of the traffic mix provided as input to the dimensioning algorithm define the result. Accordingly, if the traffic the network has to serve is known/predicted/forecasted in detail, meaning the components (applications and services) of the traffic mix, the volume of the demand those are generating and the time-wise distribution of this demand matches the reality, the accuracy of dimensioning will mostly depend only on the selected dimensioning methods. Additionally, the result of dimensioning should be robust enough to be able to serve a wide interval of different traffic mixes within the extent of the physical limits. As is discussed later in this chapter, dimensioning against the requirements is performed mostly with analytical algorithms (that have the advantage of potentially being scalable as opposed to simulation-based mechanisms). During dimensioning, an equivalent model of the system is created that conforms to the basic modeling assumptions applied by the analytical algorithms. This equivalent model should be selected so that it is capable of capturing the relevant behavior of the traffic demand and that of the system. In this section, the models are divided into traffic and network models and discussed separately for simplicity's sake, though it has to be noted there is a strong interdependency between them, that is the traffic models must be selected in order to match the network model input formats. The interpretation of the QoS/QoE requirements and their integration into the dimensioning method are discussed as well.

4.5 Traffic Models

The analytical performance (quality) analysis of LTE MBH has to deal with two distinct modeling problems that both require special modeling approach due to the dual nature of the traffic mix served by packet based radio access networks. The traffic mix consists of the demand generated by real-time "streaming applications" using UDP as the transport layer protocol, such as VoLTE, and by the demand generated by TCP-based data applications that can be modeled as "elastic sources" (that adapt their rates to the available bandwidth; each elastic source receives a fair share of this bandwidth).

The quality of the real-time streaming applications are best characterized by delay, delay jitter and packet discard/loss rates, whereas the main performance measures of the TCP-based traffic is the throughput, latency and download times. The two require dedicated and distinct modeling approaches, which makes it problematic to develop a general model capable of capturing the characteristics of both the elastic and the streaming sources at the same level of detail and to combine them in a single framework. Accordingly, a possible classification of the traffic types from a modeling point of view can be given as follows:

- *Persistent streaming*: connections that send data with a fixed rate during their lifetime therefore can be modeled with a constant rate r. The traffic source can be characterized by a deterministic packet inter-arrival time, e.g. transmission time interval (*TTI*) and packet/frame size (ω). The system should be able to offer a bandwidth that equals to their rate, $r = \omega / TTI$.
- *On–off modulated streaming* (for applications like VoLTE): These connections send data with a fixed rate (r_{On}) in the On state (during the talkspurt, for example) and are inactive in the Off state (i.e. during the silent period). Alternatively, in order to model the silent frames sent during the Off state, a second rate (r_{Off}) can also be included to the traffic description. The On and Off states alternate stochastically with predefined average durations, τ_{On} and τ_{Off} respectively. The size of the frames/packets and the inter-arrival time during the On and Off states can be given with separate parameters.
- *Exhaustive elastic traffic* (such as download of large files): The users generating exhaustive elastic traffic are considered to continuously download or upload during the time of the analysis. This type of traffic is a persistent traffic, with a rate adapting to the available bandwidth, i.e. the instantaneous download rate changes continuously with the varying network conditions. The mean value of the download rate is the main parameter that describes the level of quality, and the mean QoS parameter in this case is the minimum download rate.
- *Non-exhaustive (On–Off) elastic traffic* (like Web browsing where the On period lasts until the requested Web page is downloaded): Similarly to the On–Off modulated streaming, active and inactive periods alternate stochastically. In the active period a given amount of data is downloaded with a rate that adapts to the available bandwidth (e.g. Web page download). This is followed by an inactive (reading) period. A complication is that Web pages consist of a multitude of small objects, whose significant part is downloaded during the slow-start phase of multiple TCP connections when the elastic nature of the traffic is not yet dominant. A possible set of parameters is the request size (φ) and the length of the Off period (τ_{Off}). The length of the On period (download time) depends on the throughput achieved by the end user and it is the main QoS/QoE parameter.

This classification allows the distinct modeling of individual traffic flows and assumes already detailed information on the composition of the traffic mix. In reality, this detailed information is not always available, especially not in the case of dimensioning greenfield networks. Such information might, however, be available when an existing network is extended or re-planned/optimized.

An alternative classification follows the level of aggregation of the traffic the network has to serve (discussed below). At each aggregation level, different models can be applied that are able to capture those characteristics of the traffic which are the source of diversity and can be used when multiplexing gains are calculated. The level of detail of the traffic models (which include the user behavior models as well) defines the capability to consider the multiplexing gain. The LTE MBH multiplexes the user traffic on packet level; however, modeling abstractions can be made that elevate the multiplexing to application, connection, etc. levels. The multiplexing provides gain when certain circumstances, such as load, buffering space, are favorable.

A possible definition of the multiplexing gain can be given as follows: the multiplexing gain is the amount of transport link bandwidth that can be saved by considering the diversity of the traffic demand, i.e. the variation of the traffic load in time and the available buffering space. Multiplexing gain can be achieved only if the traffic demand is high enough (both in terms of volume and number of connections), that is there is statistical diversity; if QoS/QoE requirements allow packets to be stored in the multiplexing buffers without being served until the packet burst is over and, finally, if there is enough buffering space available, then transient overload can be handled without significantly increasing the packet discard rate. If the buffering space is limited and (or) there are strict QoS/QoE requirements, extra transport link bandwidth that can handle the fluctuation of the traffic load without storing the incoming packets in the buffers will be necessary. In general, the capacity gain, which can be achieved on a transport link, is the result of both the aggregation of traffic sources and the buffering. These two components are in an intricate relationship and cannot be clearly separated.

Traffic variation, which has deterministic and stochastic components as well, occurs on several distinct timescales, and the different components of the multiplexing gain can be associated with the different timescales. According to this, the multiplexing gain is the result of the following:

- Mobility and the daily routine of the mobile subscribers, which induces a deterministic variation of the traffic on a long timescale (typically a day). Due to geographic diversity, this traffic variation depends on the geographic location.
- Intensity and the length of the sessions (a stochastic component) referred to as "session level user behavior."
- Stochastic user behavior during the sessions referred to as "burst level user behavior."
- Packet level diversity caused by the behavior of the traffic sources and the communication protocols used in different technical solutions adopted within the system.

The estimation of the achievable multiplexing gain at different levels (as above) requires different levels of traffic modeling.

Based on these considerations, a valid classification of the traffic models/traffic description is as follows:

- Peak load or BH load: the amount of traffic the system, a given set of nodes or an eNB has to serve. This is a rather high-level traffic definition where no detail except the sheer load in Kbps is given.
- Geographic diversity and daily load profile/distribution: this modeling allows the consideration of the multiplexing gain based on the observation that distinct eNBs are not reaching their peak load at the same time of the day. The simplest traffic description of this kind is to divide the day into equal periods and define a separate load value for each period.
- Session level models are used to capture the session level dynamics of the system, i.e. the session arrival process and the session holding time. The detailed behavior of the sources

during a session is not necessarily modeled. At least, the arrival process and session holding time during peak times should be given per traffic type in addition to the traffic type specific parameters. If these parameters are given for the whole day according to the load profile of the eNB, both the geographic diversity and the session level behavior can be considered during dimensioning. The holding time in this context defines the total time an application is active on an established connection (bearer). The holding time of the OTT applications accessed by a particular user depends on the system performance and the available bandwidth; however, in this context the statistical parameters derived from measurements can be used.

- Burst level modeling captures the way of using the application: a burst is the On period of the On–Off modulated streaming or elastic traffic generated by the user.
- Packet level models are the most detailed models that describe the packet level behavior of the connection during the On or Off periods. In the case of streaming traffic, this behavior depends only on the application, whereas in the case of elastic traffic this depends on the actual networking conditions and on the application, content server, etc.
- The elastic traffic model that captures the TCP performance on session level is an alternative modeling approach which should be considered. Often, the elastic models include both the traffic and the system behavior as well.

The traffic model classes described above are not exclusive but complementary. A complete model can be created so that it allows the capturing the traffic behavior at each level of detail.

Figure 4.13 presents the relationship between the above models and the dimensioning algorithm. For the peak load or peak time load based modeling the traffic of an eNB can be given with rough parameters referring to the aggregate traffic (i.e. no individual sessions are

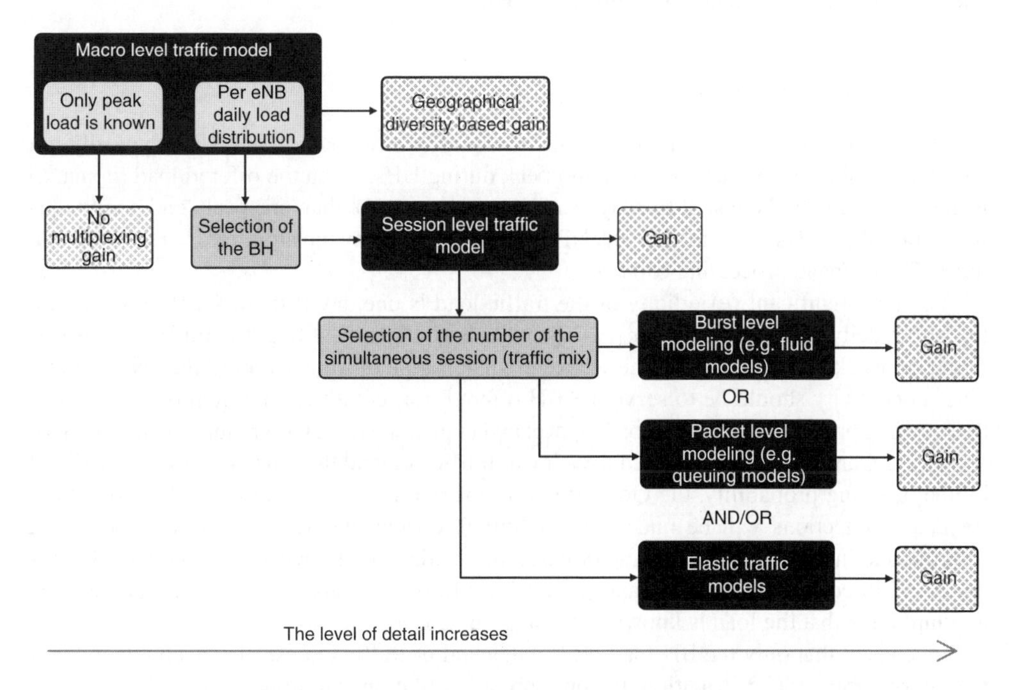

Figure 4.13 Relationship between the modeling levels

considered); this is a macro level description of the traffic. If no additional information is available on the traffic, no multiplexing gain can be assumed during dimensioning. If the daily variation of the load is known per eNB level, the geographic diversity can be exploited and multiplexing gain can be modeled. The busy hour (BH), on which a more accurate dimensioning can be based, can be identified by evaluating the daily variation of the traffic load. The traffic mix during the BH can be described with session level models that give the number of simultaneous sessions to be served by the network. A more detailed model that captures the burst and packet level behavior of the traffic which, on a macro level, reproduces the BH traffic load of an eNB or of an aggregating link can be created by including one of the following modeling approaches. Traffic parameters that capture the burst or packet level behavior of the user sessions allow detailed, QoS/QoE requirement-based dimensioning. Note that, in general, in order to select the proper BH (which, by definition, should give the maximal capacity requirement for the links of the corresponding aggregating level), the whole dimensioning process should be performed for several BH candidates. The multiplexing gain can be considered at each modeling level.

It is important to note that the achievable multiplexing gain, meaning the bandwidth saving on the aggregating links which can be considered during the dimensioning process, may strongly depend on the specific traffic mix that the dimensioning is based on.

If the prediction of the expected traffic is not fully reliable, then the resulting link capacities should be robust enough for the possible changes in the traffic mix. On the other hand, dimensioning should provide a robust system that is not specially designed for one and only one traffic mix but can deal with various traffic mixes. This in turn means that in order to achieve this robustness the resulting multiplexing gains can decrease, since the possible gain based on one specific traffic mix cannot be fully exploited.

4.5.1 Peak Load or Busy Hour Load

Due to the daily routine and mobility of the users, the traffic load of an eNB is typically not constant in time but reaches its maximum/peak during BHs, when the offered load (demand) is much higher than the rest of the day. An eNB can have more than one peak load period during a normal day. The peak load of an eNB is upper bounded by the air interface resources and the eNBs baseband processing capacity.

The most significant periodicity of the traffic load is one day (Figure 4.14); however, on top of the daily periodicity, longer periods may also be observed (e.g. the traffic can be different on weekdays than during the weekend). The scope of dimensioning the eNB's transport connectivity should be to serve the BH offered traffic with certain quality (i.e. to keep the blocking probability at an acceptable level and to guarantee the QoS requirements are met). If the eNB transport is dimensioned for a lower traffic demand than the one resulting defined by the blocking probability, the QoS offered to the admitted connections or the throughput for data connections will be unacceptably limited during the BH. The dimensioning and traffic modeling aspects of the network based on the detailed knowledge of the peak load is discussed in context of the burst and packet level traffic models. In the current context the assumption is that the load is known at the aggregate level.

In the event that only the BH load (traffic demand or traffic intensity) is available without the information on its distribution, the only possible solution is to dimension the aggregating

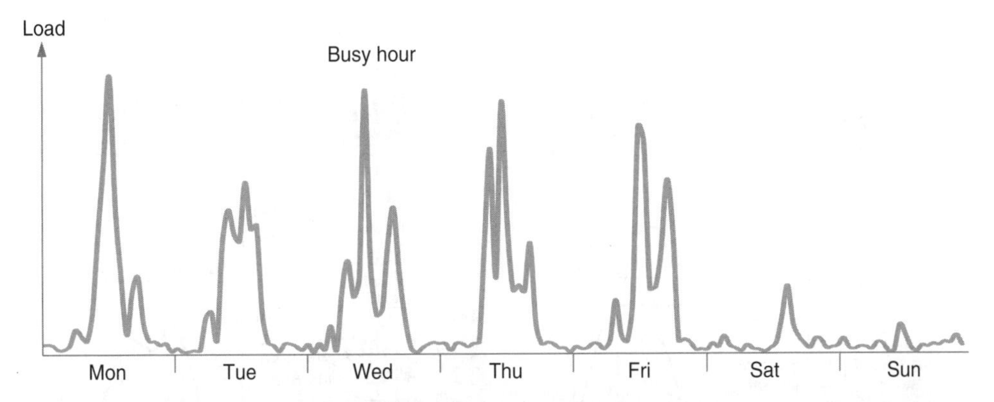

Figure 4.14 Example of the daily load profile

links by summing up the peak load of the eNBs whose traffic is aggregated, with the implication that the potential multiplexing gain is not considered. In this case the multiplexing gain can be introduced only via corrective coefficients, such as overbooking ratios.

4.5.2 Geographic Diversity and Daily Load Profile/Distribution

The daily distribution of the offered load (demand) of eNBs that are located in the proximity of each other can be very similar (two suburban eNBs serving the users of a given neighborhood will most likely have BH at the same time). As a result, the aggregated traffic intensity of these eNBs would show a similar daily distribution as the traffic intensity of the individual eNBs. However, eNBs located in different areas (e.g. urban and suburban) may have significantly different daily traffic distribution with BHs at distinct times of the day (e.g. during weekdays people are commuting from suburbs to downtown offices). Therefore, if the demand originated from two or more significantly different areas (e.g. urban and suburban) is aggregated by the underlying transport infrastructure, the resulting traffic load on the higher-order links will have a different daily distribution and the peak load on the aggregating links will not reach the sum of the BH traffic that the corresponding eNBs have to serve (Figure 4.15).

In a simple but valid dimensioning approach, the last mile links within an MBH network are assumed to be dimensioned for the peak load of the eNB (or for the nominal capacity of the eNB considering the available physical resources and baseband processing capacity) they are connected to, considering the transport overhead (including the IPsec overhead) as well. This approach ensures that the last mile links are not blocking the traffic of a given eNB and the blocking is only possible on the aggregation links. If the traffic of two eNBs from the same area (with the same BHs) is aggregated in a router or switch, then a multiplexing gain is not possible as the traffic on these eNBs will culminate at the same time. However, if, for example, at the eNBs located in suburban and urban areas the load reaches its peak at a different time of the day, the peak load on the link that aggregates the traffic of these eNBs will not reach the sum of the peak loads measured at the individual eNBs. In case this is considered during the dimensioning process, the multiplexing gain can be assumed and the overall network cost can be reduced.

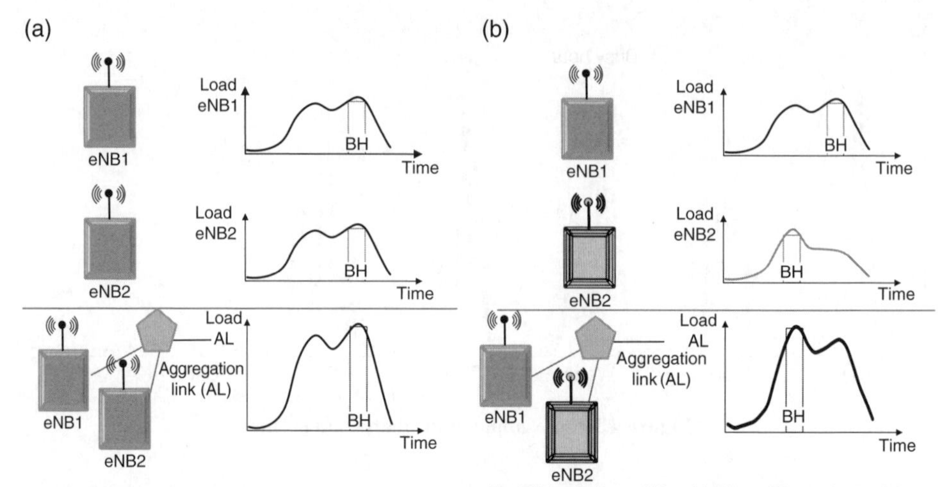
(a)
Load eNB1
BH
Time
eNB1
Load eNB2
BH
Time
eNB2
Load AL
Aggregation link (AL)
BH
Time
eNB1
eNB2
When two eNBs have BH at the same time of the day, the resulting load on the aggregation link is the sum of the BH loads measured at the two eNBs. No multiplexing gain is possible.
(b)
Load eNB1
BH
Time
eNB1
Load eNB2
BH
Time
eNB2
Load AL
Aggregation link (AL)
BH
Time
eNB1
eNB2
The BH load of two eNBs with BH at different times of the day will not sum up on the aggregation links: the BH load on the aggregation link will be less than the sum of the individual BH loads. This is the basis of the multiplexing gain.

Figure 4.15 The source of multiplexing gain due to geographic diversity

This is a simplified approach, where the links are dimensioned according to the offered load and different types of applications and services are not distinguished. The main drawback of the method is that burst level and packet level multiplexing gain and QoS/QoE requirements are not taken into account. As the traffic served by a given eNB is not homogeneous but is generated by various real time and non-real-time applications and by services that have different QoS/QoE requirements, it is no easy task to identify the period of the day the BH dimensioning should be based on. Dimensioning should identify the time period during which the traffic mix requires the highest bandwidth for the BH in order to meet the QoS/QoE requirements (Figure 4.16).

During the dimensioning process, the links are dimensioned one by one starting from the last mile links to the core network. For each link the peak load including the differences due to geographic diversity is considered when the required bandwidth is calculated. Accordingly, the network is decomposed/clustered into sub-networks, based on the time of the BH that has to be identified in some way. The transport links of the eNBs that have the BH at the same time are dimensioned separately up to the first link that aggregates traffic from separate sub-networks. The links which aggregate the traffic of several areas are dimensioned in a separate step as in general their BH (which should be identified as well) can be different from the BHs of the areas whose traffic they are aggregating (e.g. urban and suburban).

This is a valid approach also in those cases when in addition to the daily load distribution more detailed information on the traffic mix/distribution is available. For a more advanced dimensioning (along these lines), detailed information is needed about the traffic demand, which should be given separately for each application and service type. This might be possible with an existing network, which is re-planned/optimized based on the available demand/load measurements.

It is important to note that the OTT traffic is usually elastic in nature (except applications such as Skype) and it adapts to the available resources. Accordingly, the measured load is

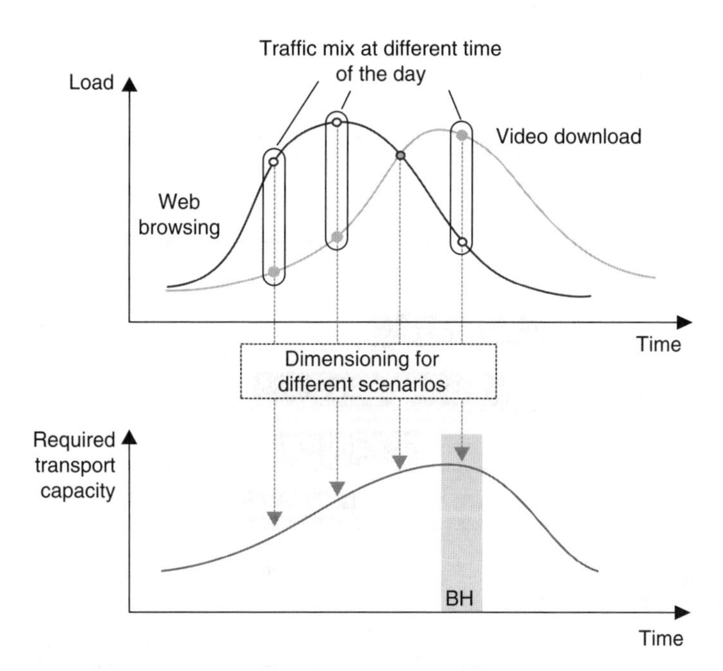

Figure 4.16 Identifying the busy hour: the busy hour is the time of the day when the highest amount of transport resources are needed. Due to the complexity of the traffic mix, the identification of the busy hour is difficult as the required amount of transport resources largely depends on the traffic mix

a result of the interaction of the traffic sources and the network. Therefore in this context "demand" is defined as the amount of data requested by the users, and this is used as the basis for dimensioning. The very same demand over the same network with a higher amount of resources would result in a different load. Having the user demand available per eNB, there are still factors that must be considered.

4.5.3 Session Level User Behavior

The traffic mix during the BH (or a potential BH) of an eNB, a sub-network (network domain) or the whole network consists of individual users' sessions. The number of simultaneous sessions is not constant but changes stochastically in time as users initiate and terminate services, as shown in Figure 4.17. Note that "session" in this context can refer to a particular application session a user is running through an already established bearer or to the bearer itself. In the former case, the model captures the way a user launches various applications, that is the way the user uses the connectivity provided by the LTE system. In the second case, the model captures the dynamics of the PDP activations and deactivations. Combined, hierarchical modeling is also possible where both the PDP activation and the application usage pattern are captured by one model. Additionally, there are cases when the PDP context and the application session coincide, for example with a native LTE service such as VoLTE. Note that in either case there is a bearer establishment process before the user can access the application or can use the native LTE service itself.

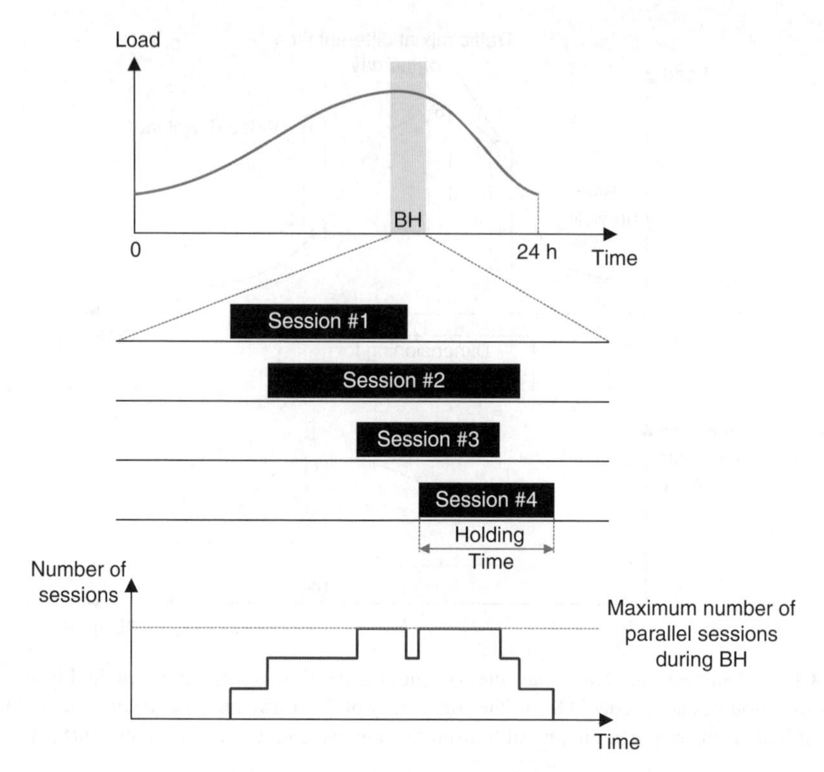

Figure 4.17 Busy hour load and session level models. The session level models capture the dynamics of the session arrival process and allow the calculation of the number of simultaneous sessions during the busy hour

In general, one bearer is used to serve multiple applications of the same user; however, there are cases when launching an application is the trigger for establishing a secondary bearer for the same user.

The dynamics of the session arrival and termination process should be considered during the dimensioning of both the user and the C-plane. When the U-plane is dimensioned, the session arrival and termination process allows the calculation of the maximum number of simultaneous sessions arriving at the system that together with the maximum allowed blocking rate or with the bearer establishment failure rate defines the maximum number of parallel sessions the system has to serve. Each PDP context establishment and termination triggers a signaling procedure, and thus the arrival and termination process allows the calculation of the signaling load. Note that during the lifetime of a PDP context additional signaling can take place (e.g. bearer QoS demotion/promotion) that increases the signaling load, and therefore a detailed C-plane dimensioning should consider these events as well as the handover intensity, the number of location updates, etc., in addition to the session arrival and termination process.

Traditionally, the traffic generated by the users is given by the traffic intensity (Erlang in case of voice traffic) or by raw data rates/load per application type (data traffic). The latter is more frequent when dimensioning is based on load measurements. In case the traffic is given in data rates/load, it is not easy to find the number of simultaneous connections that is required in order to model the individual connections. Modeling the individual connections is needed for proper QoS/QoE based dimensioning. Note that, as discussed, the demand is a more appropriate traffic parameter than the measured load.

As mentioned, the models describing the session level user behavior and those describing the geographic diversity are correlated and can be combined in a common model that captures both aspects. Accordingly, on the macroscopic level (see previous section), the traffic is described as an aggregate: often the population size (number of subscribers served by a given eNB) is given as a fixed input parameter and the offered traffic load is given in Erlang/subscriber for voice or Kbps/subscriber for data connections. In contrast, on session level the offered traffic consists of individual sessions originated by the users. A session can be, for example, a voice call, a Web browsing session, a software update or publishing some personal content. An adequate session level traffic model reproduces the average aggregated BH traffic load but, being a stochastic model, allows for some fluctuation around these values. For a correct dimensioning it is crucial to correctly estimate the maximum number of simultaneous sessions (for each application) allowing a certain blocking probability (for those services to which it is applicable). Dimensioning should be based on a traffic mix, in terms of simultaneous sessions, that represents correctly the offered BH traffic for each eNB.

At the timescale of the user sessions the traffic variation is caused by the session initiation and termination, which can be modeled as a stochastic process. A simple way of modeling this stochastic behavior at eNB is to assume that sessions start independently at a given rate r, that is in an infinitesimal time interval of length dt the probability that a new session starts is $r \times dt$, which means that the number of subscribers is not upper bounded (infinite population). In addition, it can be assumed that the duration of the sessions (referred to as "holding time") are independent with an average value of T. Note that for elastic traffic (TCP based OTT traffic, for example) the session holding time may depend on the available bandwidth. For a bulk download (software update, file download) there is a simple relationship as the download time is inversely proportional to the available throughput: *HoldingTime* = *FileSize*/*Throughput*. However, for a Web browsing session, the relationship is more complicated due to the periods between the download of two distinct Web pages. It is important to note that this model does not set an upper limit for the number of simultaneous sessions. However, in reality an upper bound is given by a finite population or by the finite capacity of the air interface. The parameters of the connection level model (r and T in this simple case) are to be derived from the macroscopic level traffic parameters with the help of some auxiliary parameters if needed (e.g. BH call attempts, target throughput for data connections, average file size).

Assuming a session requires a constant effective bandwidth, which is additive, and that blocked call attempts are lost (i.e. these attempts are not repeated or queued), dimensioning could be based on the well-known Erlang B formula. In a more general case, where different session types require different bandwidth, the Kaufman–Roberts recursion can be used. Note, however, that while these methods are adequate in a circuit switched network with resource reservation they do not take into account buffering and lower-level multiplexing gains. Therefore, a correct dimensioning of a packet-based network requires other methods.

A possible way for dimensioning is to ignore the detailed dynamics of session initiation and termination and to assume a sufficiently large but constant number of simultaneous sessions. Dimensioning can then be based on this snapshot, where the number of sessions is fixed to their "peak values" for each session type. The task is then to estimate the maximal number of

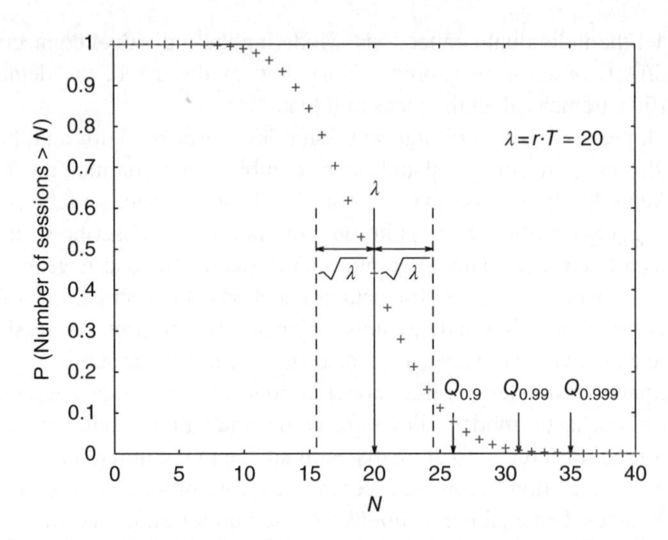

Figure 4.18 The typical tail distribution of the number of simultaneous sessions. In this example, the average number of simultaneous sessions is 20 and the quantiles Q_p are shown for $p = 0.9$, $p = 0.99$, and $p = 0.999$

sessions for each session type (with certain blocking in case blocking is relevant), which is an input of the method.

Assuming independent sessions (which is a reasonable approximation but not always correct) with no blocking, it can be shown that the number of simultaneous sessions follows a Poisson distribution with mean $\lambda_t = r \times T$ (as shown in Figure 4.18). Although this naive approach does not give a strict upper limit for the number of sessions, an upper bound (referred to as "p-quantile Q_p") can be given with any probability $p < 1$, meaning that the number of simultaneous sessions is below Q_p with probability p. (In reality, even assuming infinite transport resources, the number of simultaneous sessions is strictly limited by the maximal air interface throughput or the population size, which the above approach does not take into account.)

The number of sessions, which dimensioning should be based on, can be estimated by Q_p with sufficiently large p. For high values of λ (average number of simultaneous sessions in BHs) the quantiles Q_p can be approximated (normal approximation) as Equation 4.1:

$$Q_{0.9} \approx \lambda + 1.2\sqrt{\lambda}, Q_{0.99} \approx \lambda + 2.3\sqrt{\lambda}, Q_{0.999} \approx \lambda + 3.1\sqrt{\lambda}. \tag{4.1}$$

Note that this approach does not take into account the limited throughput of the air interface. Realistic demands have to be defined in conformance with the maximal air interface throughput.

The multiplexing gain, which can be exploited on the macroscopic level, is manifested in the fact shown in Equation 4.2:

$$Q_p(\lambda_1) + Q_p(\lambda_2) > Q_p(\lambda_1 + \lambda_2). \tag{4.2}$$

Example 4.1

As an illustration of this approach, assume two eNBs (eNB_1 and eNB_2) with average number of simultaneous sessions $\lambda_1 = 20$ and $\lambda_2 = 10$. Assuming a Poisson arrival process of identical and independent sessions, the maximum number of sessions the eNB_1 and eNB_2 have to serve with probability of 0.99 is $N_1 = 30$ and $N_2 = 17$ (calculated by using Equation 4.1). The last mile links serving these eNBs should be dimensioned by considering the maximum number of simultaneous connections they have to serve. On the other hand, an aggregation link serving the two eNBs could be dimensioned by considering the multiplexing gain. Accordingly, the average number of connections on the aggregation link is $\lambda_3 = \lambda_1 + \lambda_2 = 20 + 10 = 30$, whereas the maximum number of connections with probability of 0.99 calculated with (4.1) is $N_3 = 43 < N_1 + N_2 = 30 + 17 = 47$, thus gain can be included in dimensioning. This is a reasonable approach, and as the probability that both eNBs will simultaneously reach their peak traffic rate is small, according to Equation 4.1, the maximum number of connections on the aggregation link is 47 with a probability of 0.999.

4.5.4 Burst Level User Behavior

In general, the rate at which the traffic source of a session generates data packets is not constant in time but forms bursts on a lower timescale. In the simplest and most typical case a user session consists of On and Off periods (see Figure 4.19). For example, in the case of a voice call, the frames carrying encoded voice samples are sent during the On period. The length of the On period equals the time the user is talking, whereas the Off period is the silence period between two consecutive talkspurts. In the case of an OTT application, the On period can be defined as the time when there is active data transfer between the communication entities through one or more data (TCP and/or UDP) connections, like the download of a Web page or an upload of a photo taken by a mobile phone. The download of a given file can be, however, modeled as a session having a single On period. The burst level models are capturing the user behavior/activity as well. The mean talk/silent time of the voice codecs largely depend on the user, whereas the download time (On period) of the content depends on the size and structure of the requested Web page (assuming Web browsing connection) and on the context (network conditions, Web browser behavior, radio channel quality of the user, simultaneous connections, the QoS settings, etc.). In a similar way, the Off period depends on the user's own activity, reading time, etc. Accordingly, if the traffic is modeled at this level of detail, additional empirical parameters or assumptions have to be introduced.

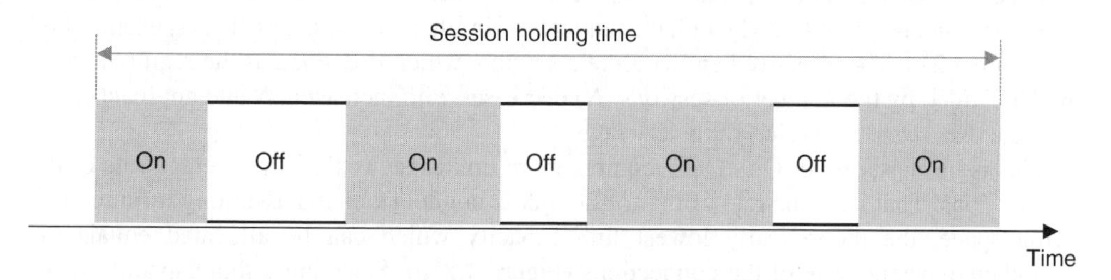

Figure 4.19 Microscopic level behavior of the On–Off traffic source

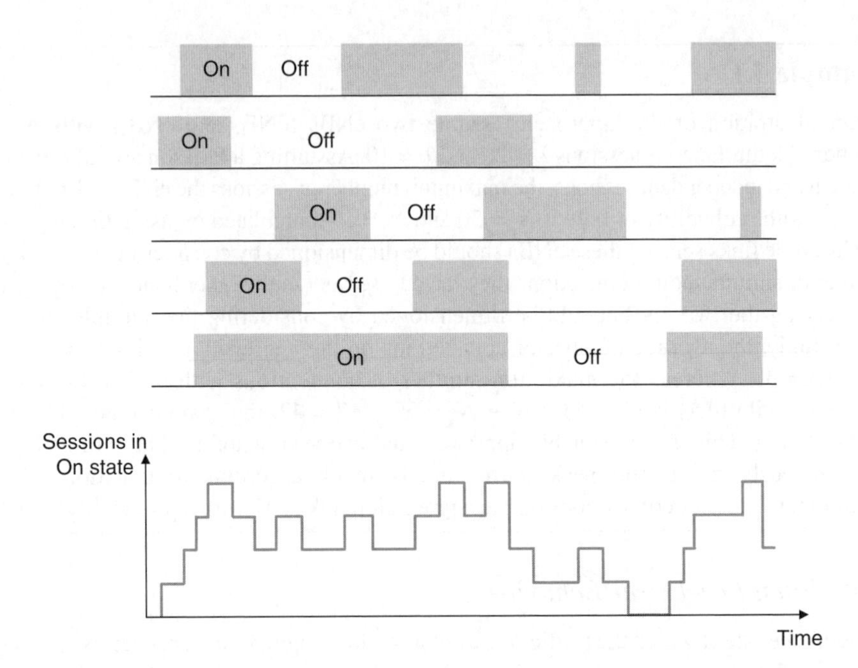

Figure 4.20 The number of sessions in the 'On' state changes dynamically

There is a difference between the behavior of streaming and OTT traffic during the On period: voice codecs are generating frames with constant rate (equal sized frames with fixed inter-arrival time), whereas the packet rate of the OTT applications over TCP is dynamically changing according to the actual network conditions.

If the number of On–Off sessions is large enough, the probability of all connections being in the On state is negligible, which leads to a multiplexing gain on the burst level (Figure 4.20).

Using a similar argument as in the previous section, the number of simultaneous sessions in the On period can be obtained as a binomial distribution. Assuming a fixed number of sessions and by calculating the given quantile, the number of simultaneous sessions in the On periods can be estimated and the On–Off dynamics of the sources can be ignored. The difference in this case is that On–Off dynamics of the sources happens on a much shorter timescale. Furthermore, and most importantly, on this timescale the buffering cannot be neglected anymore; therefore, the explicit modeling of the On–Off behavior is needed in dimensioning.

In general, owing to the achievable burst level multiplexing gain, the required bandwidth of a link does not increase linearly with the number of simultaneous sessions. This is exemplified in Figure 4.21a. The effective bandwidth of a session, which is defined as the required bandwidth divided by the number of sessions (N) decreases with increasing N and approaches the average data rate of a session for a very large N.

This is because proper QoS/QoE requires more link capacity than the average rate of the connections. That is, if the rate of the traffic mix is not constant and assuming infinite buffering space, the theoretically lowest link capacity which can be allocated equals the cumulative average rate of the connections (Figure 4.21b). Selecting a link capacity below this value would lead to an unstable system and the queue length would have a growing

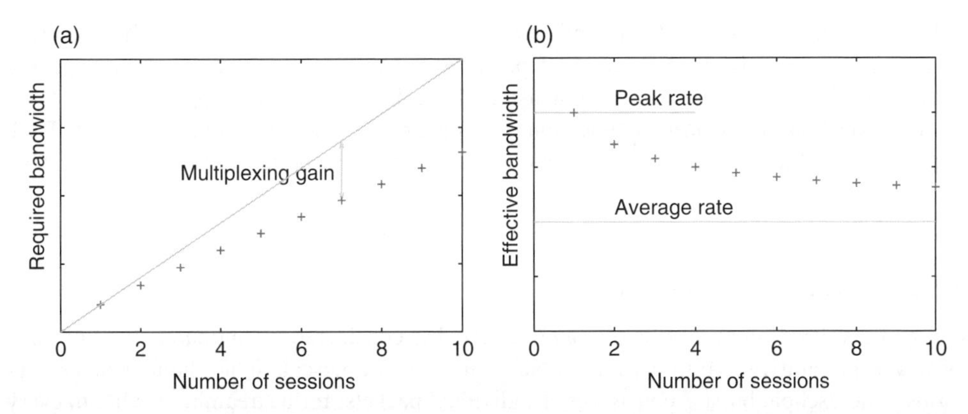

Figure 4.21 Typical dependence of the required and effective bandwidth on the number of simultaneous sessions

tendency. In reality the buffering space is not infinite, thus—depending on the dynamics of the aggregated traffic—buffers will be eventually overloaded resulting in discards and poor performance. The available buffer size gives an upper bound on the delay; therefore, stricter delay/latency requirements mandate shorter buffer sizes. To avoid high packet discard rates, the link capacity or the effective bandwidth has to be increased. The stricter the delay requirement, the higher the minimum link capacity that guarantees proper quality. This is the reason why the dimensioned link capacity should always be kept above the average rate of the traffic. As the number of multiplexed connections is increased, the required link capacity approaches the average rate.

The fact that the required bandwidth is not proportional to the number of sessions indicates, on the one hand, that the Erlang B formula and the Kaufman–Roberts iteration cannot be used, as mentioned in the previous section. The achievable multiplexing gain could be taken into account by introducing a constant activity factor; however, this approach also assumes a constant effective bandwidth (which is less than the peak rate of the connection). The maximum multiplexing gain can be achieved by setting the transport link capacity to the average rate of the aggregated traffic and setting the size of the transport buffers so that they never overflow. The problem with this approach is that it allows extremely high delay.

The estimation of the burst level multiplexing gain in packet switched networks is a highly complex task requiring a suitable modeling of the dynamics of traffic sources, the scheduling architecture and buffering. Fluid models and queuing models are suitable for modeling the burst level multiplexing (as described later).

A reasonable stochastic model for the dynamics of On and Off periods within a session is that the duration of each period is independent and follows an exponential distribution with different mean for On and Off periods (τ_{On} and τ_{Off}). For elastic traffic it can be assumed that the downloadable file sizes are independent and exponentially distributed with mean λ. The average length of the On period can then be obtained as $\tau_{On} = \lambda / Throughput$. On this level of fluid modeling (discussed later), the traffic within an On period is considered a continuous flow of data with a given rate r_{on}, and the schedulers follow an ideal fluid scheduling policy. By modeling the scheduling architecture explicitly, on top of the throughput, it is possible to capture the delay and the loss (fluid model with finite buffers) as well, which is also an important measure of quality.

In the case of a TCP-based OTT application (e.g. Web browsing) the length of the On period (e.g. Web page download time) largely depends on the capacity of the network that is being dimensioned. Therefore an iterative dimensioning mechanism has to be applied.

The throughput of an elastic connection is defined by the bottleneck link on its path. Each connection can face one, and only one, bottleneck link (narrow point) during its journey from source to destination.

4.5.5 Packet Level Behavior

On the burst level, the traffic within an On period is considered a continuous flow of data. However, in reality, the traffic within a burst on a lower timescale is not homogeneous, but forms "microscopic bursts" consisting of individual packets: for a streaming traffic, in every TTI one packet arrives, whereas for TCP connections one window size full of data (usually several segments) arrives in a short burst and the period is the RTT, as shown in Figure 4.22 (slow-start and other TCP specific issues are ignored).

Modeling the packet and burst level behavior together is, mathematically speaking, a difficult task, because the burst level behavior is typically stochastic, whereas the packet arrival pattern within a burst is deterministic (streaming sources) or can be approximated as being deterministic (TCP sources). Queuing models are typically well suited for packet level modeling, though they have limited capabilities with regards to the modeling of elastic traffic.

Suppose the time packets belonging to a streaming source (e.g. voice call) are allowed to spend at a network node that is less than their TTI. In this case the required link capacity, which guarantees that no packet will spend more time at the node than the allowed one, is bounded by the delay requirement and the aggregated packet arrival process rather than the rate of the connections (Figure 4.23).

Example 4.2

Assume a traffic mix consisting of identical and independent sources that send a 72-byte long packet every 20 ms (such as a narrow-band AMR codec). If the delay requirement (maximum allowed time the packet can spend in a router) is set to 5 ms, the minimum link is 115.2 Kbps. This link capacity is sufficient to serve three more sources given they are sending packets in a phased way, that is the first source is sending its packets at $t_1 = \{t_0; t_0 + 20$ ms; $t_0 + 40$ ms; etc.}, the second source at $t_2 = \{t_0 + 5$ ms; $t_0 + 25$ ms; $t_0 + 45$ ms; etc.;}, the third source at $t_3 = \{t_0 + 10$ ms; $t_0 + 30$ ms; $t_0 + 50$ ms, etc.} and finally the fourth source at $t_4 = \{t_0 + 15$ ms; $t_0 + 35$ ms; $t_0 + 55$ ms, etc.}. This is the minimum link capacity that can serve four simultaneous connections according to the delay requirement. The maximum link (assuming the same 5 ms delay requirement) capacity can be calculated by assuming that the four sources are sending one 72-byte packet exactly at the same time. In this case the link should be able to transfer all four packets within 5 ms and the link capacity is 4 × 115.2 Kbps = 460.8 Kbps. It should be noted, however, that as the sources are independent the likelihood of this is extremely low. In fact, any arrival pattern that depends on the relative phase of the four sources has the same probability. The required link capacity that guarantees the delay requirement in a particular arrival pattern takes a value of between 115.2 and 460.8 Kbps.

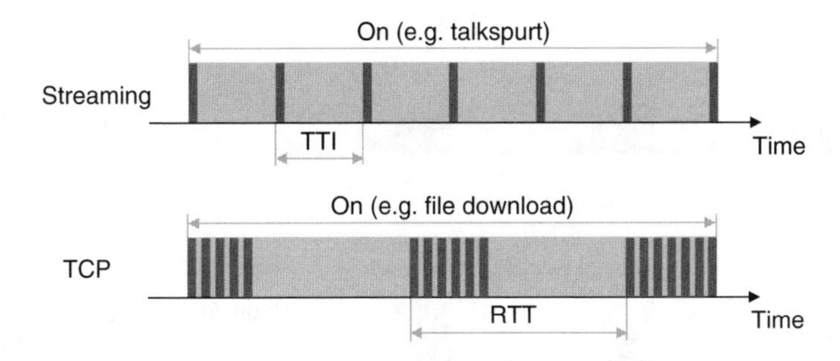

Figure 4.22 Packet level behavior within an On period

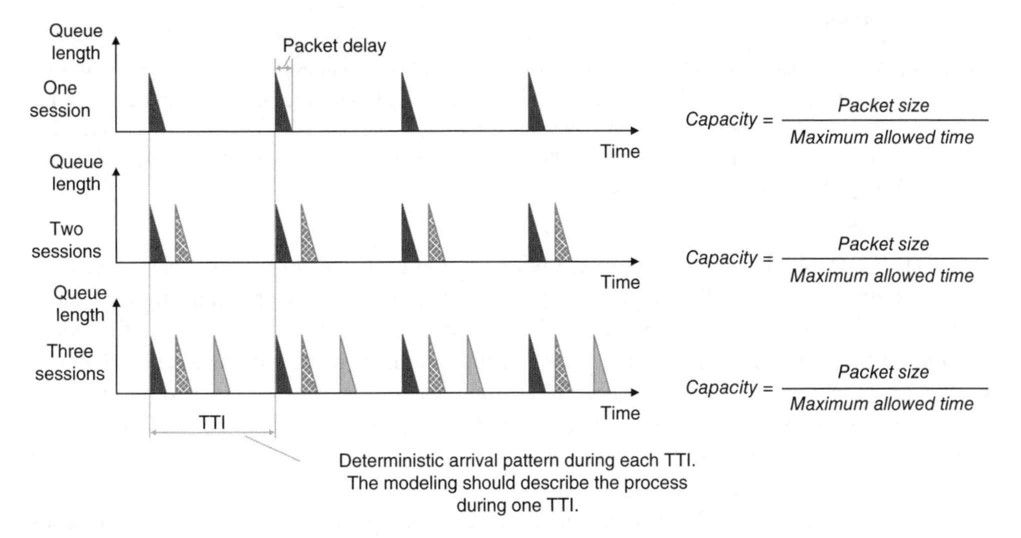

Figure 4.23 Link capacity and the arrival process: if identical streaming sources with deterministic TTI are aggregated, the required link capacity will be a function of the aggregated traffic process. In this example, the packets are always served before a new packet arrives; therefore, the required link capacity is not increasing with the number of sessions

As a packet arrives, the buffer load (queue length) jumps to a value corresponding to the packet size. Due to a finite outgoing link capacity (which is assumed to be the bottleneck resource), the queue length decreases linearly with the service rate (link capacity). Figure 4.23 shows three cases with one, two and three simultaneous identical streaming sessions (from top to bottom). If each packet can be served before a new one (from another source) arrives, the required link capacity depends only on the packet size and the maximum time the packet is allowed to spend at the network node. This can happen if the inter-arrival time of packets from two distinct sessions is larger than the maximum allowed time the packet is allowed to spend at the node.

Obviously, the aggregated arrival process of deterministic streaming sources is a stochastic process where the stochastic component is defined by the session arrivals that are independent. In the worst-case scenario, sessions are originated simultaneously, resulting in a batch arrival. In the best-case scenario, the sessions are arriving evenly distributed over

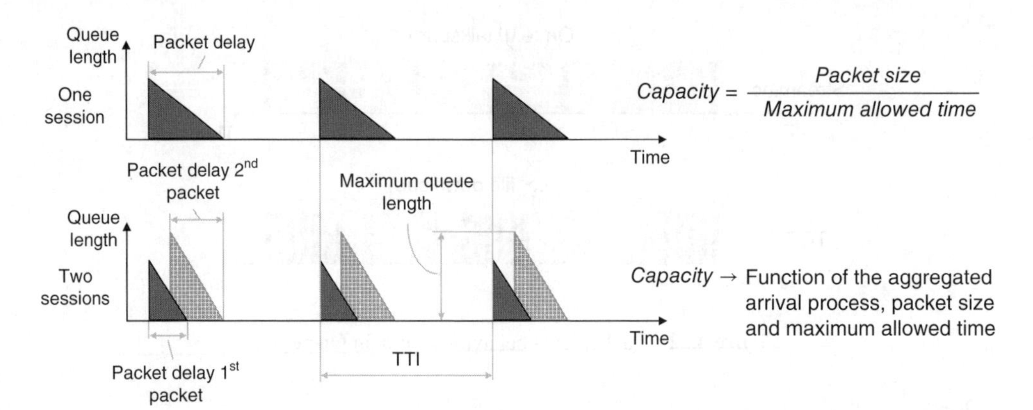

Figure 4.24 Link capacity and the arrival process: if identical streaming sources with deterministic TTI are aggregated, the required link capacity will be a function of the aggregated traffic process. In this example, the packets are not served before a new packet arrives; therefore, the link capacity is increasing with the number of sessions. Dimensioning should guarantee that the delay of the second, third, etc. packet is below the maximum allowed time

one TTI. Figure 4.24 shows an example where the aggregated arrival process is between these two extreme cases. The required capacity is the one that ensures that the maximum queue is served during the maximum allowed time any packet can spend at the node. Accordingly the link capacity will increase with the number of connections and it depends on the arrival pattern of packets, for which a stochastic description can be given (a reasonable model describing this behavior is the N*D/D/1 queuing model). In general, the resulting required capacity does not depend linearly on the number of connections, which can be interpreted as a packet level multiplexing gain.

In contrast, if the link is dimensioned so that the amount of traffic is served without losses and no strict delay requirement is given, the resulting link capacity necessarily equals the offered traffic load (and it is the minimum link capacity that can guarantee system stability). The link in this case is fully utilized and no statistical multiplexing is possible.

This scenario on the packet level is illustrated in Figure 4.25. It can be seen that the required link capacity is strictly proportional to the number of connections and the link capacity is always fully utilized (i.e. there is no unutilized time slot with a zero buffer level).

In a more realistic case, both streaming and elastic traffic are carried by a link. The strict delay requirement of the streaming traffic can be provided with higher priority queues and the rest of the bandwidth can be utilized by the elastic traffic. Similarly to the case presented in Figure 4.25, if the link capacities are fully utilized (either by streaming or elastic traffic) then there is no multiplexing gain.

The discussion so far assumed an infinite buffer. In real systems the buffers are finite. Even a small variation in the traffic load on a fully utilized link leads to a very large fluctuation in the queue length (occupied buffer space), which implies that a finite buffer would often go to saturation, which would lead to packet drops. Therefore strictly speaking, full utilization is not possible for finite buffer space.

In reality, the packet delay and the multiplexing gain are determined by both the burst and the packet level behavior. In a period of time when the rate of the aggregated traffic at a given

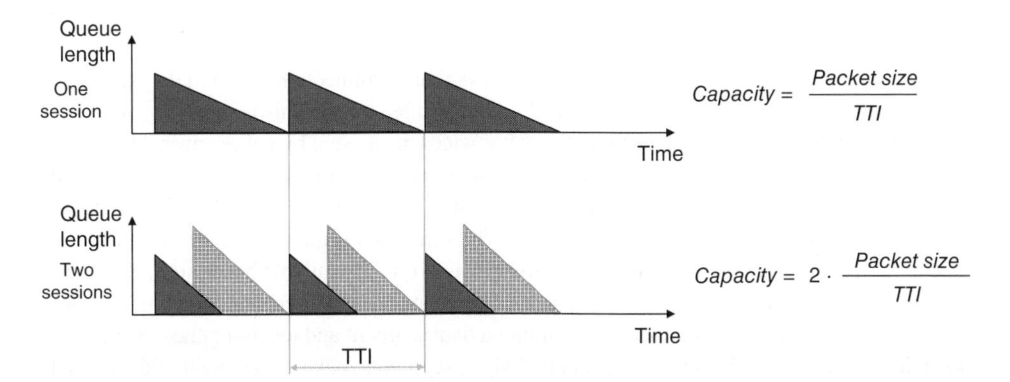

Figure 4.25 Link capacity and the arrival process: if identical streaming sources with deterministic TTI are aggregated, the required link capacity will be a function of the aggregated traffic process. If there is no strict delay requirement, dimensioning should guarantee that the buffer content is served during one TTI. The required link capacity in this case is linearly increasing with the number of connections

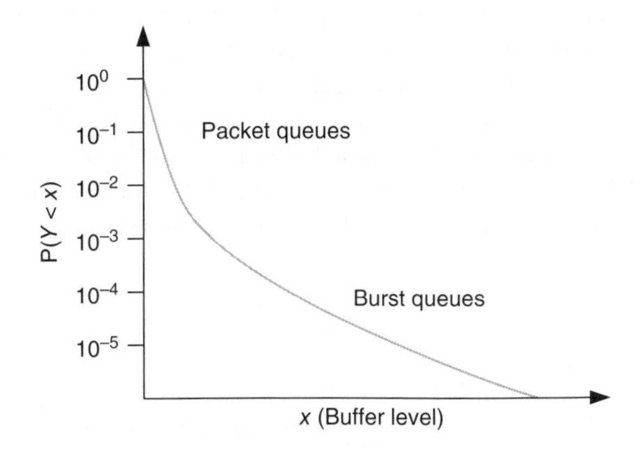

Figure 4.26 Packet and burst level queues in the PDF of the buffer occupancy Y

node is below the link capacity the queue length (buffer load) remains low. Although, owing to the burstiness of the traffic on the packet level, there is some fluctuation in the queue length, this fluctuation is limited. The buffer level can grow significantly only in the case of a large burst, with an arrival rate higher than the link capacity. The resulting (usually long) queue is referred to as a "burst queue." Therefore, the behavior on the two distinct levels (packet and burst) is typically well separated also in the probability distribution function (PDF) of the buffer load Y. The PDF for a low buffer level is dominated by the packet queues, whereas for high buffer occupancy the relevant contribution comes from the burst queues, as shown in Figure 4.26.

This also implies that the delay which is a result of a packet queue is usually low: typically the service time of a few packets. In contrast, the queuing delay resulting from a burst queue can be much longer. This suggests that packet level modeling would be necessary only if the maximal queuing delay allocated to the buffer was very low.

4.5.6 Transmission Control Protocol Models

TCP, the dominant transport layer protocol, provides reliable connection-oriented data transfer. At the TCP source (content server), the data transfer is controlled by a sophisticated flow and congestion control mechanism that is clocked by the acknowledgments sent by the receiver. These mechanisms allow the TCP to adapt its rate to the available bandwidth with some latency due to two factors. First, the TCP source extracts information about the actual network conditions from the received acknowledgments sent by the receiver when the data segments are successfully transferred to it. Second, as the TCP source has no information on the available bandwidth, it is probing the system by increasing or decreasing the rate of the connection depending on the acknowledgments it receives. The time spent between sending a data segment and receiving the corresponding acknowledgment (i.e. the RTT) and the acknowledged segments make up the main information the TCP source can rely on for flow or congestion control. Accordingly, as shown in Figure 4.27, there are periods (phases) in the TCP connection's lifetime when it is not able to utilize the available bandwidth such as the slow-start phase, which occurs at the beginning of the connection or after a timeout when the TCP source reduces the transmission rate to zero and executes a slow-start. There are cases when the TCP rate is limited by the TCP source itself, that is when the TCP maximum window size (communicated by the receiver during the connection setup) is reached. In this case the amount of in-transit segments is not increased according to the congestion avoidance algorithm (additive increase or CUBIC function) but is kept constant. Otherwise, whenever the TCP rate is increased according to the congestion avoidance algorithm and the maximum window size is not reached, the TCP is able to adjust the data rate to the available bandwidth in a timely manner.

Whenever the TCP is able to adapt to the available bandwidth, it can be regarded as an exhaustive elastic traffic source and modeled accordingly. This is the case when the congestion

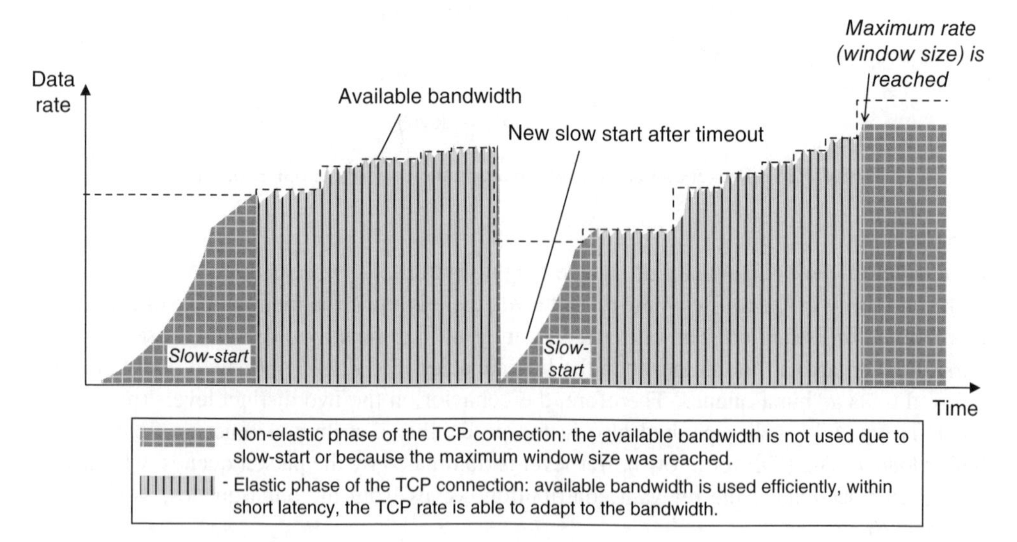

Figure 4.27 The TCP source is probing the system for the available bandwidth. As at the connection setup it has no knowledge on the system conditions, data transfer is started at a low rate, which is increased exponentially (slow-start). During this period, the source is not using the system efficiently. In a similar way, a slow-start is executed after a timeout or a longer idle period, during which the resources are not fully utilized

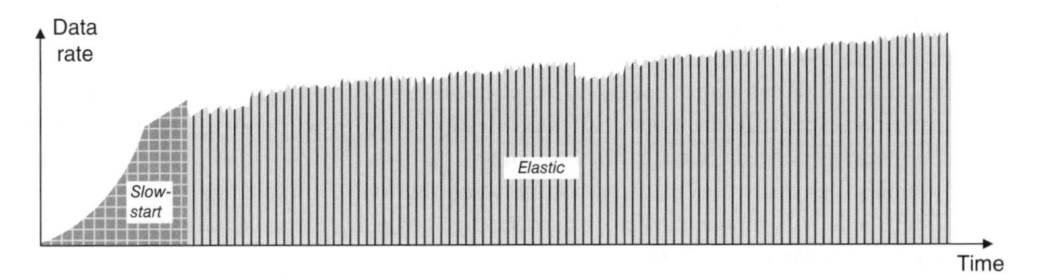

Figure 4.28 When the requested object (file, software update, etc.) is large enough, the TCP connection spends the majority of its lifetime as an elastic connection

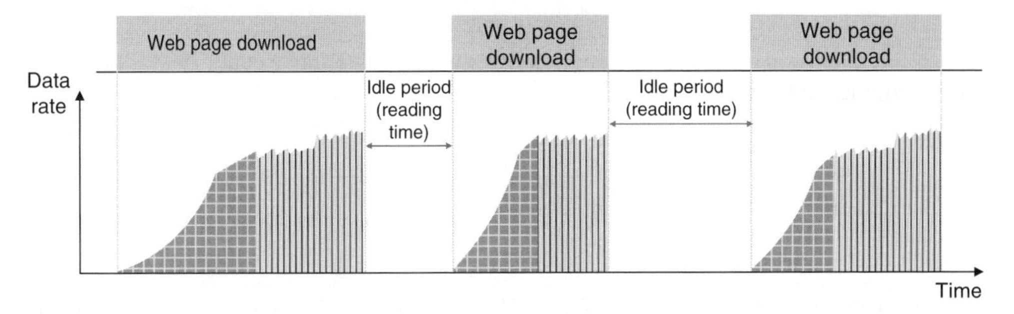

Figure 4.29 The TCP connections established during a Web browsing session are not acting as elastic sources as they spend significant time within the slow-start phase

avoidance is dominating during the connection's lifetime. Except the cases when the network conditions are causing frequent timeouts or when the maximum window size is reached, a TCP connection is in congestion avoidance. Under these circumstances, provided that the amount of data to be transferred is large enough, the major part of the TCP lifetime is spent in the elastic phase. It is reasonable to assume that in cases of proper dimensioning all TCP connections that are carrying large data (file downloads/uploads, software updates, etc.) are acting as an elastic source (Figure 4.28).

The main quality indicator of these sessions is the TCP throughput. There are, however, sessions when the profile of the applications run by the users consists of the transmission of only a small amount of data, such as Web surfing, chat, tweets, etc. During these sessions, the TCP connections providing the data transfer spend significant time in slow-start, that is the elastic nature of the connection is not dominant (Figure 4.29). Latency and content download time are the main quality parameters in these cases.

The TCP connections can be modeled with packet level models or flow level elastic models.

4.5.6.1 Packet Level TCP Models

The scope of the packet level TCP models is the estimation of the long-term average throughput of a single TCP connection/flow by using packet level parameters as round trip time (*RTT*), packet discard rate (p), number of segments acknowledged by a single ACK (acknowledgment signal) (b), packet size (s), retransmission timeout timer (RTO),value etc. Packet level dynamics of the flows is captured at some degree assuming stochastic discard procedures. As there are several TCP variants in use by different operating systems that significantly differ on their operation (especially the flow control/congestion control mechanisms are different), a dedicated model is needed for each TCP variant.

The simplest generally accepted relation between the long-term average throughput of a flow (*THP*) and the average packet discard rate is given by the inverse square root law (Equation 4.3; Altman et al., 2000):

$$THP \approx \frac{\sqrt{3/2} \times s}{RTT \times \sqrt{b \times p}}. \tag{4.3}$$

A possible extension of the formula is given by Equation 4.4 (Padhye et al., 1998; Floyd et al., 2000; Widmer, 2000):

$$THP \approx \frac{s}{RTT\sqrt{\frac{2}{3}b \cdot p} + 3RTO \cdot p \cdot \sqrt{\frac{3}{8}b \cdot p} \cdot \left(1 + 32p^2\right)}. \tag{4.4}$$

The applicability of these models in LTE MBH dimensioning is limited as they a priori require input parameters (*RTO*, *RTT*, p) that in fact depend on the link capacities selected during the dimensioning process itself. The user applications are establishing TCP connections from various locations with servers being located at different data centers spread around the globe. Accordingly, the parallel TCP connections sharing the same link will have different RTTs, thus their share of the common bottleneck link will not be equal. Note that it is reasonable to assume that identical TCP connections sharing the same link will experience the same packet discard rates due to the operation of the buffer management algorithms such as RED or CoDel. The TCP throughput formula given by these equations is used by TCP-friendly rate control (TFRC) mechanisms in order to achieve a fair bandwidth sharing between adaptive streaming and TCP flows. Another valid use of these formulas is the congestion window estimation of some advanced TCP variants. Although there are examples of a successful adaptation of these formulas in the context of radio access network backhaul dimensioning (Bodrog et al., 2009), the use of the packet level TCP models in the dimensioning of complex multilink networks, such as LTE MBH, is problematic not only from a modeling but also from a computational complexity point of view.

4.5.6.2 Flow Level TCP Models

The main assumption of the flow level TCP models is the ideal and instant rate adaptation capability of each flow meaning the ideal elasticity (Figure 4.30). The flow level models do not consider the packet and burst level dynamics of the TCP connections, whereas the buffering is ignored. Accordingly, the TCP behavior during slow-start is not captured and delay is not modeled. This makes the flow level models more suitable for bulk data transfers than for interactive connections.

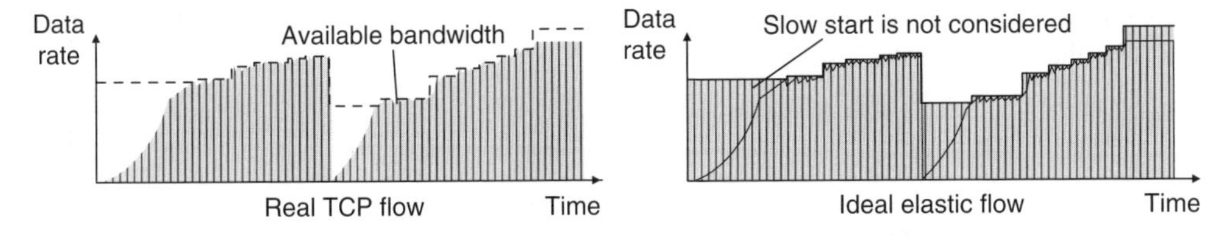

Figure 4.30 The difference between the real and ideal elastic sources

The flow level models consider that independent flows are initiated stochastically according to a random process. In this approach, there is no distinction between the sessions and bursts, that is consecutive TCP connections originated by the same user within the same session are modeled as independent flows. This is equivalent to the assumption of having one TCP flow within one session; therefore, these models are not suitable for modeling Web browsing not only because the slow-start is not captured but also because their assumption of the TCP sessions.

The amount of data to be transferred by a flow, the flow size, is given by a distribution; the instant throughput of a flow (*THP*) is assumed to be determined by a given resource sharing scheme. Simultaneous flows share the available resources according to this scheme.

The average throughput of a flow (γ) is given by the ratio of the average flow size and the average flow holding time, that is the time to transfer the whole data (Equation 4.5):

$$\gamma = \frac{E(\mathit{flow\,size})}{E(\mathit{flow\,holding\,time})}. \tag{4.5}$$

The flow level models can be categorized into single-link and multilink models.

The single-link models are suitable for initial interface capacity calculations without considering the topology. This information can be utilized as input of the link capacity calculations. It should be noted that the elastic models are not capturing the fact that the TCP is two-directional traffic, that is at least there is a flow of ACK segments in the reverse direction to the data flow. The rate of the ACKs depends on the rate of the data segments, that is it is proportional to the instant throughput of the connection. As users are not only content consumers but also producers, it is common that the data flows share transport links with ACK flows in the same direction.

The single-link models can be further classified into infinite and finite population models. The infinite population models assume that the flow start is a Poisson process with intensity λ (Núñez Queija et al., 1999; Benameur et al., 2002; Bonald and Proutière, 2004a). New flows are initiated regardless of the actual number of parallel flows. This might cause overload whenever the offered load is above the link capacity.

The flow sizes are drawn from a given distribution with expected value of *E*(*flow size*) and the link capacity is shared equally among the flows modeled based on the theory of processor sharing queues. The infinite population models are insensitive to the flow size distribution, the stationary state and the performance depends on a single parameter, the load (ρ). This holds even when the flow arrivals are not strictly according to the Poisson processes, that is, flows arrive in sessions (Bonald and Proutière, 2002 . The average throughput is given by Equation 4.6 (C is the link capacity):

$$\gamma = C \times (1 - \rho). \tag{4.6}$$

The relative load is based on Equation 4.7:

$$\rho = \lambda \times \frac{E(flow\,size)}{C}. \tag{4.7}$$

The distribution of the number of simultaneous TCP sessions is geometrical (Equation 4.8):

$$P(number\,of\,flows = x) = (1 - \rho) \times \rho^x. \tag{4.8}$$

The infinite population elastic models provide a very simple and robust analytical tool for performance evaluation on a single link, thus simple dimensioning can be done by selecting the link capacity to a value that guarantees that the average throughput equals the target throughput defined for the flows.

The finite population elastic traffic models define an upper limit to the number of users (N). Each user may have at most one flow at a time. If the data requested by the user has been downloaded, the user will initiate the next download in a new flow after an idle period of length τ. The model allows the definition of user-specific parameters, both the flow size distribution and the idle period can be defined per user level, with a distribution with expected value of $E(flow\ size)$ and $E(\tau)$ respectively. A user with infinite long flow can be considered with $E(flow\,size) \to \infty$. This models the case when during the analysis period the user has continuous bulk data transfer.

The calculated per flow level average throughput does not depend on the detailed traffic characteristics (flow size, idle periods and their correlation) but on a parameter $(r, r = E(flow\ size) / E(\tau))$ that can be interpreted as the offered traffic load of a user, (Equation 4.9; Bonald et al., 2003):

$$\rho = r \times \frac{N - E(x)}{E(x)}. \tag{4.9}$$

$E(x)$ is the mean value of the parallel active flows.

This model allows a simple and robust dimensioning algorithm that is able to capture the differences between the various OTT applications.

In real environments, TCP flows are not identical. Connections served by a given eNB can have different routes even on the MBH as the eNBs are multi-homed (content can be reached via different S1 interfaces). Users can establish TCP connections with servers at various locations (Figure 4.31). Due to the different routes, the congestion point of the flows served by an eNB might be different. At a given link, the throughput of connections that have experienced congestion already at another link is upper limited according to the congestion at that link. The

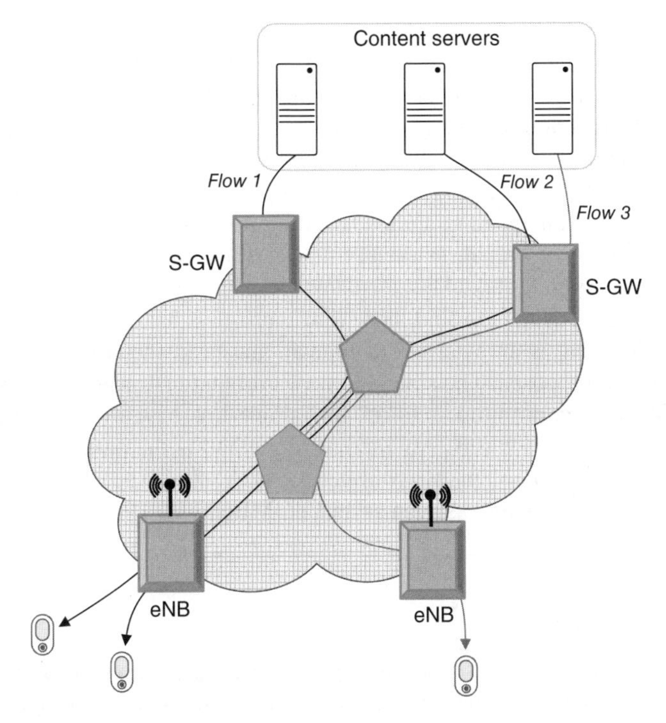

Figure 4.31 Users may establish TCP connections through several S-GW to various content servers, that is the TCP connections served by an eNB can have diverse routes

RTT of the connections is different as the content is downloaded from different servers. Accordingly, the applicability of single-link models is limited to interface capacity calculation (see Figure 4.2) or to the cases when the transport connectivity of the eNBs is provided by point-to-point transport tunnels with capacity reservation. Proper dimensioning requires that at least the topology and the diverse routes are considered.

Beside these factors, at a bottleneck link, the resource sharing of the simultaneous TCP flows is a result of the packet level congestion. The flow level models postulate the resource sharing of simultaneous flows at a bottleneck link. The scope of the resource sharing schemes (e.g. utility resource sharing schemes, balanced fairness) is to provide a reasonable good approximation of the resource sharing realized by TCP. Based on these schemes the instant rate (R) of a flow can be calculated.

The utility based resource sharing schemes approximate the resource sharing of TCP by maximizing a utility function, $U(R_1, R_2, \ldots, R_k)$, where R_k is the rate of flow k and $\sum R_k \leq \textit{link capacity}$. Example utility functions are the min/max fairness (maximizes the throughput of the "worst" flow), proportional fairness or weighted α-fairness. The latter is a general scheme that has as special case the max/min fairness, the proportional fairness and the inverse square root law. These utility functions provide slightly different results. The link capacity is calculated by an iteration where the minimum link capacity is selected to guarantee that the rate of each flow (R_k) equals or exceeds the corresponding QoS requirement (minimum throughput, THP_k), where $R_{k \geq} THP_k, \forall k$.

The problem of the utility function based resource sharing schemes is that the stationary performance proposes an insensitive scheme, the balanced fairness (Bonald and Proutière, 2002). The condition for insensitivity is given by a balance function. This method provides similar results to the utility function based mechanism.

In a multilink network model based on balanced fairness, flows are assumed to be originated and terminated stochastically (e.g. the Poisson process) according to an infinite population. Each flow is mapped to a class based on its route, flow size distribution, arrival intensity and load. For each link the stationary state can be calculated according to the balance function. The stationary state should be calculated for each link within the network. The scope of the network dimensioning is to find those link capacities that are providing for each flow the required quality, that is an extensive computational task; therefore, approximations for balanced fairness are proposed. The store and forward bound (Bonald and Proutière, 2004b) reduces the problem of a multilink network to a single-link case by assuming that data are transmitted separately on each link. An alternative method is the parking lot approximation (which decomposes the problem by considering each class separately. The solution is especially suitable for tree networks such as an LTE MBH.

4.6 Network Models

As discussed, the applicability of single-link model based dimensioning is limited to the initial interface capacity calculation as part of a more detailed dimensioning method or to the cases when the eNBs are connected via point-to-point transport services to the core with capacity allocation. In all the other cases, the topology of the network, the route of the traffic and the transport QoS architecture should be considered for efficient dimensioning in terms of network costs.

Various dimensioning approaches require dedicated models. At the different phases of the dimensioning process (reliability requirement based dimensioning, QoS driven dimensioning, initial dimensioning) different network models are needed that can capture the special aspects addressed by the dimensioning itself. The modeling technique/approach itself (elastic traffic model based dimensioning, queuing theory, fluid models, etc.) defines the applicable network model. In fact, as it was presented in the elastic traffic modeling part, the traffic and network model are strongly coupled, the separate discussion is due to the scope and structure of this section. In each case, the modeling approach-specific traffic parameters, the routing and the topology information itself are mandatory components of a network model.

In general, the network models are used for network analysis and not for dimensioning, that is to provide performance evaluation of a given network configuration (topology, resources, parameters) under the load generated by the estimated traffic. The performance evaluation is done against the quality requirements. The dimensioning algorithm itself is an iterative analysis process, where the optimal system resources that are still able to provide the required quality are identified. Therefore, the scalability and the computational complexity are important aspects of the modeling process that limit the feasible alternatives, the level of complexity of the models and thus the level of details. At the selection of the modeling technique several aspects should be examined:

- The relevant components of the network and the network behavior to be modeled. If the scope is to address a specific part of the network, such as the radio interface or transport network, or given technology layer only, or the C-plane only, there is no need to model the whole e2e in all its details. It is a reasonable approach to apply a high-level model to the rest of the network and to analyze the targeted network segment with the required accuracy level.

- Network state to be captured: steady state behavior, transient events, rare events, etc. The scope of the MBH dimensioning is to calculate the required network resources that provide the required services in steady state. Transient performance degradation caused by a node or link failure is beyond the scope of dimensioning. If possible, these failures are resolved by the transport network through a path switch. During the path switch, the service quality might degrade. The role of dimensioning in this context is to take the right technology choices that can guarantee operation according to the requirements in these cases.
- The level of details of the requirements. For example, when there is only rough traffic mix information, such as raw BH load, there is no need for fine-grained network modeling.
- The feasibility of the analysis method of the problem to be modeled. As an example, rare events are difficult to capture with simulation methods, as collecting enough statistical samples with simulations is challenging.
- The available time and processing power. Detailed analytical models are numerically complex and, similarly to the simulation methods, require extensive processing; therefore, reasonable simplifications/assumptions should be considered.
- The size of the problem: a complex, large network that cannot be decomposed into smaller sub-networks to be handled more easily requires simplifications even if there is enough detailed information available.
- The performance parameters of interest. If only the target throughput is of interest, there is no need for detailed delay or loss modeling.
- The required accuracy of the result. Different modeling approaches result in different accuracy.

4.6.1 Queuing Methods

As queues are intrinsic to packet switched networks such as LTE MBH, queuing models are suitable for delay and loss analysis. These two parameters quantify the level of QoS. If delay, delay variation (which is calculated as part of the delay analysis) and loss are below the target values defined for the S1 and/or X2 interface, the system meets the expectations. If it does not, the link capacities must be increased.

In a general packet switched network, packets (generated by the users) arrive at the nodes (routers) asynchronously and are processed and forwarded according to a service protocol, for example strict priority scheduling (SP), weighted fair queuing (WFQ) or FIFO or hierarchical schedulers that are a combination of these. If a node is not able to handle/forward each packet coming to it simultaneously, it uses its buffering space to store the excess packets until they can be served. In other words, the packets are waiting service from the node, which is referred to as the "server." Buffer management mechanisms—such as RED or CoDel, and its improvements—are applied to prevent TCP flow synchronization and bufferbloat. Queuing models are capturing (with exact or approximate solutions) the packet level behavior of the scheduling mechanisms.

Figure 4.32 shows the basic terms and parameters used in the queuing models.

The queuing delay on the nodes is influenced by:

- The packet arrival process (arrival statistics) at the node: the packet arrival can be defined as a stochastic process where the packet inter-arrival time (τ) is given by a distribution. The main parameter of the inter-arrival time is the arrival intensity λ, which is the mean number

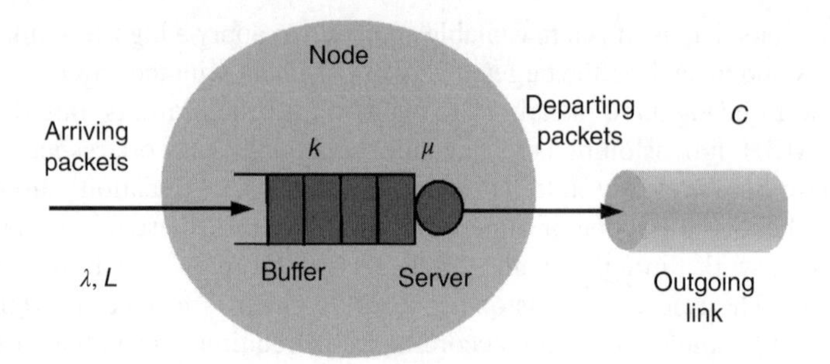

Figure 4.32 Queuing model of a transport node. The figure assumes that the outgoing link is the bottleneck link; therefore, the node is modeled by its output scheduler

of packets arriving at the node during the time unit. In addition to this parameter, further parameters can be used to describe the variability of the arrival process in order to increase the accuracy.

- Buffer size k, for example finite buffering system ($k < \infty$) in which the packets that arrive at a full buffer are dropped or infinite buffering system ($k = \infty$) when the packets are always accepted into the node. The buffer size is measured in packets.
- The distribution of the service time (service statistics): the main parameter of the service time is the mean service time $1/\mu$. The service time varies as it depends on the size of the packets served. The service time is a parameter of the server. If the server capacity is the outgoing link capacity (C), as is the case in packet switched networks, and the mean packet length is L, then the mean service time is $1/\mu = L/C$.

A queue is considered stable if $\lambda/\mu < 1$. It should be noted that when the load approaches 1, the delay increases significantly and in case of a finite buffer the drop rates too.

The delay (d) in queuing is a random variable. The main parameter (calculated by the queuing models) of the delay is its mean value. Some models can provide other parameters of the delay distribution as well. The queuing delay on a node has two components:

- The waiting time in the buffer (W), i.e. the time the packet has to wait until it is served (Figure 4.33).
- The service time of the packet (Figure 4.33).

If the server is free when a packet arrives, it receives service immediately and its delay is equal to its service time. If the server is busy then the packet is buffered and has to wait until the server is ready to serve it. The total delay will be the sum of the waiting time and the service time of the packets.

Packet loss probability can be calculated or approximated by assuming infinite buffer space and calculating the probability of buffer states above k (the size of the buffer). In the event of packet discard, the intensity of the packet flow of the departing packets will be decreased.

As the figure shows, for the arrival process in the example, the buffering space (if no packet loss is allowed) must be at least enough for four packets.

In the event of transient overflow, when packets are dropped, the intensity of the departure process will be lower than the arrival intensity of the traffic.

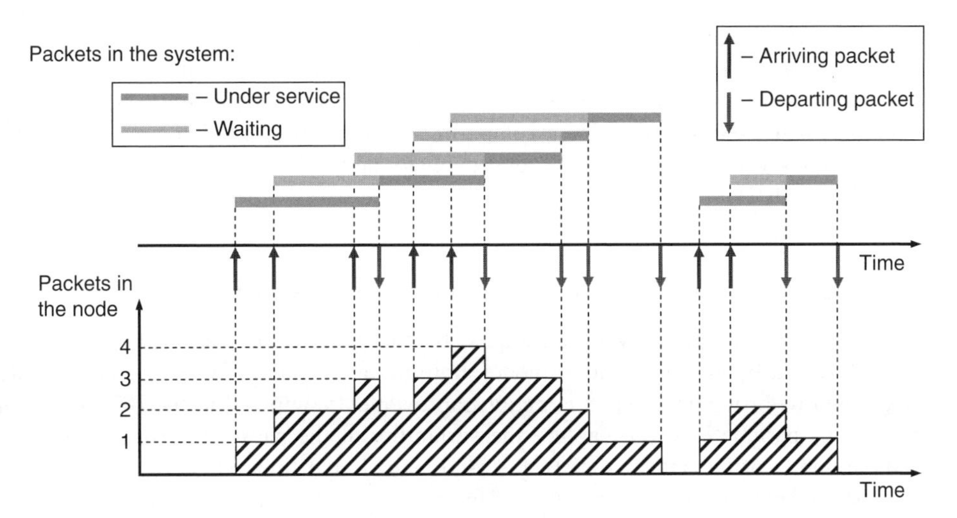

Figure 4.33 An example of delay and waiting packets in a node

The literature concerning the queuing models based node and network analysis is extensive. Each of the available models provides an exact or approximate analysis of a queuing node based on simplifying assumptions on the arrival and service processes. The application of any queuing model requires the conversion of the traffic parameters to the arrival process parameters which the given queuing model uses (i.e. the creation of an equivalent model that captures the traffic behavior). The modeling of the streaming sources is relatively straightforward as there are models that can describe On–Off traffic sources with different rates during the ON and Off periods. However, modeling elastic sources is problematic as due to the TCP flow and congestion control their packet level behavior depends on the actual circumstances. A possible approach is to introduce an iterative analysis method in which the TCP sources are modeled as streaming sources with a fixed packet rate during the On periods. The initial length of the On period is defined based on the amount of data to be transferred and the data rate derived from the QoS requirement. The analysis consists of several iterations. After the first step of the analysis, during which the QoS parameters (delay, loss) and the throughput are calculated based on the initial traffic parameters of the TCP connections, the second step of the iteration is executed with elastic traffic parameters corrected based on the results of the first analysis.

A second aspect which limits the applicability of the queuing models is that proper modeling requires that the applied model is able to capture the behavior of the hierarchical scheduling discipline of today's transport nodes. The rest of this section provides an overview of the popular queuing models.

4.6.1.1 Exact Solutions

BCMP (Baskett, Chandy, Muntz and Palacios) (Bascett et al., 1975; Bolch et al., 1998) is a widely known and applied queuing model with an exact solution. The arrival process is assumed to be a Poisson process, which means that the packet inter-arrival time is exponentially distributed. There are several service disciplines that can be modeled with BCMP, such as the FCFS (first come, first served) that is the most relevant in case of a packet transport network. The analysis and the QoS parameter calculation is the result of solving the corresponding M/M/1-FCFS queue that assumes exponential service times (packet sizes exponentially distributed). Due to the assumptions it introduces (i.e. Poisson arrival process and exponential service time), BCMP is not suitable for detailed network analysis and dimensioning of LTE MBH.

4.6.1.2 Two-Parameter Approach: The Queuing Network Analyzer

The two-parameter approach was introduced by the queuing network analyzer (QNA; Whitt, 1983a and b). In this queuing model, both the arrival and the service process are approximated with two parameters, a major improvement compared to the BCMP, which uses only one parameter, λ. To capture the variability of the packet inter-arrival time, a second parameter of the arrival statistics is used, namely the squared coefficient of variation of the packet inter-arrival time (c_a^2). This parameter describes the burstiness of the traffic: in the event of a deterministic arrival it is 0, whereas in the event of a Poisson arrival it is 1. In a similar way, the packet length variation is captured by a second parameter (in addition to a mean service time, τ), which describes the variability of the service time (c_s^2). The model supports infinite buffer, and so loss is not modeled. The other QoS parameter, the mean waiting time, can be calculated with Kramer–Langenbach-Belz formula (Kramer and Langenbach-Belz, 1976).

The QNA is a framework that allows the analysis of a whole queuing network. The main assumptions and modeling mechanisms introduced by the QNA can be easily extended or used as network models of alternative analytical modeling mechanisms (queuing models, fluid models). Based on the traffic routes, a packet network can be decomposed into its equivalent queuing network model using the following assumptions:

1. The output link capacity is the bottleneck, i.e. a node is modeled with as many queuing nodes as outputs it has.
2. The individual queues are independent.
3. The UL and DL traffic is not correlated.
4. The system is an open system, i.e. the packets generated by the traffic sources are served by one or more nodes and leave the system.
5. The waiting space is not limited.
6. The service discipline is FCFS. The QNA network consists of a network of directed links and nodes.

The nodes represent outgoing links from the routers, the links represent packet flows. Packets enter the network on directed links, move from one node to the other and leave the system on directed links. The packet flow on the links is considered random. If the server of the queuing node is busy (is serving a higher priority buffer or there are packets in the buffer that have the same priority as the arriving packet) when a packet arrives then the packet is buffered. If the queue is empty, the arriving packet is served immediately.

The QNA can be easily extended by replacing the node model with queuing models that are able to calculate the parameters of alternative scheduling disciplines, that is by replacing (5) and (6) with new assumptions. The extension of the QNA to finite buffers was provided by Heindl (1998). Queuing models for the strict priority (Horváth, 2005) and weighted for queuing (Horváth and Telek, 2003) scheduling disciplines are also available.

4.6.1.3 Markovian Arrival Processes Based Decomposition

The two-parameter approach that models the inter-arrival time as an independent random process is not able to capture the true nature of real traffic sources that follow a pattern, for example a streaming On–Off source generating a data frame periodically during the On period, whereas no data (or smaller data frames with longer periodicity) are transmitted during the Off period. This can be modeled with MAPs (Markovian arrival processes) (Latouche and Ramaswami, 1999; Heindl, 2001). The generator matrix of the underlying Markov chain is shown in Equation 4.10:

$$\begin{bmatrix} -\dfrac{1}{E(On)} & \dfrac{1}{E(On)} \\ \dfrac{1}{E(Off)} & -\dfrac{1}{E(Off)} \end{bmatrix}. \tag{4.10}$$

Both On and Off periods are assumed to be exponentially distributed with mean *E(On)* and *E(Off)* respectively.

The advantage of the model is that the aggregated arrival process of simultaneous connections (described by their own MAP) can be described by a MAP as well. The state space of the aggregate is the direct product of the state spaces of the individual connections. State space explosion can be prevented by applying state-space compression.

Batch Markovian arrival processes (BMAPs) allow a more general traffic description where the On and Off periods of the arrival processes are not exponentially distributed.

The buffer is modeled by a Markov chain that follows the number of packets in the buffer. The inter-arrival time of the incoming traffic to the buffer is described by the MAP of the traffic aggregate and the service time by the distribution of the packet sizes and the link capacity. In the event of an FCFS queue, the Markov chain can be solved with matrix geometric methods. This queuing model allows calculating not only the mean waiting time but also higher-order moments of the waiting time distribution, the mean queue length and the higher-order moments of the queue length distribution and finally the packet loss probability.

The extension of the model to an SP discipline is possible, there are computationally complex exacts solutions (Bai et al., 1997) and faster and robust approximations (Horváth, 2005) as well. Approximate analysis of a WFQ discipline is provided by Horváth and Telek (2003).

4.6.2 Fluid Network Models

Fluid modeling is a framework that allows burst level modeling of a packet switched network with a mathematical model that approximates a network node as a container that is filled by a fluid with a given inflow rate $\alpha(t)$ and drained with a service rate $\beta(t)$. The fluid level in the container at t is the queue length, $x(t)$ increases if $\alpha(t) > \beta(t)$ and decreases if $\alpha(t) < \beta(t)$. The container overflows if the fluid level $x(t)$ reaches the maximum fill level, meaning the queue length grows above the buffer size (c). The amount of fluid (data) lost during a unit of time due to overflow, $\gamma(t)$, when the buffer is full is $\gamma(t) = \alpha(t) - \beta(t)$, otherwise $\gamma(t) = 0$. The outflow rate of the fluid, $\delta(t)$, meaning the amount of served data during a unit of time when the buffer is empty, equals the fill rate ($\delta(t) = \alpha(t)$), otherwise it equals the service rate ($\delta(t) = \beta(t)$). The fluid models calculate $x(t)$, $\gamma(t)$ and $\delta(t)$. In the context of a fluid model, the traffic demand is given by the $\alpha(t)$, whereas the service process is given by $\beta(t)$, both being stochastic functions. The complexity of the fluid analysis depends on the complexity of these two stochastic functions. When the inflow and service rate are Markov modulated, the system is referred to as a "Markov fluid model." The inflow and service rate are Markov modulated when they are piece-wise constant. They can vary only when the modulating Markov chain changes its state and their values are determined by the state of the Markov chain. An On–Off source with constant data rate during the On period, with On and Off periods being independent and exponentially distributed, is one example of a Markov fluid model. In case there are N simultaneous (and identical) connections, the corresponding Markov chain consists of $N + 1$ states, one state corresponding to the number of active connections ($0, 1, \ldots N$). The analysis of a Markov fluid model is given by (Kulkarni, 1997; Bai et al., 1997; Gribaudo and German, 2001).

The basic fluid model assumes identical connections. In real systems this assumption is not valid as there are many traffic classes the network has to serve (Elwalid and Mitra, 1994).

The scope of the analysis of the fluid model is to calculate the stationary distribution of the fluid level ($x(t)$) within the container (buffer), which, when there is a constant rate of traffic sources modulated by a Markov chain, requires solving a set of first order ordinary differential equations. Fluid networks can be analyzed by decomposing the system to independent fluid node models, which are then analyzed independently.

Similar to the queuing models, fluid models can provide the waiting time and the loss probability of the On–Off modulated streaming traffic, but they have no direct methods to capture the behavior of the elastic flows. Fluid models have the advantage over the queuing models that they are capable of modeling the deterministic manner of the data transmission on the links, while queuing network models capture the packet level behavior of the traffic better.

4.6.3 Network Model

The LTE MBH is a network of queues. The QNA approach (to create an equivalent queuing network model that consists of directed links and nodes) allows an e2e performance analysis of the system and the system level QoS parameters to be calculated. The nodes within the queuing network model are modeling the egress schedulers of the transport nodes, whereas the directed links are representing the direction of the traffic. The conversion requires the prior knowledge of the physical topology and the traffic routes, which means that the QoS-based network analysis and dimensioning can be performed only after the topology has been designed. The analysis is performed hop by hop following the directed links. The output traffic parameters of a node (or the aggregated traffic parameters of several nodes) are the input parameters of the next node during the node analysis. This approach can be applied to fluid models as well, by decomposing the system to independent fluid queues. In addition to the output traffic parameters, the QoS parameters are calculated for each node separately. The e2e QoS parameters can be calculated by aggregating the per node level parameters along the traffic route.

Accordingly, the equivalent queuing network is created by identifying the transport schedulers and the queues these are serving that are relevant from QoS perspective. In the case of an MBH, such relevant buffers are the transport buffers associated with the physical ports and buffers of the transport schedulers located in the S-GWs and in the eNBs. The equivalent queuing network is created by considering the physical topology, the traffic demand, the configuration of the schedulers and the route of the traffic over the transport network. The modeling assumption is that transport links are the bottleneck links, which implies that the output schedulers are the relevant ones. The UL and DL traffic is modeled separately (as the model assumes that these are not correlated); therefore, the equivalent queuing topology might consist of several distinct topologies. Additionally, the traffic routes, and the redundancy of the transport topology, might result in further independent sub-networks when the equivalent queuing network model is created.

The initial step of the quality analysis is the creation of the equivalent queuing network model based on the considerations described above, the physical network topology consisting of transport network devices connected with transport links or internal data buses and the traffic routes. For example, the simple transport network shown in Figure 4.34 results in the equivalent queuing network model shown in Figure 4.35. Note that buffers corresponding to the DL and UL directions are treated as separate nodes on the queuing network model. The structure of the eNB, transport and S-GW schedulers (shown in green, orange, and blue respectively in Figure 4.35) defines the queuing node model.

4.6.4 *Routing and Requirement Allocations*

The conversion of the transport network into its equivalent queuing network is possible if the traffic routes are known as within the equivalent queuing network the links represent the

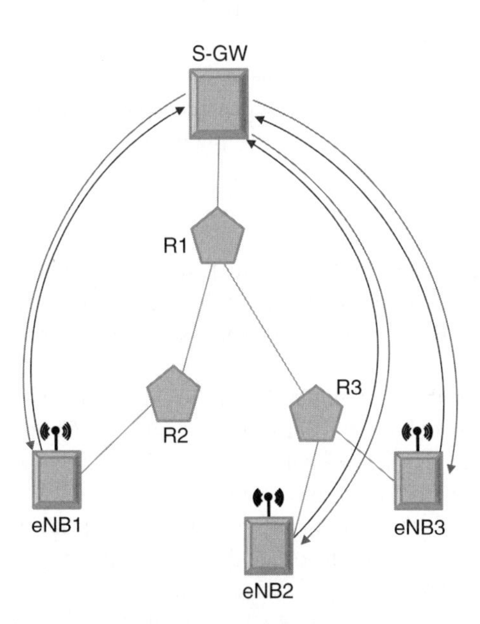

Figure 4.34 Example of a transport network topology with uplink and downlink traffic

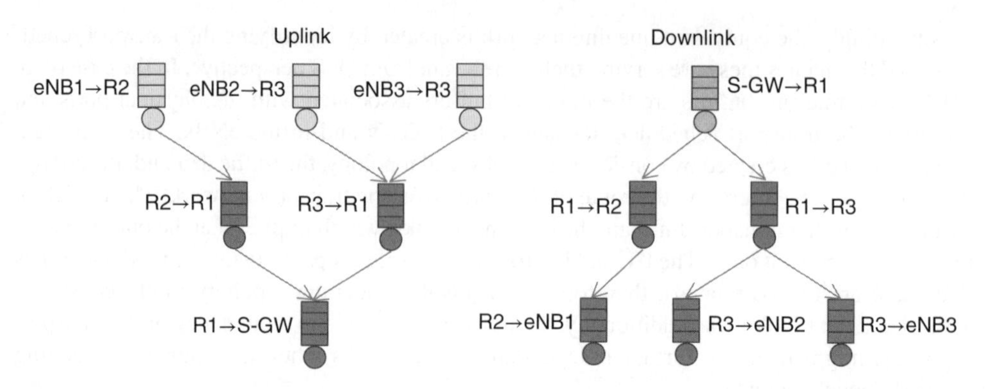

Figure 4.35 Queueing network model of the reference transport topology. The topology is converted into two independent queuing networks with directed links corresponding to the uplink and downlink traffic considered not correlated

path of the traffic from one queuing node to the next. The routing information is the result of the logical and physical topology planning: the primary (protected) and the secondary (protecting) paths of transport tunnels give the route of the traffic they are serving. If the MBH is a pure IP-routed network, the definition of the routes requires a routing algorithm, such as Dijkstra's shortest path algorithm. Having the transport topology, the traffic endpoints and the routes of the connections, the equivalent queuing network topology can be created.

Dimensioning is an iterative process, which by following the directed links within the equivalent queuing network topology calculates the required link capacity hop by hop. At a given hop, the link capacity is calculated that meets the QoS requirements defined for that particular node/link. Requirement allocation is the process of converting the QoS requirements defined for the e2e to node/link/network segment level requirements. This process requires the consideration of the following aspects.

The e2e transport delay (D_{TN}) consists of the media delay (d_m), processing delay (d_p), service time (d_s) and queuing delay (d_q), as shown in Equation 4.11:

$$D_{TN} = d_m + d_p + d_s + d_q. \tag{4.11}$$

The media delay is the latency on the physical links (copper, fiber or microwave) that is physical media specific and proportional to the distance between two transport devices. The total media delay on the MBH experienced by each packet depends on the number and medium of the intermediate links.

The processing delay is the time required at each node to process the packet (policy check, routing or switching table lookup, transfer from the ingress buffer to the egress buffer, time to live and checksum update, etc.). The total processing time on the MBH experienced by each packet depends on the number of intermediate hops.

In the context of an eNB, the media and processing delays are the intrinsic delay components (latency) specific to each LTE interface that each and every packet experiences. As the link capacities must be calculated so that the e2e delay requirement is met, the intrinsic delay must be subtracted from the delay budget (maximum allowed e2e delay). The resulting delay requirement/budget (d) is the one that was reduced with the intrinsic delay component, as shown in Equation 4.12:

$$d = D_{TN} - \left(\sum d_m + \sum d_p\right). \tag{4.12}$$

The service time is the time required by the scheduler to transfer the head of line packet over the transport link and it is proportional to the packet size.

Finally, the queuing delay or waiting time is the time spent by the packet within the buffers of the transport node waiting for the scheduling. This component depends on the load, the priority of the packet (connection it belongs to), the scheduling discipline, the scheduling parameters (weights, etc.) and the available link capacity. The queuing delay can be decreased by increasing the link capacity. Accordingly, the scope of the dimensioning of a given link is to find the smallest capacity that guarantees the reduced delay requirement is met.

As discussed, the delay requirement should be converted into node, link or network segment level requirements first. In general, the delay requirement is defined as the maximum delay a packet mapped to a given QoS class is allowed to experience. In packet switched networks the delay at a given transport node is upper bounded by the buffer size (often defined as a maximum delay target). A delay requirement violation probability is also given in case the maximum allowed delay is less than this upper bound; as in packet switched networks, no hard guarantee can be given on the delay except the upper bound. When packets belonging to different QoS classes are mapped to the same buffer, for example, there is a single buffer to serve the assured forwarding behavior groups AF11, AF12, AF13 and AF14 per-hop behaviors (PHBs) and there is a separate delay requirement for each class, the dimensioning should consider the stricter requirement in order to guarantee that each individual delay requirement is met.

In LTE, the U-plane connections served by a given eNB can have different routes within the MBH and those can be established to provide connectivity to different public data networks (PDNs). Accordingly, two simultaneous connections belonging to the same QoS class having the same delay requirement can have different routes. This implies that the connection with the longer route can spend less time at a given node. If the node is shared by the two connections, in order to meet the delay requirements, the link capacity should be selected so that QoS expectations of the flow with the longer route are met and a larger link capacity must be selected. Similar considerations are needed, in case the traffic served by two distinct eNBs share the same transport links, resulting in different delay requirements. During dimensioning, the stricter requirement must be considered when the corresponding link capacity is calculated.

Figure 4.36 shows an example of when the traffic of two eNBs is sharing the same transport link. The connection with the longer route (to eNB_3) imposes stricter delay requirements, and the maximum value of d_l is bounded by the delay requirement of the flows served by eNB_3.

Even if the connections have the same route to the eNB, those that are having handovers can spend less time on the S1 interface as the latency caused by the X2 forwarding must be taken into account.

Considering all the restrictions listed above during dimensioning, the connections served by an eNB with a short transport path (few intermediate hops) would experience considerably shorter delay than the corresponding delay requirement. Compensating this would result in

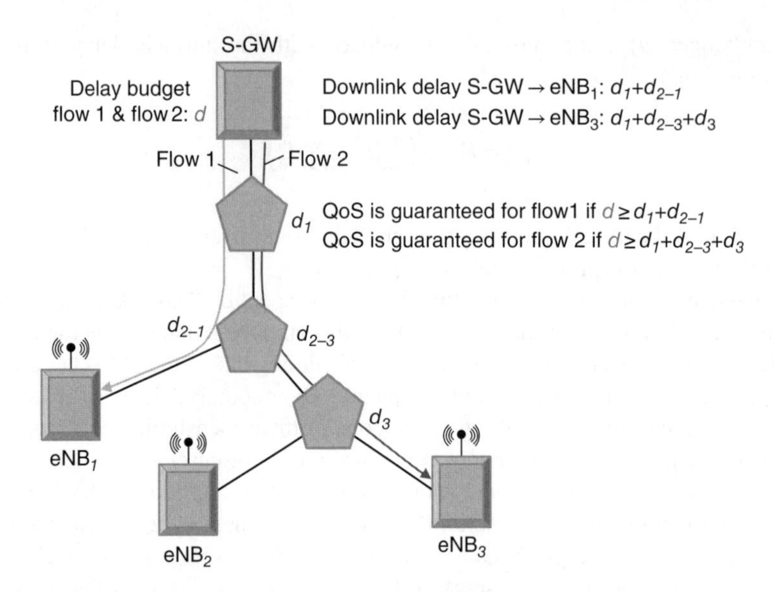

Figure 4.36 Example of delay requirements derived from the expectations of different connections sharing the same link

reduced network costs. Possible compensation would be to reallocate the remaining delay budget to the links that were dedicated only to the given eNBs. In the example shown in Figure 4.36, the allowed delay on the shared link (i.e. d_1) should be derived from the delay budget of flow 2. To compensate the stricter delay requirement of the common link, the allowed maximum delay on the last mile link of eNB$_1$ (d_{2-1}) can be defined as: $d_{2-1} = d - d_1$.

The maximum allowed delay is a meaningful parameter in case of delay or latency-sensitive applications when the transfer time of a single or a few packets has a measurable impact on the overall experience.

The delay requirement violation probability and the discard rate are relevant to real-time applications. As the volume of the OTT data traffic dominates the traffic mix—and, on the other hand, the real-time streaming applications are by default mapped to the EF class on the transport network—the likelihood that these requirements are not met in case of proper planning is low. However, the dimensioning algorithm should be executed in such a way that these requirements are guaranteed in e2e.

The throughput requirement should be interpreted as the minimum bandwidth reserved for the given connection (for which the requirement was defined) at each and every node along its path.

4.7 Dimensioning

4.7.1 QoS-Driven Dimensioning

One possible way of implementing a dimensioning algorithm based on analytical methods is to execute an iteration on the link capacity until the QoS requirements are met. The scope of this mechanism is to find the minimum capacity that is able to guarantee the level of QoS to the amount of traffic that the network is dimensioned to serve. In order to limit the interval where the algorithm is searching and thus to reduce the number of iterations for the optimal value, a minimum and a maximum capacity should be defined. A possible approach is to set the minimum capacity to the average throughput the link has to carry and the maximum capacity to the peak rate of traffic demand. The optimal capacity value is between these two extreme values.

The capacity of each link (resource) has to be adjusted during the optimization until the QoS requirements are satisfied. This means the number of variables involved in the optimization is the number of links in the network. With such a large number of variables the classical optimization algorithms (like hill climbing or simulated annealing) need to perform a large number of steps until a solution is found, and this solution is hardly the global optimum.

One way to handle this problem is to separate the traffic classes during the capacity dimensioning.

First, the capacity dimensioning should be performed by considering only the real-time traffic sources as to be in the network. The link capacities are then optimized such that the delay and loss requirements of these traffics are satisfied. After this step the mean throughput of the elastic traffics can be checked (e.g. by using the processor sharing model based analysis). If the mean throughput is lower than intended, the capacities can be increased while all demands are met.

A possible approach to decrease the number of variables during the optimization is to fix the ratio of the link capacities according to the ratio of their traffic load. Thus, the capacity of link i should be set to $\alpha_i \times C$, where α_i is the sum of the traffic loads of all the traffics crossing link i. The traffic load is the product of the mean arrival intensity of the packets and the mean packet sizes. This way the only single variable to optimize is C.

When the optimal value for C has been found, the network with capacities $\alpha_i \times C$ can be considered an initial guess for a refining optimization phase with all the link capacities involved.

In general, there are several alternatives to how the dimensioning of an MBH can be approached. One valid method is to start with the output of the radio planning, meaning the cellular site location, the access technology, the peak rate of the radio interface at each cell site and the traffic mix forecast used in the radio planning. The location of the cell sites represents an important input of the planning of the physical topology. The first step is to define the physical topology by considering the existing infrastructure, if such exists. The transport technology—that is microwave, optical and electrical links and the transport protocol stack (Ethernet, IP, MPLS, MPLS-TP)—and the related choices on the logical topology—that is the use of point-to-point connectivity (E-Line, MPLS tunnels) or VLANs, L2VPNs, L3VPNs, etc.—are mandatory inputs of the capacity calculations. Once the physical and logical topology options are known, the dimensioning against the QoS requirements can be started with the definition of the e2e routes (assuming "shortest path routes" is a valid approach in this step when the routes of the transport tunnels is not known). There are several modeling methods as listed in this chapter that can be used in dimensioning.

The bandwidth of the logical links (E-Line, MPLS) or PE–PE (provider edge to provider edge) LSPs (L3VPNs) can also be calculated with the same methods based on the traffic mapping information and QoS requirements.

4.7.2 Reliability Requirement Based Dimensioning

Throughout this section, a reference MBH is considered that consists of Ethernet access (VLAN domain) and IP/MPLS aggregation. This consideration simplifies the discussion as the dimensioning problem and approach is valid in case of other packet-based MBH technologies which are able to provide protected transport services as well; however, discussion of the different technology-specific aspects can be avoided. As opposed to having a homogeneous transport solution—for example when the whole MBH is IP/MPLS, which allows the creation of e2e label switched paths (LSPs) and e2e path protection—this MBH architecture introduces additional challenges with regard to topology design and reliability requirements. In order to provide the necessary level of protection the dimensioning should consider the relevant failure cases. In what follows, a failure case is associated with one or several links or node failures in the MBH, that is a failure case is defined as a set of broken links and nodes. A failure in the IP/MPLS domain triggers the protection mechanism. As a result, the traffic which is carried by LSPs routed along a broken link or node is rerouted. Local node or link protection provides fast recovery, while intra-domain 1:1 path protection (e2e within a domain) provides more optimal operation in a failure case. 1:1 protection can be used also in the VLAN domain.

The dimensioning of the MBH (that consists of two separate domains)—the aggregation domain, which is based on IP/MPLS (possibly Layer 3 VPN), and an Ethernet-based access domain referred to as the "VLAN domain" (which may be a carrier Ethernet network)—should ensure that the LTE services are provided with the right level of quality even in the event of link or node failures. The number of consecutive failures the transport services are expected to survive are given as input parameters (reliability requirements). Accordingly, up to the specified amount of failures, the end user will experience no change in the level of service. This has implications for the topology and the overall cost, as the redundant resources must be created in the MBH network.

Due to different technologies in the two domains, an e2e protection is problematic (if not impossible) to establish. A possible solution, if e2e protection is provided within both domains independently, is shown schematically in Figure 4.37a. As an alternative, it is possible that the VLAN domain has a tree topology that doesn't allow the creation of disjoint VLAN tunnels. Therefore, protection is not an option, which is shown in Figure 4.37b. Optionally, site protection can also be provided for the S-GW, in which case the core elements are connected to two edge routers of the MPLS domain (shown in Figure 4.37c and d). This solution ensures uninterrupted traffic flow even in the event of an edge router breakdown or the failure of the link connecting the gateway and the MPLS domain. Switching between the working and protecting path requires that the two routers are communicating with each other using a dedicated mechanism such as the Virtual Router Redundancy Protocol (VRRP).

On top of the above protection scenarios local link or node protection may also be used in the IP/MPLS domain. In this case, the input of the dimensioning should include the list of LSPs of the MPLS domain for which local protection is required.

During the dimensioning, shortest path routing algorithms can be used to establish shared risk link group (SRLG) independent working and protection paths on the link topology of the MPLS and VLAN domains. If bandwidth reservation is used, the bandwidth of each LSP/tunnel has to be calculated as a first step by using one of the QoS/QoE-based dimensioning methods described above. A plausible approach is to model each tunnel as a point-to-point

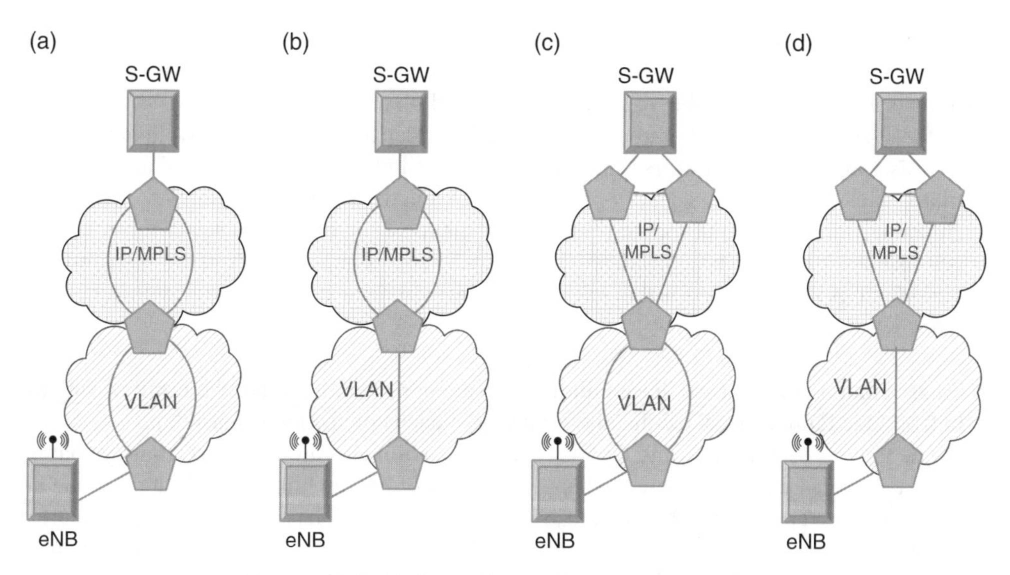

Figure 4.37 End-to-end protection over two domains

link connecting the eNBs to the S-GW and calculate the bandwidth accordingly. This bandwidth will only include the gain of multiplexing the traffic mapped to a given VLAN tunnel (e.g. the whole or part of the traffic served by the eNB). When the aggregation domain provides a L2VPN or L3VPN service, the bandwidth of the PE–PE LSPs can be calculated by summing up the individual tunnel bandwidths of the inner tunnels mapped to the PE–PE LSPs. Once the tunnel bandwidth is available, optimization algorithms such as MLO should be applied to calculate the path of each tunnel so that at a system level the optimal resource utilization is achieved. During this calculation, the resource sharing options (given as input) should be considered. The details of this step are beyond the scope of this chapter. In what follows it is assumed that, regardless of the level of protection (local or intra-domain 1:1), the routing information (working and protection paths) is given as an input to the QoS/QoE requirement based dimensioning algorithm. It is to be noted that the logical interfaces (S1 and X2) of the eNBs are first mapped into LSPs/tunnels which are in turn routed on the links. However, from the QoS/QoE requirement based dimensioning point of view the detailed modeling of LSPs can be ignored; therefore, it can be assumed that the connections are directly routed on the links of the MPLS and VLAN domain according to the path routing information given as input. During dimensioning, the protocol overheads of the LSPs/tunnels should be considered only.

The scope of dimensioning is to calculate the physical link capacities that guarantees the required level of QoS/QoE is provided even in the event of link/node failures when the traffic affected by the failure is switched from the working to the protecting path. From the modeling point of view, each failure results in a new topology where the network elements (links/nodes) affected by the failure are removed from the original physical topology, whereas the traffic is routed according the route of the protecting tunnels. For each new topology dimensioning must be executed in order to ensure that the QoS/QoE requirements are always met. For each link, the largest calculated capacity must be configured to the system. With a large network, the amount of alternative topologies that must be evaluated (considering the potential combinations) can be extensively high, requiring significant computational effort. A possible simplification is to reduce the number of failure states based on their relevance.

In the event of link or node failure the connections mapped in the transport tunnels (LSPs of VLAN tunnels) which traverse the broken node or link are rerouted. In what follows a "failure state" is defined as a set of broken facilities together with the routing information of each connection on the link topology of both domains. The scope is to select those failure states that are going to be considered in the QoS/QoE requirement based dimensioning. Several options are possible for this selection process. In what follows, a few approaches are listed.

- Single failures: In this case those failure states are considered that are the result of single-link or node failure (provided that the rerouting of each connection is feasible).
- Single failures in both domains: In this case the two domains are treated separately, that is those failure cases are considered, where the number of broken links/nodes is at most one in both domains. This approach is less restrictive than the previous one, since double failures can also be considered, provided that the two failed links/nodes are in different domains. The number of failure states in this case is N1 × N2, where N1 and N2 are the number of failure states within the MPLS and the VLAN domain respectively.
- Probabilistic approach: With this method a probability is assigned to each failure state, and the aim is to cover a certain fraction of the cases, i.e. the net probability covered by the considered states must be above P, which is assumed to be given in the input (e.g. $P = 0.999$). This approach requires a weight (w_i) to be assigned to each facility, which represents the probability (ratio of time) for that facility to be broken. This approach tries to capture the fact that some links/nodes are more prone to failures. In this case the probability of a particular failure can be calculated (assuming independence of individual link failures) as Equation 4.13:

$$\text{Prob}(\text{Failure case}) = \left(\prod_{i \in \text{broken facilities}} w_i\right)\left(\prod_{i \in \text{working facilities}} (1 - w_i)\right). \tag{4.13}$$

In order to assign a weight (probability) to a failure state corresponding to a particular failure case some auxiliary input parameters are needed, such as the (expected) life time of the local protection (τ_{local}) and the expected duration of a failure (τ_{failure}). The probability of a failure state corresponding to a single facility failure can then be estimated as Equations 4.14 and 4.15:

$$\text{Prob}(\text{Failure state with local protection for facility } k) = \frac{\tau_{local}}{\tau_{failure}}\left(\prod_{i \in \text{working facilities}} (1 - w_i)\right) w_k. \tag{4.14}$$

$$\text{Prob}(\text{Failure state with intradomain protection for facility } k)$$
$$= \left(1 - \frac{\tau_{local}}{\tau_{failure}}\right)\left(\prod_{i \in \text{working facilities}} (1 - w_i)\right) w_k. \tag{4.15}$$

After a probability is assigned to each failure state, the next step is to decrease the number of considered states by considering that the net probability of considered states must be greater than P. This can be achieved by ordering the failure states by their probability in descending order and selecting the first n of them (Equation 4.16):

$$\sum_{k=1}^{n-1} \mathrm{Prob}\left(\text{Failure state}\, k\right) < P < \sum_{k=1}^{n} \mathrm{Prob}\left(\text{Failure state}\, k\right). \tag{4.16}$$

Those failure states which are not in the selected set are not going to be considered in the rest of the process.

For simplicity's sake, the local protection that guarantees that availability of the transport services until the e2e path protection detects the failure and executes the path switch to the protecting path was not considered in the discussion above.

The net probability, P, can be set at any arbitrary value.

References

Altman E., Avrachenkov K. and Barakat C. (2000) A stochastic model of TCP/IP with stationary random losses. *ACM SIGCOMM Computer Communication Review* **30**: 231–42.

Bai Z., Demmel J. and Gu M. (1997) Inverse free parallel spectral divide and conquer algorithms for nonsymmetric eigen problems. *Numer. Mathematics* **76**: 279–308.

Bascett F., Chandy K., Muntz R. and Palacios F. (1975) Open, closed and mixed networks of queues with different classes of customers. *Journal of the ACM* **22**: 248–60.

Benameur N., Ben Fredj S., Oueslati-Boulahia S. and Roberts J. W. (2002) Quality of service and flow level admission control in the Internet. *Computer Networks* **40**: 57–71.

Bodrog L., Horváth G. and Vulkán C. (2009) Analytical TCP throughput model for HSDPA. *IET Software* **3**(6):480–494.

Bolch G., Greiner S., de Meer H. and Trivedi K. (1998) *Queueing Networks and Markov Chains: Modelling and performance evaluation with computer science applications*. Wiley-Interscience, New York.

Bonald T. and Proutière A. (2004a) On performance bounds for the integration of elastic and adaptive streaming flows. *ACM SIGMETRICS Performance Evaluation Review* **32**(1): 235–45.

Bonald T. and Proutière A. (2004b) On performance bounds for balanced fairness. *Performance Evaluation* **55**: 25–50.

Bonald T. and Proutière A. (2002) *Insensitive bandwidth sharing in data networks*, http://perso.telecom-paristech.fr/~bonald/Publications_files/questa.pdf, accessed 1st June 2015.

Bonald T., Olivier P. and Roberts J. (2003) *Dimensioning high speed IP access networks*, http://perso.telecom-paristech.fr/~bonald/Publications_files/itc18-dim.pdf, accessed 1st June 2015.

DOMO (2015) *Data never sleeps 2.0*, http://www.domo.com/learn/data-never-sleeps-2, accessed 1st June 2015.

Elwalid A. I. and Mitra D. (1994) Statistical multiplexing with loss priorities in rate-based congestion control of high-speed networks. *IEEE Transaction on Communications* **42**: 2989–3002.

Floyd S., Handley M., Padhye J. and Widmer J. (2000) Equation-based congestion control for unicast applications. *Proceedings of the conference on Applications, Technologies, Architectures, and Protocols for Computer Communication SIGCOMM '00*: 43–56.

Gribaudo M. and German R. (2001) Numerical solution of bounded fluid models using matrix exponentiation. *Proceedings of 11th GI/ITG Conference on Measuring, Modelling and Evaluation of Computer and Communication Systems (MMB)*. VDE Verlag, Aachen, Germany, 2001.

Heindl A. (2001) *Traffic-based Decomposition of General Queueing Networks with Correlated Input Processes*. Shaker Verlag, Aachen, Germany, 2001.

Heindl A. (1998) *Approximate analysis of queueing networks with finite buffers and losses by decomposition*. Technical report. Technische Universitat Berlin, Institut fur Technische Informatik.

Horváth G. (2005) A fast matrix-analytic approximation for the two class Gi/G/1 nonpreemptive priority queue. *12th International Conference on Analytical and Stochastic Modelling Techniques and Applications*, Riga, Latvia, 1st–4th June 2005.

Horváth G. and Telek M. (2003) An approximate analysis of two class WFQ systems. *Workshop on Performability Modelling of Computer and Communication Systems (PMCCS)*, Monticello, IL, September.

HTTP Archive (2014) *The HTTP archive tracks how the Web is built*, http://httparchive.org/index.php, accessed 1st June 2015.

KISSmetrics (2015) *How loading time affects your bottom line*, https://blog.kissmetrics.com/loading-time/, accessed 1st June 2015.

Kramer W. and Langenbach-Belz M. (1976) Approximate formulae for general single systems with single and bulk arrivals. *Proceedings of 8th International Teletraffic Congress (ITC)*: 235–43.

Kulkarni V. G. (1997) Fluid models for single buffer systems. In: J. H. Dshalalow (ed.) *Models and Applications in Science and Engineering: Frontiers in queueing*. CRC Press, Boca Raton, FL: 321–38.

Latouche G. and Ramaswami V. (1999) *Introduction to Matrix Analytic Methods in Stochastic Modelling*. American Statistical Association and the Society for Industrial and Applied Mathematics, Alexandria, VA.

Núñez Queija R., van den Berg J. L. and Mandjes M. R. H. (1999) Performance evaluation of strategies for integration of elastic and stream traffic. In: D. Smith and P. Key (eds) *Teletraffic Engineering in a Competitive World*. Proc. ITC, Edinburgh, UK: 1039–50.

Padhye J., Firoiu V., Towsley D. and Kurose J. (1998) Modeling TCP throughput: A simple model and its empirical validation. *ACM SIGCOMM Computer Communication Review* **28**(4): 303–14.

Szilágyi P. and Vulkán C. (2014) Two-way TCP performance issues and solutions in asymmetric LTE radio access. *IEEE 25th International Symposium on Personal, Indoor and Mobile Radio Communications (PIMRC)*, Washington, DC, September 2014.

Website Optimization (2015) *The psychology of Web performance*, http://www.websiteoptimization.com/speed/tweak/psychology-Web-performance, accessed 1st June 2015.

Whitt W. (1983a) The queueing network analyzer. *The Bell System Technical Journal* **62**: 2779–815.

Whitt W. (1983b) Performance of the queueing network analyzer. *The Bell System Technical Journal* **62**: 2817–2843.

Widmer J. (2000) "Equation-based congestion control," Diploma thesis.

5

Planning and Optimizing Mobile Backhaul for LTE

Raija Lilius[1], Jari Salo[2], José Manuel Tapia Pérez[1] and Esa Markus Metsälä[1]
[1]*Nokia Networks, Espoo, Finland*
[2]*Nokia Networks, Doha, Qatar*

5.1 Introduction

Planning the mobile backhaul network for LTE is rarely done without considering the available backhaul infrastructure or other radio access technologies. As the proportion of the LTE capable end user equipment is increasing and LTE usage increases, there will likely come a time when the backhaul network needs to be modernized to support the increasing traffic. This is referred to as "strategic planning" in Chapter 2 and Chapter 3, and as a result of this strategic planning the L1/L2 technologies are selected, and the topology of the network is planned taking into account the resiliency methods available for use. Strategic planning defines the cost of the backhaul network as it defines the transport media, bandwidth requirements and resiliency to be used that have a major impact on the total cost of implementing the backhaul network, as discussed in the examples in Chapter 3. Also the schedule outlining the phases in which the network is to be upgraded or replaced to meet the requirements of the increased traffic is a result of strategic planning. These results define the boundary conditions for high-level planning.

In addition to LTE, many backhaul networks support other radio access technologies such as Global System for Mobile communications (GSM) and Wideband Code Division Multiple Access (W-CDMA). In this case LTE mobile backhaul cannot be properly planned without taking this into account, as the transport for those technologies primarily utilizes packet transport. In practice other technologies need to be taken into account at least in the backhaul dimensioning, for example QoS settings to guarantee appropriate priority to different traffic types as well as IP subnet planning.

LTE Backhaul: Planning and Optimization, First Edition. Edited by Esa Markus Metsälä and Juha T.T. Salmelin.

Network sharing is attractive to mobile operators to share the cost of transport and mobile network equipment in the RA network. It also affects the LTE backhaul network design, as the operators want to separate their traffic to different virtual local area networks (VLANs) and use IP addresses from dedicated address spaces.

LTE backhaul planning is also affected by network implementation: operators often have lists of friendly sites that may need to be utilized to speed up the implementation of the network. There are vendors that have equipment which introduces multiple technologies in the same base station equipment, resulting in backhaul sharing starting from the first mile link.

Small cell planning is causing new challenges to backhaul design. The cost of planning per eNB needs to be radically reduced as the number of eNBs increases. The small cell design may also be driven by the availability of the backhaul or power supply as the decision criterion. The small cell design is briefly examined in Chapter 2.

5.1.1 Planning and Optimization Process

Strategic planning sets the boundary conditions for further planning steps: high-level and detailed planning. The following sections will examine the implementation of strategic planning in greater detail. So-called detailed planning defines the configurations and parameters of the various elements in the network. Figure 2.30 presented in Chapter 2 shows how the strategic, high-level and the more detailed planning processes interact. One can also separate it into a phase of upfront planning (e.g. a greenfield case or some other larger network change) and ongoing operational planning, where the larger changes are driven more by strategies. In the end, the detailed planning phase is needed before the network can become operational.

Detailed planning is itself a continuous process. The inputs to the planning process could be simplified, the traffic in the mobile network is forever changing and affected by the continuously changing environment of end-user equipment and applications. There are many places in the process where assumptions are made and only once the network is implemented can we know how well the assumptions matched the actual network conditions. This means that the backhaul for mobile network is never ready or complete in the sense that there can never be a time when no additional changes will need to be made. The backhaul network requires a constant monitoring of traffic, for example in terms of capacity utilization of the network, analysis of the network statistics, definition of the optimization actions to improve the performance and once again re-planning parameters and configuration for the backhaul network or the mobile network equipment like eNBs.

This continuous process—which involves parameter planning, implementation in the network, performance monitoring and analysis and optimization to improve the performance—is illustrated in Figure 5.1.

This chapter concentrates on the high-level planning tasks that define the overall guidance, for example, for resiliency planning, QoS, dimensioning, routing and subnetting. High-level planning defines the overall guidelines for detailed planning and puts those guidelines into practice by defining the actual parameters for the different equipment in the backhaul and eNBs. How to apply these principles in practice is presented as practical planning examples for leased line and microwave backhaul in Chapter 6.

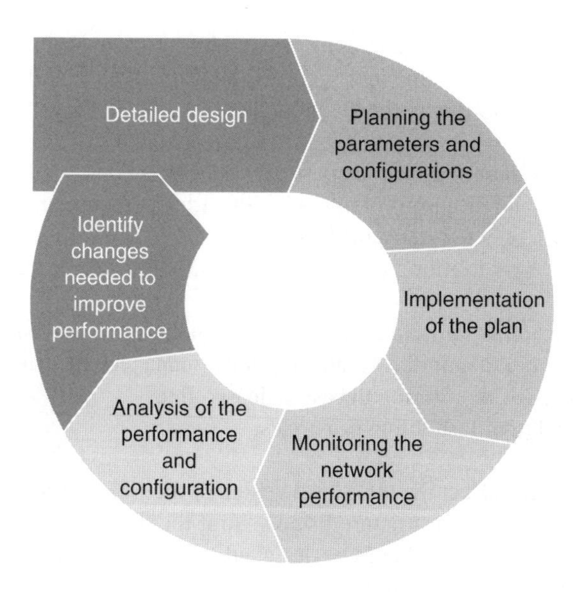

Figure 5.1 Detailed planning is a continuous process

5.1.2 High-Level Design Overview

As the output of the high-level design, the overall technical solution, basic design rules, technologies and concepts used should be defined. The high-level planning process consists of the following tasks:

- network topology and selection of transport media such as Ethernet, copper, fiber (Ethernet, SDH), microwave Ethernet radio
- selection of transport resiliency mechanisms: equipment protection, link protection, routing/path resiliency, geo-redundancy
- definitions of QoS and availability target per traffic type
- dimensioning backhaul bandwidths based on information of the mobile traffic and network topology
- planning the site solution based on information of the different radio access technologies sharing the base station site
- logical networking: flat network for all different traffic types or separated routing, e.g. for user and control plane traffic and O&M traffic
- IP subnetting and routing design taking into account necessary traffic differentiation and possible extension of the network
- security design: IPsec, PKI solutions, centralized user account control, O&M security
- selection of synchronization method and design
- management system connectivity or use of self-organizing network (SON) functionality like auto-connection.

The list is long and the tasks are challenging having dependencies between them. Topics are described in more detail in the following sections. Additionally, the following details should be taken into account when making the planning decisions:

- management network planning for transport network equipment
- design for scalability and future-proof upgradability, transport equipment selection to match and interoperate with selected transport features
- migration planning, how to migrate from hybrid TDM/IP to all-IP transport network in a cost-efficient way
- planning for SON features, self-configuration functionality such as ANR (3GPP TS 32.511)
- planning of performance monitoring and capacity management principles for transport network, including SLA monitoring in case of leased lines
- fault monitoring and troubleshooting principles.

All of the above factors affect the high-level design of the LTE access network. Constraints set by the existing network infrastructure usually limit the degree of freedom in the design. In practice, the transport network should be designed jointly for all radio access technologies, making the task even more complicated.

5.2 Backhaul Network Deployment Scenarios

Operators can implement the backhaul network in many ways. Considering the definition of network areas given in Chapter 2 (Figure 2.2), the backhaul network from the eNB to the core network consists of access, pre-aggregation and aggregation areas. These can have different topologies and media, redundancy and protection mechanisms and logical network architectures. There are many alternatives, and Section 5.2 should be considered as an introduction to backhaul network deployment scenarios using various technologies, presenting a couple of examples.

5.2.1 Connectivity Requirements

Each eNB requires IP layer connectivity to the core network for the S1 U-plane (to the S-GW) and for the S1 C-plane (to the MME) and additionally for network management and, potentially, synchronization. All these traffic types are forwarded from the access through the network tiers to the core.

X2 connectivity to neighbor eNBs is, instead of direct X2 links, today mostly arranged via a higher network tier hub point back to the neighbor eNB (known as "hair-pinning"). This hub point is often a SEG. Thus all traffic types from the eNBs typically travel together at least until this hub point. Potential evolution and needs of LTE-A is discussed in Chapter 2.

IP connectivity may be provided by any underlying technology; the technology for wide area connectivity usually relies on native IP or MPLS with an IP C-plane. IP supports those services that are needed for backhaul resiliency, robustness, troubleshooting and security. It is expected that at least the aggregation and backbone tiers use IP and potentially MPLS and on the services offered by IP and MPLS.

Differences between Ethernet MAC layer bridging and IP routing are discussed in many networking titles and are revisited briefly in the next section.

5.2.2 Differences Between Ethernet and IP Connectivity

5.2.2.1 Bridging and Routing

The basic operation of the Ethernet switch is that of Ethernet "bridging," meaning flooding frames to every port except where the frame originates. Only if a filtering rule exists (i.e. MAC address is found in the forwarding table of the switch), can the frame be forwarded (only) to the correct port, and not to all ports. Frames with an unknown address as well as any broadcast frames are flooded to all ports in the same broadcast domain, a broadcast domain equaling a VLAN. The bridge then gradually learns the MAC addresses and the ports behind which the addresses reside by observing the source MAC addresses of the frames as they pass through it. After this, flooding can be avoided as the forwarding table is populated.

Once a MAC address is learned and appears on the forwarding table of the switch, it ages out after a certain time (e.g. several minutes) if no frames are passing through with that specific source address. If stations remain silent for a while there will again be a temporary flooding of frames until the MAC addresses are again learned.

One well-known concern with Ethernet bridging is the possibility of an L2 loop, where frames loop endlessly until the network operator resolves the situation by disabling one active path or by otherwise changing the configuration and breaking the loop. Because the network device and the whole bridged domain (VLAN) are usually overloaded in forwarding frames endlessly, it is very difficult to troubleshoot the situation and find out the root cause of it. It could be due to a single site which then disrupts the traffic of all the other sites on the same VLAN. The risk for extended downtime in such a situation is high; therefore, the risk should be mitigated by avoiding large broadcast domains.

IP routing avoids this situation. If a router does not know how to reach the destination, the packet is dropped rather than being flooded to every direction. With Ethernet bridging the frame with unknown destination is instead flooded to all ports. With this mechanism, Ethernet bridging tends to spread the problem. With IP, a missing route to a destination only affects that single destination, while other destinations are not affected and thus IP isolates the problem.

If a routing loop occurs due to a networking error, the IP layer packet will not circulate endlessly, as each router decrements the time to live (TTL) field of the IP packet, and so the packet will be dropped when TTL reaches zero. There is no similar field in the Ethernet frame, and this allows frames to circulate in the system indefinitely if the problem persists.[1]

5.2.2.2 Security

Since all unknown unicast frames as well as broadcast frames are flooded by Ethernet bridges, inherently any station can reach any other station in the same broadcast domain, with no configuration required in the bridge. A broadcast domain (VLAN) is then vulnerable to bogus stations, since if they can physically connect to the switch, Ethernet bridging arranges connectivity to all other stations in that VLAN.

[1] The Spanning Tree Protocol (STP) is intended to block any active paths in L2 topologies except the one that is deemed best, and so L2 loops can occur only when there is some configuration failure in the protocol or a network problem, e.g. if spanning tree control messages do not reach all nodes correctly.

When this connectivity is not needed for any legitimate purpose, it should be restricted. IP traffic gets forwarded by routers only when the address in the IP packet header matches an entry in the routing table. It is useful to impose some control to the Ethernet bridge, so that not everyone has immediate connectivity by L2 bridging function.

With IP, additionally access controls are commonly implemented based on IP addresses. To control access at the Ethernet MAC layer, Ethernet switches may be configured to accept only certain MAC addresses or a certain amount of MAC addresses, in order to control which and how many stations may access the VLAN. Also segregating traffic to different VLANs helps. The downside of this is the added configuration effort related to VLANs and in maintaining legitimate MAC addresses of individual end nodes in the switches.

Access control and security with Ethernet can be further improved using IEEE802.1X port-based authentication. This supports strong cryptographic authentication of a station before it is admitted to use services of the MAC bridge. These and other Ethernet layer security topics are discussed in (Metsälä and Salmelin, 2012) with further references listed there.

5.2.3 *Implications to Backhaul Scenarios*

Simply because of the nature of native Ethernet bridging, for connectivity it is recommended to rely on IP layer forwarding for site-to-site connectivity instead of bridging. The Ethernet still often serves as the underlying layer, as a point-to-point link, and as a physical port. The tendency of IP to limit and isolate problems generally reduces the impact of failures on the network operation. Also, troubleshooting is simpler.

Security is an important consideration in the LTE backhaul. If the layer underneath IP is not fully secured, the backhaul network is compromised to an extent. However, when IPsec is used to protect the traffic, the IPsec cryptographic protection itself remains and the threats that exist are limited. Still, for security reasons (see below), it is recommended that instead of relying on Ethernet bridging for connectivity the IP layer should be used whenever possible, with IP layer access controls and cryptographic protection of the IP layer with IPsec.

5.2.4 *Ethernet Services*

Metro Ethernet Forum (MEF) has defined abstract services (e.g. E-Line for point-to-point and E-LAN for multipoint). The benefit of the service is that customers (e.g. mobile operators) can now use it between sites over a physical distance greater than that of a single site or LAN.

MEF is agnostic to the technology being deployed to deliver the service, and only defines the characteristics the user of the service sees. So the service can be implemented by any technology that is capable of delivering Ethernet frames and connectivity and other characteristics as defined in the service. Both port-based (non-VLAN-aware) and VLAN-based services are supported.

MPLS is commonly used by service providers to deliver different types of L2 and L3 services. When MPLS is used, Ethernet connectivity is emulated, and the service provider can offer the Ethernet service by MPLS instead of native Ethernet bridging.

5.2.4.1 Point-to-Point (E-Line)

Point-to-point service could be used (e.g. by Ethernet over SDH, Ethernet over wireless or as MPLS pseudowires). Point-to-point service over a MPLS pseudowire is described in Figure 5.2. It uses pseudowire encapsulation for transporting Ethernet traffic over an MPLS tunnel.

The point-to-point (E-Line) service is also referred to as "virtual private wire service" (VPWS) or "virtual leased line" (VLL).

5.2.4.2 Multipoint (E-LAN) with MPLS L2 VPN

The basic idea of the MPLS L2 VPN is that it acts as a virtual bridge, and the virtual private LAN service (VPLS) emulates a bridge with the characteristics of an Ethernet bridge, as described earlier: unknown unicast and broadcast frame flooding and learning of MAC addresses.

If all base stations in Figure 5.3 are attached to the same VPLS service (via customer edge [CE] as in the figure), they all share a broadcast domain, and also have MAC layer connectivity. Thus also limitations that exist in using Ethernet bridging for connectivity are relevant for the use of MPLS L2 VPN when looking from the viewpoint of the customer. From the service provider's perspective, the benefit of VPLS is that the service provider can rely on MPLS as a wide area network technology rather than on native Ethernet, while being able to offer a multipoint Ethernet LAN service to the customer.

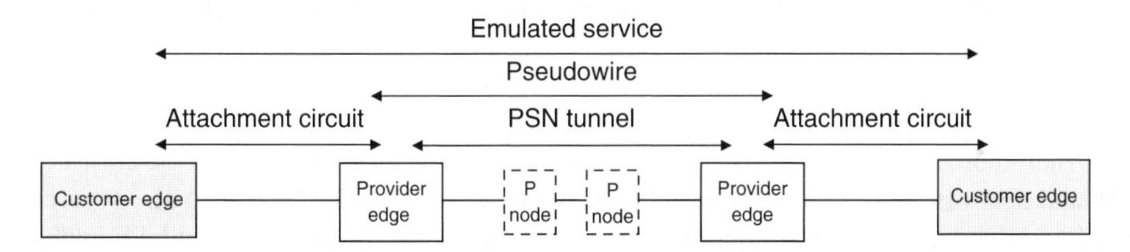

Figure 5.2 MPLS pseudowire

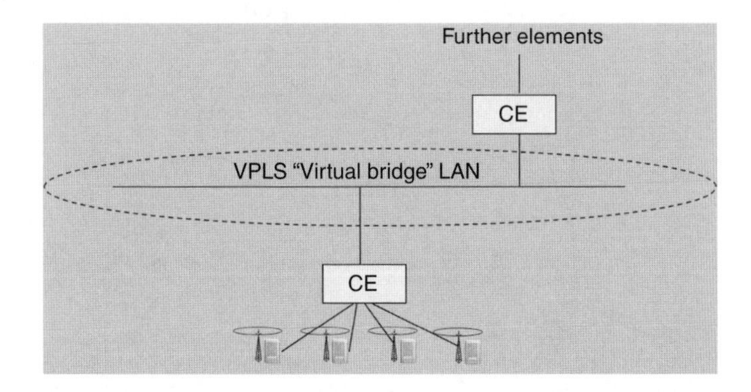

Figure 5.3 MPLS L2 VPN supports multipoint-to-multipoint connectivity

5.2.4.3 Comparison

E-Line suits the needs of the eNB backhaul better than E-LAN does, in the sense that only a point-to-point service is needed and there is no direct Ethernet layer connectivity needed between multiple sites.

5.2.5 L3 VPN Service

5.2.5.1 MPLS L3 VPN

MPLS is also used to implement L3 VPN (IP) services, and an example logical architecture is shown in Figure 5.4. Here, each eNB peers directly with the MPLS Provider Edge (PE) node, or via CE. MPLS L3 VPN is a "cloud" providing IP connectivity between all the customer sites, as defined by the service. Each customer's routing information is kept separate by the use of virtual routing and forwarding (VRF) instances, and traffic can be mapped to the VRFs based on, for example, IP addresses and/or VLANs.

In the LTE backhaul, VRFs can be configured to have U-, C- and M-plane traffic as separate "customers" of the L3 VPN service having either dedicated VRF per traffic type or sharing, for example, a VRF with user and control plane traffic. VRF separation has direct implications for VLAN and routing planning in the eNB, as is described in Section 5.10 in more detail.

Using a separate VRF for U-plane helps in troubleshooting as the routing information and connectivity related to the U-plane can be separately tracked. Similarly, management traffic is often required to be kept separate from both the user and the control plane to increase security against hostile attacks using the management connectivity.

VRFs could also be divided into one for IPsec and another one for non-IPsec traffic.

5.2.5.2 Comparison to L2 Services

The benefits of L3 VPN are that the architecture is scalable as more sites can be added to the backhaul network and connectivity can be traced at the IP layer. L3 VPN also allows overlapping IP address spaces at customer sites. This may be useful in some cases in LTE backhaul, although IP addresses should of course not be overlapping to begin with.

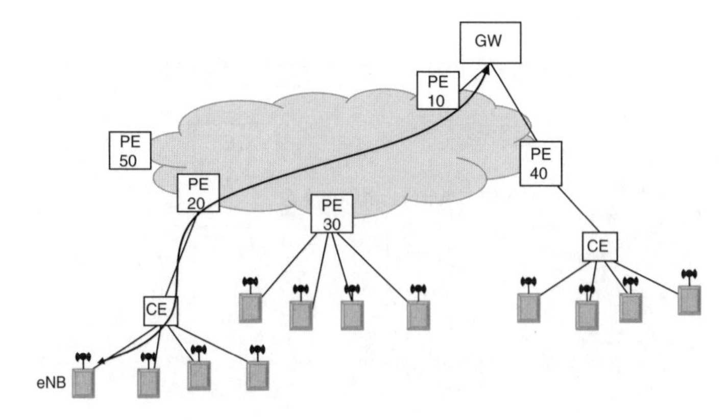

Figure 5.4 L3 VPN

With Ethernet services a benefit exists in that there is no need for the customer to peer at the IP layer with the service provider but rather use a lower-layer service at the Ethernet layer. This way the IP layer operation of the mobile operator is completely separated from that of the service provider.

Specific security aspects with Ethernet bridging and VPLS were discussed above.

5.2.6 Scenario 1: IP Access

In Figure 5.5 eNBs connect via point-to-point links to the access hub and the access hub acts as a router. Each access port in the access hub is an L3 port and the router is forwarding traffic based on the IP. An IP connectivity service exists between each eNB and the EPC.

Aggregation devices can be duplicated and the link from access hubs to the aggregation devices can be duplicated for resiliency. IP routing is used, or alternatively Ethernet link aggregation.

An eNB does not necessarily require routing capability since there is only a single link and a single interface to use for forwarding traffic UL.

5.2.7 Scenario 2: Ethernet Service in the Access

In this network deployment scenario presented in Figure 5.6, L2 service is used in the access (e.g. E-Line or E-LAN type of MEF service). This service is terminated at the access hub, where traffic is routed, as in Scenario 1, toward the EPC.

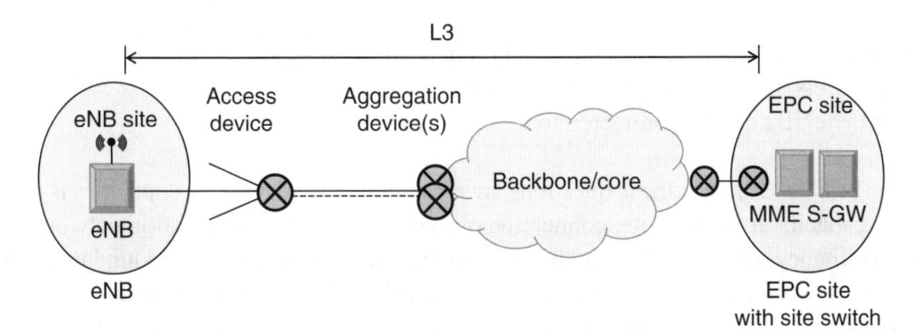

Figure 5.5 Network deployment example using IP access

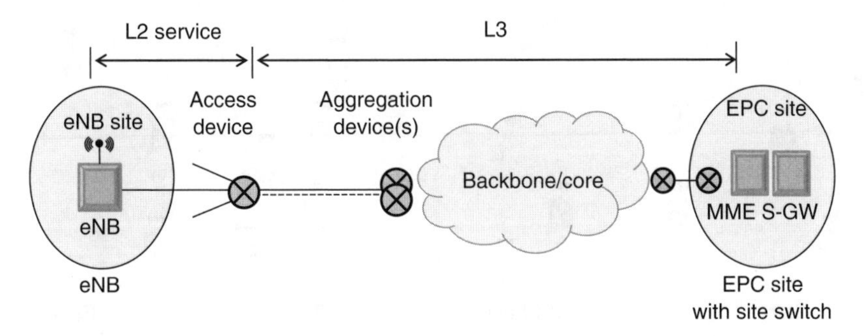

Figure 5.6 Network deployment example using Ethernet in the access

As a variant of this scenario, the Ethernet service may extend further to the network (e.g. to the aggregation edge), in which case the L3 demarcation starts from this point. It is recommended that the L2 domain be as small as possible and that IP is used for connectivity.

Dual attachment of the Ethernet service to the aggregation device, depending on the case, can be managed e.g. by link aggregation over multiple chassis.

5.3 Network Topology and Transport Media

In this section on network topology and transport media the objective is not to concentrate on the transport media selection in detail but to describe its effect on the network topologies.

5.3.1 Access Network Topologies and Media

Wireless transport is often used in the access part of the mobile backhaul network, when the capacities supported by the microwave equipment are enough for the expected and forecasted traffic, and fiber is not available cost-efficiently.

Topology is typically tree, chain, star or ring, or a combination of these. Selected topology has an impact on the feasible resiliency methods; protection methods for a ring topology are different from those for a chain topology. Figure 5.7 presents these traditional network topologies for the mobile network.

High-level design answers the following questions related to topology planning:

- How many eNBs can be connected to a hub in a star topology?
- How many eNBs can be connected to a chain?
- How many eNBs can be connected to a ring?

In order to be able to answer these questions, investigation of available link capacities is required, as well as capacity at the hub site, connection of the hub site to the aggregation network and the planned resiliency methods. When link capacities are known and the redundancy method selected, the maximum number of hops that the capacity allows can be defined. The higher the capacity per eNB, the fewer hops can be deployed or fewer eNBs can be included in the ring.

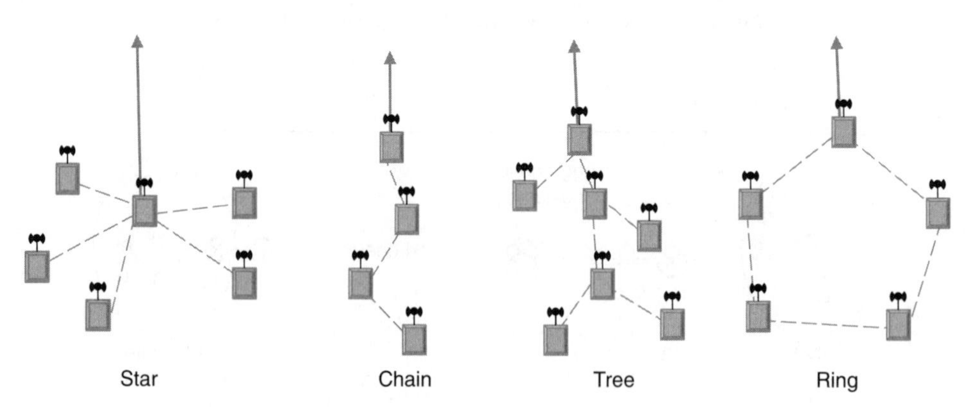

Figure 5.7 Simple network topologies

For wireless access, when both delay and bandwidth is taken into account, it does not make sense to chain more than a few eNBs. The star configuration is also the most simple to operate and maintain. Any failure in the first mile link will cause transmission failure, but that is easy to locate and link failure has an effect to one site only. Some of the links can be only partially filled that can cause higher transmission costs.

Ring topology provides protection against single link failure in the ring, and provides path diversity and therefore improves the resiliency compared to star, chain or tree topologies, without the need to implement a double link capacity for link protection.

Fiber access is available in urban areas, but availability is greatly dependent of the country in question. If fiber is available, there is no reason to use other media to connect the eNB to the aggregation point.

Leased lines can be used to connect eNBs to aggregation sites, when they are available, and this is a feasible access method. If the delivery time for leased line is good and cost is reasonable compared to quality, then leased lines can be a fast way to deploy remove the backhaul.

5.3.2 Aggregation Network Topologies

Aggregation network means further concentrating the eNB traffic toward the core network sites. S1-flex connects the eNB to multiple core sites with MME, S-GWs and often security gateways (SEGs), enabling load sharing and geo-redundancy for the core network elements. Therefore, the aggregation network design must support connectivity from eNB to multiple core network sites. The topology is typically ring structure, enabling redundancy and offering resilient connectivity to the core network. An example of this topology is presented in Chapter 2 (Figure 2.2) and is referenced throughout this chapter.

An aggregation network provides connectivity for the eNBs, and an availability requirement for the aggregation network is higher than the aggregate availability of the first mile. This can be achieved using protection methods like routing gateway redundancy and Ethernet ring protection (ERP), described further in the next section.

The capacity required for the aggregation network, and QoS for mobile traffic, needs to be flexible enough to accommodate the growth and upgrade of the network.

The core sites with MME and S-GW are connected together using a high-capacity core backbone network.

5.4 Availability and Resiliency Schemes

Resiliency in the backhaul network is needed to ensure mobile service availability in the event of failure. Failure can be due to a link or equipment failure, a power outage, weather conditions or human error in configuring the network. Links and devices can be also either completely unavailable or partially unavailable. Partial availability may mean that only a part of the traffic is affected or that the service is partly degraded (e.g. in terms of QoS).

Depending on the Stream Control Transmission Protocol (SCTP) and General packet radio service Tunneling Protocol (GTP) parameters, the maximum transport impairment time on the S1 bearer to drop a call can be up to 20 seconds, that is a voice call is likely to be ended by the subscriber before eNB/S-GW/MME drops it. The maximum tolerable S1 interruption time affects resiliency requirement for the transport network.

Section 5.4.1 describes common terms for the calculation of availability. Section 5.9. describes the security solutions and more details on protection and redundancy mechanism related to S-GWs and CAs. The theory on availability serves as an introduction to the different resiliency and redundancy methods available to improve the availability of the backhaul network.

5.4.1 Availability Calculation

Availability is calculated using commonly known formulas for mean time between failures (MTBF) and mean time to repair (MTTR).

MTBF is the predicted time between failures of a system during operation. MTBF can be calculated as the arithmetic average time between failures of a system. The MTBF considers failures which place the system out of service and assumes that the failed system is immediately repaired. Actions like scheduled maintenance are not considered within the definition of failure. Equipment vendors provide MTBF values for equipment and hardware modules.

MTTR is the time taken to repair a failed system. Systems may recover from failures automatically or by manual repair. With dynamic recovery (e.g. with dynamic routing around link or node failures), MTTR depends on the time needed for detection and for the actual recovery of traffic via the new path. SDH-like protection switching can act as fast as 50 ms, but even slower recovery (on the seconds range) is much faster than any manual action.

Repair may also mean replacing the hardware module or manually rebooting the network element in case the problem is caused by a software failure. MTTR varies from operator to operator, depending, for example, on the availability of the replacement hardware and the time it takes to deliver the replacement to the site and change it.

Availability (A) is calculated from the MTBF and MTTR in following manner in Equation 5.1:

$$A = \frac{MTBF}{MTBF + MTTR}. \tag{5.1}$$

Availability is presented in Table 5.1 as so-called nines, "3-nines" means 99.9% and "5-nines" means 99.999% availability. Downtime per year is a more intuitive representation of the availability and in Table 5.1 one finds the availability percentage and the corresponding downtime per year.

The relationship between availability and unavailability is presented in Equation 5.2.

$$U = 1 - A. \tag{5.2}$$

Table 5.1 Availability and corresponding downtime per year

Availability	Downtime per year
99%	3.65 days
99.9%	8.76 hr
99.99%	52 min
99.999%	5 min

Any problems that damage physical equipment will degrade availability dramatically. To be able to reach high availability in the backhaul network, equipment protection within single chassis or redundant equipment as well as link level protection are required. The following section discusses this further.

5.4.2 Link Resiliency and its Impact on Availability

Transport node failure is considered here as a failure causing an outage of the whole equipment and affects all traffic carried through the equipment. The equipment designed for access, such as small site switches and routers, is cost and feature optimized and therefore often provides fewer protection mechanisms than equipment that is designed for more traffic and aimed at the aggregation network.

Depending on the equipment architecture, certain functions that can affect traffic can be made redundant, like the power supply, interface and control cards in a single chassis. It is also possible to have redundant (duplicated) equipment to prevent failures, requiring additional features within the equipment.

Link failure is likely to happen more frequently than equipment failure (Metsälä and Salmelin, 2012) and there are multiple possibilities to detect and protect the network from link failures. Whether to protect a link from failures using link protection depends on how much traffic is carried over the link. A single eNB connection is not likely to require link protection, but if that link carries other radio access technologies as well and provides coverage for large areas, the situation is different. In this latter case, one needs to consider especially how to maintain voice service and coverage in the event of link failures. Often the availability of the first mile link to the single eNB dominates the availability calculation of the backhaul connectivity of the eNB.

In a tree or chain topology, the links carrying traffic for multiple eNBs are likely to be protected. Ring topology is useful for protection since the topology protects against any single link failure. The availability of the transport connection from an eNB perspective can be calculated using the formulas for link, chain and ring unavailability below, and then Equation 5.2 can be used to calculate the resultant availability.

5.4.2.1 Unavailability of a Single Link

With wireless links, the total link unavailability (Equation 5.3) can be calculated by summing up first the unavailability of the indoor (U_i) and outdoor unit (U_o) unavailability and the link media (U_m) unavailability that depends on rain intensity and propagation anomalies. Note that in Equation 5.3 the unavailability for the indoor and outdoor units for a link are considered at both ends of the link, that is the near end as well as the far end.

$$U_{link} \cong U_i + U_o + U_m + U_o + U_i. \tag{5.3}$$

For fiber connection the indoor and outdoor equipment can be replaced with relevant termination equipment, and use the related "unavailabilities."

The equation is accurate when unavailability is low, since it assumes that unavailabilities of components in the chain do not coincide. If the unavailability of any of the components in the chain is high, there is a possibility that two or more components in the chain are unavailable at the same time; in this case, the equation gives pessimistic results. Unavailability can be calculated more accurately by using the serial availability in Equations 5.4 and 5.5.

$$A_{link} = A_i \times A_o \times A_m \times A_o \times A_i. \tag{5.4}$$

$$U_{link} = 1 - A_{link}. \tag{5.5}$$

5.4.2.2 Unavailability of a Chain

When calculating the unavailability of a tree or chain topology, the above single link unavailability can be used to calculate the unavailability of chains of links by summing up the unavailability of the links with Equation 5.6, as presented in Equation 5.3:

$$U_{chain} \cong \#links \times \left(U_i + U_o + U_m + U_o\right) + U_i. \tag{5.6}$$

Again, the equation gives pessimistic values if non-negligible probability and more than one link at a time fails.

5.4.2.3 Unavailability of a Parallel System

Assuming that two systems in parallel have unavailability U_1 and U_2, then the unavailability of the system is calculated in Equation 5.7:

$$U_{parallel} = U_1 \times U_2. \tag{5.7}$$

In other words, the parallel system becomes unavailable only if both branches fail.

5.4.2.4 Unavailability of a Ring

A ring can be split to a parallel system and then the parallel system unavailability formula can be used. Figure 5.8 illustrates this.

Considering the availably from a specific eNB point of view that is located in a ring, the availability can be calculated based on the link availabilities. In Figure 5.8 the ring is presented as an equivalent parallel system having two parallel connections and their availabilities A_1 and A_2.

The unavailability U1 is the unavailability of the three links in series and the unavailability of the U2 is the unavailability of the two links in series and those can be estimated using Equation 5.6 for the U_{chain} considering the individual unavailability of the links. Generally, the availability of x links in series (each with availability A) can be calculated as A^x.

Unavailability of two systems in parallel U_{ring} is calculated in Equation 5.8:

$$U_{ring} = U_a \times U_b. \tag{5.8}$$

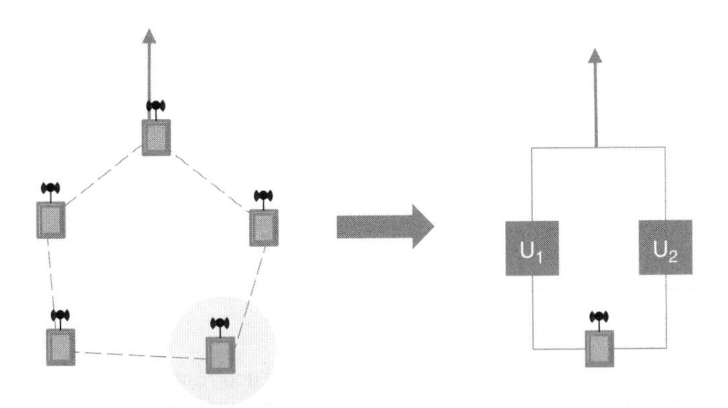

Figure 5.8 Ring availability from an eNB point of view presented as an equivalent parallel system

And finally the availability of the ring transport from an eNB point of view is calculated in Equation 5.9:

$$A_{ring} = 1 - U_{ring}. \tag{5.9}$$

The calculation can be repeated for all eNBs in the ring and the worst-case (or average) unavailability can be used as the total ring availability.

Using the equations presented above, the availability of the transport from an eNB point of view can be calculated using the specific formulas for a single link, chain or ring. The following example is calculated for a microwave using different protection methods and numbers of links in a chain.

5.4.2.5 Availability Calculation Example

Availability calculations can be done to compare different network topologies. The first task is to calculate the availability of the different equipment taking into account the replacement time. In case there is hot stand-by or 1+1 protection, the availability can be considered 100%. The following example shows how availability can be calculated for chained microwave links using hardware protection options. The MTBF values vary dependent on the vendor, and are typically somewhere between 20 and 100 years. The values in Table 5.2 are examples.

Unavailability U is calculated as 1 – A, and for links the unavailability is calculated in Table 5.3 summing up the different equipment, indoor and outdoor units.

Summing up the indoor unit unavailability and the number of links provides the unavailability for the chain of links. As described earlier, due to delay and capacity requirements, only a few eNBs should be chained in microwave access; therefore, the calculation in Table 5.4 has been done up to three links. Note that the MWR link availability calculation takes into account only the indoor unit failure; additionally the weather conditions have an effect on the availability.

Availability for the link is 100%—the unavailability and values for the link chains have been translated to availability in Table 5.5. Depending on the availability targets set for the backhaul, the maximum number of links in a chain based on availability and the redundancy can be selected.

Table 5.2 Availability calculations for different hardware configurations

Equipment	MTBF (years)	MTTR (hours)	Availability (A)	Unavailability (U)
Microwave outdoor unit	80	6	99.9991%	0.0009%
Microwave indoor unit, 1+0	60	4	99.9992%	0.0008%
Microwave outdoor unit, hot stand-by	—	—	100.000%	0.0000%
Microwave indoor unit, 1+1	—	—	100.0000%	0.0000%

Table 5.3 Unavailability for different types of links is calculated from the outdoor and indoor units

Equipment and link unavailability	Unavailability (U)
Link unavailability[a] – non hot standby + 1 + 0 indoor unit	0.0008% + 2 × 0.0009% = 0.0025%
Link unavailability – hot stand-by + 1 + 0 indoor unit	0.0008%
Link unavailability – hot stand-by, 1 + 1 indoor unit	0.0000%

[a] Unavailability of link = 2 × outdoor unit unavailability + indoor unit unavailability.

Table 5.4 Unavailability comparison for different types of links[a]

# of links	Non-hot-stand-by chain	Hot-stand-by chain	Hot-stand-by chain, 1+1 indoor unit
1	0.0008%+1×0.0025%= 0.003%	0.002%	0.000%
2	0.006%	0.002%	0.000%
3	0.008%	0.003%	0.000%

[a] Unavailability of links = indoor unit unavailability + # links × indoor unit unavailability.

Table 5.5 Availability comparison for different types of links

# of links	Non-hot-stand-by chain (%)	Hot-stand-by chain (%)	Hot-stand-by chain, 1+1 indoor unit (%)
1	99.997	99.998	100.000
2	99.994	99.998	100.000
3	99.992	99.997	100.000

The section will continue by describing different redundancy and resiliency methods for improving availability, as described at the beginning of the Section 5.4.

5.4.3 Routing Gateway Redundancy

The routing gateway from the eNB point of view is the next router. As illustrated in the backhaul network topology in Chapter 2 (Figure 2.2), depending on the access network implementation, the access gateway router (the first-hop router) is often at the pre-aggregation site or aggregation site, when the access backhaul is implemented using point-to-point Ethernet or Ethernet service.

In addition to the transport link and equipment, the access router redundancy is part of the resiliency design and depends on the number of eNBs connected through the router. On telecom sites there is also battery backup normally available and it can be also provided to the routing equipment. There are different alternatives to introduce routing gateway redundancy and these are briefly described in this section.

5.4.3.1 Gateway Hardware Redundancy

When enabled by the router architecture, a simple scheme to make the access router redundant is to use redundant routing engine/switch fabric, power supplies, fans and link aggregation groups (LAGs) to two different line cards. Thus a fairly resilient router configuration can be built using a single chassis only, thereby saving the cost of second chassis. This kind of redundancy is completely transparent to the eNB.

5.4.3.2 Gateway Redundancy Using VRRP/HSRP

Router redundancy protocols Virtual Router Redundancy Protocol (VRRP) or Hot Standby Router Protocol (HSRP) can be considered a form of L3 redundancy. The operation principle of both VRRP and HSRP[2] is the same, although the protocols are not interoperable.

The VRRP/HSRP operation for a single VLAN and single VRRP/HSRP group is shown in Figure 5.9.

The eNB uses virtual IP of the redundancy group and the protocol will decide which of the real interface IP addresses and equipment will be used. The virtual IP address can be configured as the default gateway in the eNB. In this way the redundancy is transparent to the eNB.

The hello interval and failover time can be less than a second with VRRPv3 (RFC5798) and HSRPv2, if supported by all the routers in the group. Many routers also support BFD-based polling between the routers, and use of BFD can reduce failover triggering time to lower values.

For access router design, the following design aspects should be taken into account:

- The number of VRRP/HSRP groups per physical router is limited. In case each eNB is assigned its own VLAN, or more than one VLAN per eNB is needed, the group limitation is quickly reached. The number of supported instances varies from tens to hundreds, but latest product details should always be checked from router vendors.
- Most routers support creating more than one HSRP/VRRP group in one VLAN to balance router load. In this case the eNBs in the common VLAN are split equally between the two HSRP/VRRP groups, each group having a different physical router as master in order to balance load.
- UL interface tracking is supported by some routers. This allows triggering of router failover in case of UL interface failure in the router.
- Sub-second failure detection time can be achieved by using BFD. One BFD session between two routers can provide early failover notification for multiple HSRP groups.

[2] HSRP is a Cisco® proprietary protocol.

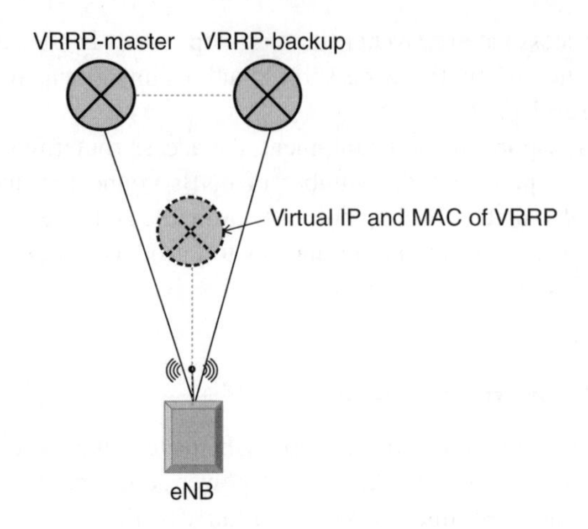

Figure 5.9 HSRP/VRRP principle

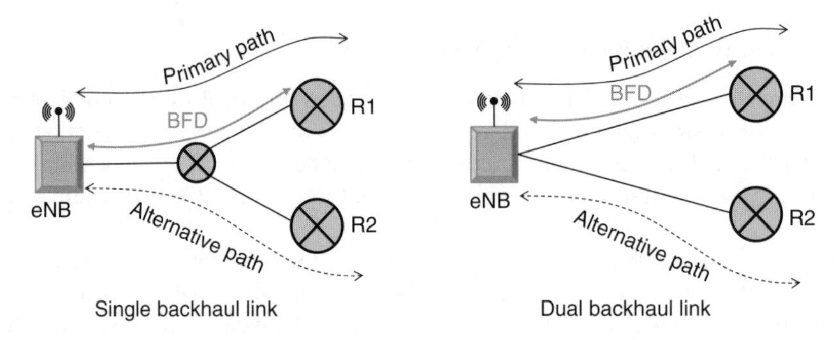

Figure 5.10 BFD-triggered gateway protection

5.4.3.3 Gateway Redundancy Using eNB-Controlled Gateway Selection

Without VRRP/HSRP, the first-hop gateway can, in special cases, be made redundant by eNB detecting gateway availability using BFD or Internet Control Message Protocol (ICMP). This requires that the eNB can change the default gateway address based on the detected availability, which solves the UL direction. For DL direction, R1 should be able to route traffic via R2 to the eNB, in case the detection fails. This can be regarded as dynamic routing without a routing protocol.

The eNB has a primary path toward router R1 and an alternative path toward router R2, either directly from eNB using two physical paths or via, for example, an L2 device as shown in the left-hand side of Figure 5.10. If first-hop gateways are located on different sites and two at least partially separate paths exist, as shown in Figure 5.10, the method also provides resiliency against link failures.

In addition to requiring support from the eNB, it should be verified that the access router supports single-hop BFD. Another design consideration is the number of BFD sessions supported by the access router. This number may be a limiting factor if BFD is processed on the router's central processing unit (CPU). Therefore, offloading BFD processing from the router's CPU to line cards is recommended, if possible. Alternatively, load balancing can be applied by using BFD in both routers with one router supporting the primary route of half of the eNBs and the other router supporting the other half of the eNBs.

5.4.3.4 Cell Site Gateway

L3/IP and lately also MPLS can also reach the eNB site. In this case there is less need to duplicate the router equipment, if the router failure only affects a single base station.

5.4.4 Ethernet Ring Protection (ERP)

Ring topology can be used for protecting the access traffic. There are multiple ways for protecting the traffic in an Ethernet ring and the section describes few concepts that can be utilized.

5.4.4.1 Ethernet Ring Protection Switching

ITU-T (2012) G.8032 provides protection and recovery switching within 50 ms for Ethernet traffic in a ring topology. One of the links in the ring is not permanently used for traffic. This avoids a loop in the topology. This can be a pre-determined link or a failed link. There are two versions of the recommendation: G.8032v1 supports a single ring topology and G.8032v2 supports multiple rings or ladder topology. Many vendors support the standard and there are many other benefits, such as:

- 50 ms protection switching
- no complicated C-plane or design
- predictable protection method regardless of the load
- support for forced switching of traffic direction, which is useful for transport network maintenance.

ITU-T G.8032 could be used to protect a microwave-based access ring. The number of eNBs in the ring is limited often by both capacity available in the ring and by the delay through the ring.

5.4.4.2 Spanning Tree Protocols

The Rapid Spanning Tree Protocol (RSTP) and Multiple Spanning Tree Protocol (MSTP) are improved versions of the Spanning Tree Protocol (STP) and avoid looping by determining the best path to the destination and disabling the other paths or links. Protection is triggered when the best path fails and the protocol resolves a new path by activating the blocked links. MSTP (which has RSTP built in for rapid convergence) enables VLANs to be grouped into a spanning-tree instance, with each instance having a spanning-tree topology independent of other spanning-tree instances. Hence MSTP is used with multiple VLANs while RSTP can be used if there is only one (V)LAN in the L2 domain. Some switch vendors have their own proprietary variants of xSTP. Compared to ERP (according to G.8032) the different versions of spanning-tree protocols can support topology resolution within seconds and the time can depend on the size of the networks and load. It is a well-known limitation for STP-based protection that it does not suit a wide-area recovery, but is more feasible in a small and limited segment.

5.4.4.3 Ethernet Automatic Protection Switching (EAPS)

Ethernet Automatic Protection Switching (EAPS; RFC3619) uses a so-called domain that includes one master node and all the other nodes are transit nodes. Loop avoidance is implemented in the master node that has two ports: primary and secondary. The secondary port is blocked in normal operation to avoid loops. If the master detects a fault in the network, it activates the secondary port. Functionality depends on the master node and it is the single point of failure in the topology. EAPS was originally developed by Extreme and even if it were standardized not many other vendors have adopted the standard.

There are also other Ethernet-based protection methods that are vendor proprietary and therefore not within the scope of this document. Many of these are intended for wireless ring protection methods for packet traffic.

5.4.5 IP and MPLS Rerouting

Backhaul protection can utilize IP re-routing in cases of link failures supported by Open Shortest Path First (OSPF) or other routing protocol. With routed access, alternative paths can be utilized by the routing decisions of the routers from the first-hop gateway onwards. With a cell site router, whole access resiliency can also be based on routing.

IP-based route load balancing mechanisms, for example equal cost multipath (ECMP) routing, could be utilized if supported by a corresponding edge router. When considering using ECMP routing, it is worth mentioning that it may not work well in cases of packet fragmentation or when synchronization traffic is carried.

OSPF protection latency has been considered higher than the SDH protection schemes that are often used as reference for other protection methods, but can be accelerated with parameter optimization and usage of BFD protocols.

Alternatively OSPFv2 Loop-Free Alternate Fast Reroute can be used to achieve faster protection times. It uses a pre-computed alternate next hop to reduce failure reaction time when the primary next hop fails. Configuration of a per-prefix loop-free alternate (LFA) path enables redirection of traffic to a next hop other than the primary neighbor without other routers' knowledge of the failure. Without LFAs, link down indication triggers a re-calculation of the routing table and selection of the resulting next hop.

MPLS fast reroute protects an LSP in an MPLS network. The protection latency can be reputedly made about the same as in SDH by using BFD (<50 ms claimed possible in some documents). The idea is similar to the LFA in that a protecting LSP is already set up which reduces switching time.

OSPF or other routing protocol general design guidance for enterprise networks is mostly valid for LTE access design as well, and for IP MPLS designs for aggregation and core tiers.

5.4.6 SCTP Multi-Homing

SCTP multi-homing is commonly used protection method in GSM and W-CDMA networks for protecting C-plane connections from controller toward core network elements. In LTE SCTP multi-homing can be considered for transport path redundancy for eNB signaling traffic. It should be noted that SCTP multi-homing supports only against C-plane path failures and is useful only if there are two physically different transport paths available in the access transport network.

As SCTP multi-homing offers no protection for U-plane traffic, some other redundancy mechanism is typically required in the transport. If there are two separate transport paths available, instead of SCTP multi-homing, it may be simpler and more efficient to use IP/MPLS rerouting, or in case of L2 connectivity ERP since this provides protection for all traffic planes—and not only the C-plane—for the dual-routed section of the transport network. Also at the MME end, there are other interface protection methods to protect signaling traffic.

When using SCTP parameters, such as maximum and minimum retransmission timers and heartbeat interval, these should be aligned with MME and the transport network in order to guarantee correct operation.

5.4.7 Connectivity Toward Multiple S-GWs and MMEs

An eNB can be connected to multiple core network elements and this is called S1-flex. The C-plane connection to the MME is done by the eNB for every S1 context request. MME addresses need to be configured in the eNB and eNB needs to have connectivity to all MMEs. If the core network element fails, for example due to power failure, the ongoing contexts are lost. New contexts can, however, be established by the other core network elements.

Multi-operator core network functionality enabling network sharing for different operators can be considered a generalized case of S1-flex, where different MMEs are from different public land mobile networks (PLMNs). PLMN and MME selection is done by eNB also in this case.

For the U-plane connection from the eNB to the core the S-GW and P-GW selection is done by MME and a domain name system (DNS) server. It is transparent to the eNB and the eNB only receives the S1 U-plane IP address during the bearer setup by S1-AP signaling from MME. The eNB routing table should be designed in such a way that all possible S-GW addresses will be routed to the U-plane subnet. Hence some coordination is typically required with EPC planners, regarding S1-U subnetting plan.

The MME and DNS can load balance S1 traffic based on S-GW selection procedure, and also use S-GW redundancy. This redundancy is transparent to the eNB, since the destination endpoint in the S-GW for the U-plane bearer is given to the eNB by MME in the S1-AP signaling.

Example 5.1

Assume that the network master IP plan states that all S1 and X2 U-plane addresses in the network are to be allocated from 10.10.0.0/16 subnet and one eNB VLAN is dedicated to the U-plane. In this case, the routing entry at eNB would state “to 10.10.0.0/16 via the UP VLAN gateway for that eNB.” In this way, all S1 and X2 UP packet are routed to UP VLAN. Of course, the UP VLAN addresses themselves in all eNBs would have to be allocated from this same subnet for X2 routing to work. Note, however, that in general it is not a good idea to dedicate a VLAN for U-plane traffic, but only for QoS and routing reasons.

5.4.8 Synchronization Protection

Synchronization protection does not have to be as fast as protection for user and control plane connections since a macro eNB can survive on-air in holdover mode for a longer time—some days—especially in the case of frequency synchronization. Synchronization protection can also be implemented by provisioning multiple synchronization sources at eNB such as the Global Positioning System (GPS), IEEE1588v2 (IEEE1588, 2008) and Synchronous Ethernet (SyncE), and also the co-sited base station can be a source of synchronization for the eNB.

"Timing over packet" implemented using PTP defined by the IEEE1588v2 provides synchronization for eNBs without support from the transport network apart for the required delay and delay variation requirements. There is a master clock that provides the synchronization to the slave clocks located in the eNBs. There can be multiple master clocks to provide synchronization redundancy, as illustrated in Figure 5.11. The slave clock exchanges messages with the master, and when the master cannot be reached a new master clock connection can be re-selected. Also route redundancy is possible to build this way if master clocks are reached over different IP routes. If master clocks are installed on different sites, also geo-redundancy is possible. eNBs should be allocated evenly among all master clocks so that no single master clock carries more than its total capacity in a failure situation.

From an eNB viewpoint, the two (or more) potential timing masters need to be reachable via their IP addresses. When the slave master's messaging detects that one master has failed, the eNB should be able to receive timing from the other master – and respond to the messaging originating from that endpoint.

IEEE1588v2 is normally used in unicast mode where eNB is connected to a single timing master that sends the synchronization massages to the slave. When using multicast mode, the redundancy is already built in.

IEEE1588v2 boundary clock (BC) can also provide synchronization redundancy using the backhaul equipment. The BC has multiple interfaces and can act as a master as well as a slave. It synchronizes itself to the best master clock using a best master clock algorithm (BMCA) and can provide synchronization onwards to eNBs.

5.4.9 OSS Resiliency

Operating support system (OSS) resiliency ensures that there are multiple network management systems that can connect to the network. It can be considered as having multiple intermediate

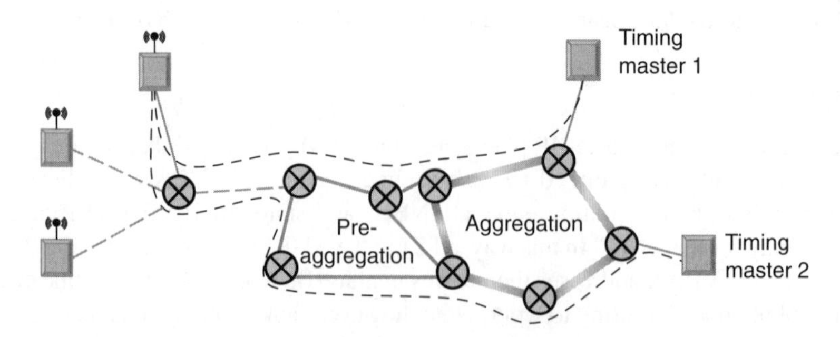

Figure 5.11 Synchronization protection using IEEE1588v2 timing over packet

points to access the network equipment, eNB. When there is some equipment between the OSS systems and the actual network equipment that translates protocols used in the network equipment and that can also introduce security for operations and maintenance traffic, this equipment should be backed up in order to avoid single point of failure. This is further elaborated on in Chapter 7.

5.4.10 End-to-End Performance of Multilayer Redundancy

When protection mechanisms are combined, it is important to ensure that the lower layer provides the protection path before the higher layer reacts, as otherwise network instability can cause "flapping" (the available path changing in quick sequence). Table 5.6 indicates typical reaction times of protection methods.

Typical problems in transport network redundancy are related to the use of multiple protection schemes and non-visibility beyond the protection boundary, for example protection in one network segment has no state information of the neighboring segment or lower-level resilience and protection is invisible to higher layers. The following lists a few potential issues that should be considered when planning redundancy for the backhaul network.

- SCTP multi-homing as protection mechanism is intended to be working when there are two physically different paths over the network between the two endpoints. If one path is not working, traffic is switched to the protecting path. In case there are lower-level protections mechanisms like IP rerouting in use, those mechanisms need to be faster than SCTP. In case the SCTP is faster to react, it will start sending traffic using an alternative path. Once the lower-level protection has repaired the primary path, the alternative path may become unavailable and the traffic is switched back, causing unnecessary flapping of the SCTP path.
- In case the SCTP timers do not match between eNB and MME, MME can declare S1 as being down and cause call drops.
- Using SCTP multi-homing together with VRRP can be difficult as the SCTP requires two physically separate paths in order to provide protection. Protection provided by VRRP and the related virtual router change can eliminate the secondary route to the destination.

Table 5.6 Protection latency times from different layers

Layer	Protocols	Typical reaction time
Transport layer	SCTP	Seconds, depends on parameter settings
Network layer	Reroute (any routing protocol, such as OSPF)	Seconds, below one second with BFD
Data layer	ERP, RSTP, MSTP	50 ms (ERP), to dozens of seconds (xSTP)
Physical layer	IEEE 802.3z, SNCP, MSP, BSHR/MS-SPRING	Below 50 ms, 700–1000 ms (IEEE 802.3z)

ERP: Ethernet ring protection; MSP: multiplex section protection; BSHR: bidirectional self-healing ring; MS-SPRING: multiplex section protection ring

- VRRP timers should be tuned so that the router in restart process (e.g. after recovering from failure) does not revert back to active state when it is not yet fully operational and able to carry traffic (e.g. still booting up).

As a summary, end-to-end resiliency when using multiple schemes should be designed and if possible tested carefully before rollout to prevent problems in the network. Keeping the design simple enough helps planning of the relevant parameters.

5.5 QoS Planning

5.5.1 QoS in an Access Transport Node

To facilitate discussion it is useful to define a QoS model for a generic access transport node. Examples of an access transport node could be:

- eNB, or a transport interface card within an eNB
- layer 2 switch connected to a microwave radio outdoor unit
- access router connected to several eNBs.

The simple model shown in Figure 5.12 is adopted in this section.

While by far not the most generic model possible, it nonetheless incorporates the basic building blocks relevant to the planning of QoS parameters for a typical access network transport node. The figure shows one physical Ethernet port hosting three VLANs, each of the VLANs having identical QoS processing. The building blocks in the model are, from ingress to egress:

- *Mapping to VLAN queues:* Mapping of packets with a given VLAN identifier into scheduler queues typically based on DSCP (differentiated services code point), MPLS TC (multiprotocol label switching traffic class) field, or VLAN 802.1p priority bits. More complex mapping rules are also possible (e.g. based on source/destination IP address or L4 protocol type), depending on the implementation.

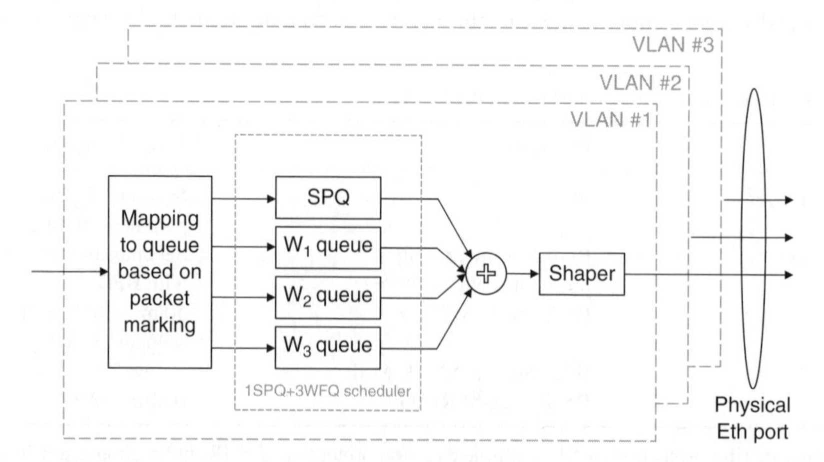

Figure 5.12 A simple QoS model for an access layer transport node. Other scheduler types could also be used

- *Scheduling:* Reordering packets in the transmit buffer based on the scheduler algorithm, and sending the reordered packets to the egress physical interface
- *Shaping:* Delaying individual packets within a burst of packets so that packets are not allowed to exit the egress port at line rate (back-to-back).

Policing, sometimes confused with shaping, is not considered in the QoS model. While shaping only delays egress packets not conforming to a traffic policy, policing in turn would discard ingress packets not conforming to the same policy. Therefore, policing is typically used by a leased line provider to discard (or mark packets for a possible later discard) packets violating an SLA, whereas for a mobile operator there is often little point in discarding his own traffic at an ingress port of his own transport node. Rather the preferred way is to let shaping handle packet dropping at the egress port when necessary. A potential usage for ingress policing is to prevent DoS attacks where the attacker floods the highest strict priority queue so that all lower-priority traffic is starved. If it is suspected that the network is vulnerable to such an attack, one option would be to police the high-priority traffic entering the node. Policing will be re-visited briefly later in this section when discussing how to plan shaper parameters to conform to a given leased line SLA.

It is assumed that the actual classification and marking is done outside the transport node at the traffic source. In the UL direction the packet marking is done at the eNB, and in the DL direction in MME, SP-GW[3], or the network management server. Especially in DL it may be more convenient to mark signaling and management traffic packets in core site routers, allowing a centralized configuration management of packet marking. This is not possible for U-plane traffic since the router has no knowledge of the QCI of the S1 bearer, and because of this the SP-GW needs to mark the U-plane traffic.

It should be noted that some routers erase DSCP marking by default, and therefore it is important to check that the intended marking is maintained throughout the transport chain. If IPsec is used, the SEG and the eNB should be configured so that the source packet DSCP is copied into the outer IP header.

Numerous scheduler implementations are possible. In the model shown in Figure 5.12, a four-queue scheduler is shown with one strict priority queue and three weighted fair queues. Most access transport equipment in the market at the time of writing support more than four queues, possibly with several scheduling schemes to choose from. Scheduling can also be performed at the physical interface level.

As with scheduling, shaping can also be performed at multiple locations within the QoS processing chain. In addition to VLAN-level shaping, a traffic shaper could also be located at the physical interface level, at the scheduler queue level or at any combination of these. A multi-stage scheduling and shaping is sometimes called "hierarchical QoS."

5.5.2 Packet Classification

Traffic differentiation in the transport domain can be based on DSCP, 802.1p bits or MPLS-TC[4] field. In the UL, eNB marks U-plane packets based on QCI values. QCI of the bearer is

[3] SP-GW means S-GW, P-GW or a combi-node implementing both.
[4] MPLS-TC field was initially called EXP (experimental bits).

received during S1 bearer setup signaling from MME. For U-plane traffic, the mapping from bearer QCI to packet priority needs to be configured in the eNB parameter. In the DL, the corresponding mapping table is in SP-GW.

The practical task of the access transport planner is to map different traffic types (identified by DSCPs, 802.1p bits or MPLS-TC field to scheduler queues in the egress ports. The usual challenge in the task is that the egress ports in the e2e transport path may have different types of schedulers and often implement different numbers of queues. Parameter design should be such that a given traffic type experiences similar treatment across all schedulers. To simplify this task, a pragmatic approach is to resort to as few traffic types as possible; at most, four traffic classes should suffice unless the operator has a very rich service offering.

Starting from some basic rules of thumb, a few general guidelines are given below.

- S1 and X2 signaling traffic should be assigned to the highest strict priority queue with the smallest delay as this guarantees shortest latency for signaling procedures, such as handovers and call setup.
- Network control traffic (e.g. IKE, OSPF) can also be assigned to the highest priority. Sometimes network control traffic is assigned even higher priority than telecom signaling, but in the case of mobile access backhaul the resulting increase in QoS design complexity may not be warranted since the volume of the Evolved Universal Terrestrial Radio Access Network (E-UTRAN) S1/X2 and IP Multimedia Subsystem (IMS) signaling traffic is at most a small percentage of the E-UTRAN U-plane traffic volume, and hence it is unlikely that telecom signaling would excessively delay network control traffic, or vice versa.
- The load of the strict priority queue should be kept low relative to the link capacity. This can be established by reserving the highest strict priority queue only for the most delay-sensitive traffic types.
- eNB management traffic should be assigned higher priority than the best effort (BE) data as otherwise congestion or abnormal network conditions could lead to remote connectivity problems. On the other hand in some cases (mass remote software download to multiple eNBs, cell tracing) the management traffic, usually carried by TCP, can temporarily consume large bandwidth and hence it is not advisable to map it to the highest priority queue either.
- Best Effort non-GBR (guaranteed bit rate) data should be assigned to the lowest priority.
- IMS voice codec specification sets fairly tolerant requirements on packet delay jitter 3GPP (2013) and hence the QCI 1 voice bearer does not necessarily have to be mapped to the highest priority strict priority queue. On the other hand, to minimize the number of traffic classes it could also be mapped to the same queue as signaling as the resulting total signaling plus voice traffic is still expected to be a small fraction of the link bandwidth. Based on testing with early commercial VoLTE UEs, the subjective voice quality is not very sensitive to the average packed delay variation. However, even a few infrequently occurring packet delay peaks—which have only a minor effect on the average packet delay variation—can cause audible degradation in speech quality. For this reason the end-to-end transport QoS design should attempt to guarantee that packet peak delays are bounded. In practice, the simplest way to do this is to use highest strict priority queue for voice traffic. It should be noted that the majority of the e2e voice packet delay jitter is caused by air interface scheduling and retransmissions.

The number of traffic classes in practical networks typically varies from four to eight. Prior to designing scheduler and shaper parameters, one has to decide how many QoS classes should be implemented. For U-plane traffic, the answer can be determined from the operator service offering, although perhaps more often the operator business plans are not known at the design phase, in which case it is advisable to leave some traffic types reserved for future services.

Table 5.7 presents an example mapping with four traffic classes, which is suitable, for example, for schedulers with at least four queues. As discussed, voice traffic could also be mapped to the highest priority queue resulting in an even simpler scheme, especially if the real-time data volume is high.

Table 5.8 presents an example mapping with eight traffic classes, which is suitable for schedulers with at least eight queues.

In this example, migrating from the four-class design to the eight-class does not require reconfiguring DSCP marking in the network (but would probably require reconfiguring queue weights). From a network management point of view, it is often convenient to base packet marking on DSCP throughout the network, if possible. In this case there is no need to configure and maintain any DSCP-802.1p mapping tables in the access network elements.

Table 5.7 A simple example of traffic type to service class mapping, with four classes

Traffic type	DSCP dec	802.1p priority	Queue
Telecom signaling and network control traffic	46	6	Queue 1, strict priority queue
Voice and real-time data	34	5	Queue 2, strict priority queue or WFQ with high weight
O&M traffic	18	2	Queue 3
Best effort non-real time data	0	0	Queue 4

Table 5.8 An example of traffic type to service class mapping, with eight classes

Traffic type	DSCP dec	802.1p priority	Queue
Telecom signaling and network control traffic	46	7	Strict priority 1 (Queue 1)
Voice	34	6	Strict priority 2 (Queue 2)
Reserved	30	5	Queue 3
Reserved	28	4	Queue 4
Real-time data, e.g. video	20	3	Queue 5
O&M traffic	18	2	Queue 6
"VIP non-real time data"	10	1	Queue 7
Best effort non-real time data	0	0	Queue 8

5.5.3 *Scheduling*

The purpose of scheduling is to prioritize important packets over less important ones when packets need to be queued at the egress port due to bandwidth limitation. For example, in a chain configuration with both incoming and outgoing ports having 1 Gbps line rate, there is no benefit in scheduling since outgoing packets can be sent out as soon as they arrive at the node. However, if the outgoing bandwidth is smaller than the instantaneous incoming traffic rate then there is a need for packet buffering (queuing), as otherwise a burst of incoming packets may need to be dropped if they cannot be sent out fast enough from the egress port. A scheduler provides a way to send the important packets first from the transmit buffer in order to minimize their delay. The question of how to choose the relative importance of packets (traffic types) is discussed in the previous section. In the following, the focus is on how to implement the desired prioritization for an egress port by means of scheduling parameter design.

Numerous types of scheduling algorithms have been proposed in engineering literature. The most common ones in access transport are the strict priority queuing (SPQ), weighted round robin (WRR) and weighted fair queuing (WFQ), or a combination of these. Perhaps the most useful scheduler type is a hybrid scheduler with one or two strict priority queues and a few weighted fair queues, for an example of 1SPQ + 3WFQ scheduler (Figure 5.12). For example, the so-called Cisco Low Latency Queuing is a combination strict priority and WFQ. Such scheduler implementation allows mapping delay-sensitive traffic to the high-priority queue, while other traffic types can be assigned to weighted fair queues with queue bandwidths designed according to service requirements.

5.5.3.1 Weighted Round Robin/Weighted Fair Queuing (WRR/WFQ)

A WRR/WFQ scheduler with four queues is shown in Figure 5.13. In the WRR algorithm the queues are checked cyclically (in a deterministic order) and if there are packets in the queue they will be sent to the egress port. The weight parameter defines the maximum number of packets per cycle that can be removed from the queue. The queue relative output rate is equal to or larger than the queue weight divided by the sum of all queue weights. For example, if the queue weights are {4,3,2,1} and the incoming rate is 1000 packets per second, then the first queue (weight 4) would be served at a rate of at least 400 packets per second (4/(4+3+2+1) = 40% of the relative capacity of the queue). It is important to note that, regardless of the packet size, for the WRR scheduler the per-queue service guarantee is given in terms of packet rate, not bit rate.

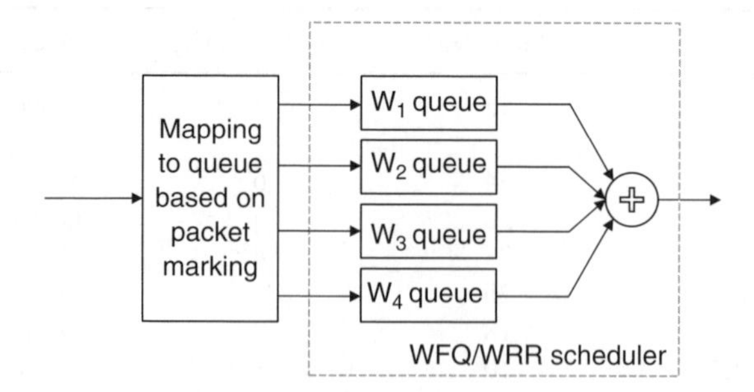

Figure 5.13 WRR/WFQ scheduler with four queues

The WRR scheduler is easy to implement in hardware but the queue weights are difficult to design if the packet average sizes of different traffic types are not known, which is typically the case in practice. For example, since S1 signaling traffic packets are relatively small with respect to DL user data packets, the WRR queue weights for signaling should be made sufficiently large relative to BE data to guarantee sufficient prioritization.

The WFQ algorithm solves the inconvenience of WRR, that is of requiring the knowledge of average packet sizes. This is essentially achieved by counting bits per queue instead of packets per queue. It follows that WFQ guarantees an average queue bit rate proportional to the relative weight of the queue. This is indeed a useful property for a transport planner as it simplifies the network design considerably. Assume again that the queue weights are {4,3,2,1} and the egress port bandwidth is 100 Mbps. Then the highest priority queue will receive at least an average of 40 Mbps bandwidth. Compared with a WRR scheduler with the same weights, if highest priority queue packet size is 1000 bits and the packets in other queues have 10,000 bit average size, then the highest priority queue would receive only 4 Mbps service rate with WRR scheduling. Depending on application, this may or may not be enough. In some WFQ implementations the guaranteed bandwidth of the queue is defined in percentage points of the egress port bandwidth.

While the WFQ scheduling principle sounds simple, hardware implementation of exact WFQ scheduler is difficult and some practically realizable approximation of the ideal WFQ scheduler is usually found in transport equipment; an example is the deficit weighted round robin (DWRR).

A remarkable theoretical property of the WFQ scheduler is that the worst-case packet delay in a transport network of several WFQ nodes can be guaranteed under some mild side conditions,[5] essentially that the maximum burst size is limited by traffic shaping at the network ingress. For example, no such delay guarantees are known for pure strict priority schedulers. This result, together with the simple design of weights, lends some justification for favoring the WFQ scheduler.

5.5.3.2 Strict Priority Queuing (SPQ)

The SPQ scheduler repeats the following two steps:

1. the scheduler (Figure 5.14) checks the transmit buffer and selects the highest priority packet
2. the scheduler sends the packet out from the egress port and moves back to the first step.

If in the first step there are several candidate packets with the same priority in the buffer then the packet that has waited in the buffer for the longest time is selected and sent out. A high-volume flow of high-priority packets can completely starve a lower-priority packet flow, causing unpredictable delay variations and bandwidth throttling for the lower-priority packets. Therefore, in sharp contrast to WFQ, SPQ offers no per-queue bandwidth or delay guarantee except for the highest priority queue which is guaranteed to be served at the egress port rate. Such a bandwidth guarantee for lower-priority queues could in principle be introduced by

[5] The worst-case delay guarantee holds if the traffic has been shaped to smoothen packet bursts. Details can be found in Stiliadis and Varma (1998). Similar, although looser, upper bound for the worst-case delay applies to the WRR scheduler as well.

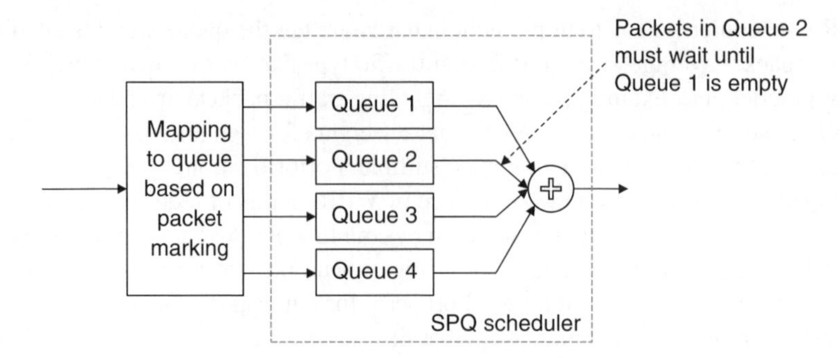

Figure 5.14 Strict priority queuing scheduler with four queues

policing high-priority queues which is possible in some scheduler implementations. On the other hand, it could also be argued that if there arises a need to artificially throttle SPQ bandwidth due to large amount of high-priority traffic, this to some extent defeats the whole idea of SPQ and in fact is a trigger for the capacity management process to optimize the QoS design itself, most likely by lowering priorities of some traffic types or by increasing the link capacity. There may be some use cases where rate limiting high-priority queues is desirable, for example prevention of DoS attacks.

Note that there are no parameters that need to be designed for a pure SPQ scheduler.

For the SPQ scheduler the average delay per queue can be calculated analytically assuming Poisson distributed packet arrivals. This is discussed briefly in Section 5.6.

5.5.4 *Traffic Shaping*

Traffic shaping limits the average transmit rate and breaks long packet bursts into shorter ones. Without shaping, packet bursts would leave the egress port at a line rate. This by itself is not a problem if the next transport node in the path is able to process the burst without losing any packets. Problems arise when the node receiving the packet burst has a small transmit buffer (relative to the burst size) and the egress port rate of the node is smaller than the incoming line rate. This scenario arises frequently in access network, especially when transitioning from a 1 Gbps Ethernet to sub-Gbps microwave links or sub-Gbps Next Generation Synchronous Digital Hierarchy (NG-SDH) links. This is illustrated in Figure 5.15, where the first node has ingress and egress line rate of 1 Gbps and the second node has an egress rate of 100 Mbps and buffer size of four packets (four-packet size is an example to illustrate the point, only). Without shaping the back-to-back packets arriving at a line rate to the first node pass the node unaltered since negligible buffering is needed. The second node, however, is unable to de-queue the buffer quickly enough and tail-drops packets after the first four packets in each burst. Traffic shaping improves the situation by increasing the packet inter-departure time from the first node by delaying the transmission of consecutive packets at the egress buffer. If the node #1 shaper maximum burst size and output rate are less than the buffer size and output rate of node #2 then node #2 is able to de-queue the packets from its output buffer without tail-drop.

Another use case for traffic shaping is when a leased line provider is policing incoming traffic to prevent SLA violations. If traffic sent to the leased line network is not scheduled and

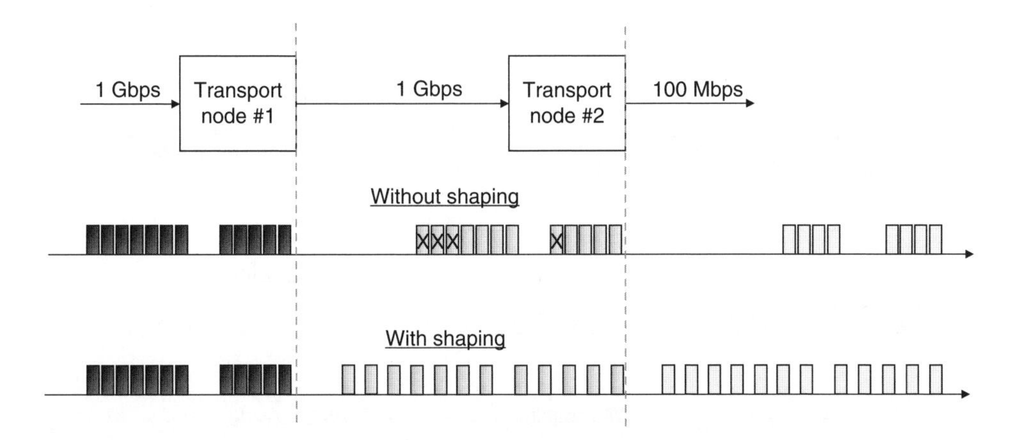

Figure 5.15 The operation of traffic shaping. Transport node #2 has a buffer size of four packets

shaped to comply with the SLA, the provider input policer may drop non-compliant packets. By designing the shaper parameters the operator may prevent this, or ensure that the less important packets (e.g. BE data) are dropped first. Figure 5.16 illustrates a scenario where the committed information rate (CIR) is 100 Mbps and the maximum burst size is two packets. Without shaping, the two bursts entering the leased network would violate the SLA and the provider would discard the non-compliant packets.

From field experience, for TCP traffic flows packet drop in the transport network is well known to degrade throughput. This is because the TCP protocol by design reacts to even minor packet loss by reducing the transmit rate.[6] Moreover, the TCP protocol slow-start and air interface scheduling tends to increase the burstiness of TCP flows, resulting in the tail-drop problem illustrated in Figure 5.9. It is not uncommon to see three to four times improvement in TCP throughput after shaping parameters in the access network have been optimized. The problem is made more severe by the introduction of LTE-A, which requires peak rates of up to 300 Mbps and beyond.

Since the purpose of the basic shaper described (also called the "single-rate shaper") is to control the average output rate and the maximum output burst size, it is natural that the shaper parameter design then involves choosing these two parameters.

Example 5.2

Using Metro Ethernet Forum terminology, the SLA states that the leased line CIR is 100 Mbps and committed burst size (CBS) is 9132 bytes. For this SLA the average output rate of the shaper should be set to, at most, 100 Mbps and maximum burst size to not more than 9132 bytes (e.g. six back-to-back frames of 1522 bytes each). The SLA may also define the non-zero excess information rate (EIR) and excess burst size (EBS). Designing shaper parameters for this kind of "dual-rate SLA" is left as an exercise.

[6] Several variants of the TCP protocol exist and the specific implementation in use depends on the server IP stack implementation and is in some cases configurable (e.g. Linux). Some TCP variants recover from packet loss faster than others.

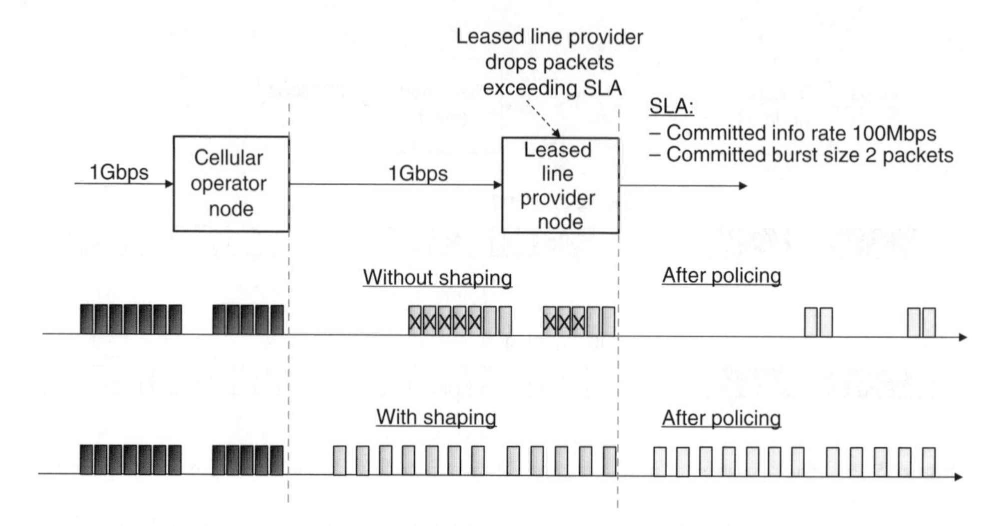

Figure 5.16 Traffic shaping and policing for the leased line provider use case. Service provider polices packets entering its network

It should be noticed that since traffic shaping modifies inter-packet delays it may also increase packet jitter, which is undesirable for some applications. For example, IEEE1588 (2008) Precision Time Protocol (PTP) packets should not be delayed since the protocol operation is based on a constant delay difference between consecutive packets. For this reason it is advantageous if the transport equipment supports shaping also at the queue level, instead of only at the physical interface of the VLAN level. This allows leaving the highest-priority low-delay queue unshaped, while the BE data queue (TCP traffic) can still be shaped.

The two most common shaper implementations are the leaky bucket shaper and the token bucket shaper. However, the algorithmic difference between various shaper implementations is not crucial for transport planning purposes and details are omitted here.

5.5.5 *Active Queue Management and Bufferbloat*

As discussed earlier, by default an egress port starts dropping packets when the egress buffer fills up and incoming packets cannot fit in. Transport equipment nowadays can house very large egress buffers (e.g. dozens of megabytes per output port), which may result in excessive queuing delays in the order of hundreds of milliseconds if the egress link is congested. When downloading a large file, the increased packet-level queuing delay cannot be noticed by the end user since latest TCP implementations typically have very large transmit and receive windows and are therefore able to operate at 100+ Mbps speeds even when the RTT is several hundreds of milliseconds. The drawback is that small TCP transfers such as Web page downloads are also affected (if sharing the same queue) since the long TCP RTT resulting from the queuing delay can slow down the Web page fetch time enormously. The problem of long queuing delays due to large transmit buffers is sometimes called "bufferbloat."

To combat these problems one suggestion has been to use smaller buffer sizes, since this will directly reduce the maximum possible queuing delay. A more advanced approach is to resort to active queue management (AQM) schemes such as RED or one of its many variants. (See Chapter 4 for more information on RED.) The underlying idea of RED is to start randomly dropping packets before the egress buffer fills up completely. This is in contrast to dropping packets only when the queue overflows (tail-drop). Dropping TCP packets will then result in the TCP flows crossing the link, reducing their transmission rate, which in turn lowers the buffer fill factor and queuing delay. The drawback is that high-rate TCP flows may suffer in the process, thus preventing achieving the peak throughput offered by LTE and LTE-A. The TCP peak throughput reduction is not a problem for Web browsing, but could limit the performance of large file download and "speed tests," which, depending on the operator, may or may not be more important than Web browsing. Therefore, whether to apply AQM techniques (such as xRED) depends on not only technical aspects but also the strategic decision of which type of BE data the transport network is optimized for.

Conventionally, AQM techniques have been proposed for high-capacity backbone links that carry thousands of TCP flows. For access links where the number of flows is typically in the order of hundreds, the deployment and optimization of AQM is likely to be more challenging. Access layer average link utilization also tends to be lower than in backbone links, making the bufferbloat problem less severe.

In engineering literature, practical guidelines for planning and tuning AQM algorithms in wireless backhaul are scarce, even non-existent, and very little field experience has been gathered. The interested reader can find more details and pointers to further references in Adams (2013).

Finally, it is worth noting that the packet delay caused by bufferbloat should primarily be solved by prioritizing the delay-sensitive traffic in network schedulers, or increasing the link bandwidth. After all, if the egress buffer of a link is continuously full, this should in most cases be a trigger point for upgrading the link capacity, rather than reducing buffer size.

5.5.6 Connection Admission Control

Connection admission control (CAC) decides whether to accept a new bearer setup into the system. The decision depends on the available resources and the amount of resources that the new bearer would consume if admitted to the system. Usually CAC is applied only to GBR bearers, such as voice. In E-UTRAN the admission control can be applied only at the S-GW and eNB since the transport network has no information of the call setup process at a higher layer. The actual CAC implementation is not specified in 3GPP and left to the network element vendor.

There are two components in CAC: radio interface admission control and transport admission control. The radio interface voice capacity of an LTE cell may be quite large, as simulation predictions state well over 200 simultaneous voice calls for a 20 MHz bandwidth cell, albeit the various assumptions made will have a strong impact. Even so, usually it should be the radio capacity limiting the number of admitted GBR bearers instead of the transport GBR bandwidth.

While there are as many CAC implementations as there are eNB vendors, a simple transport admission control algorithm would work as follows:

1. A new GBR bearer setup request is received from MME. Within the call setup S1-AP signaling message there are information elements for UL and DL GBR bandwidth, without any transport overheads.

2. eNB determines the transport bit rate used by all active GBR bearers in the UL and DL. This may be the actual total GBR measured by eNB, or in a simpler algorithm it may be the current number of GBR bearers times the pre-configured bit rate of one bearer. Transport overheads are taken into account and they depend, for example, on whether IPsec is used, VLAN header overhead, etc.
3. eNB sums the current GBR plus the GBR of the new bearer (with transport overheads). If the value exceeds a pre-configured threshold, the bearer setup is rejected. Otherwise, it is accepted. The check is made for both the UL and DL separately. Even if the bearer is accepted from a transport capacity viewpoint, it could still be rejected by radio interface admission control, though (again) this depends on eNB implementation.

If the total bit rate of the current GBR bearers in step 2 is not the actual measured one, it may make sense to apply "overbooking," for instance, by simply increasing the acceptance threshold used in step 3. For a voice bearer, the typical activity factor is about 0.6 and hence the acceptance threshold could be scaled by upwards of this amount. The e2e design should be coordinated with EPC planners so that the GBR values received in S1 signaling do not already contain overbooking.[7]

Example 5.3

Assume narrow-band adaptive multi-rate coding (AMR) codec with a source bit rate of 32 bytes per voice packet transmitted at 20 ms intervals. The UE adds a 40-byte RTP/UDP/IP header to the voice packet bringing the total S1 payload size to 72 bytes. The S1 transport overhead without IPsec consists of a GTP/UDP/IP header of 36 bytes (assuming no GTP extension header) and an Ethernet header of 42 bytes (including one VLAN header and Ethernet L1). Adding up, the total S1 PHY bit rate becomes 62.4 kilobits per second per voice call, with 100% voice activity. Suppose the eNB should support 400 simultaneous voice calls at transport layer. Then the admission control threshold should be set to 400 × 62.4 Kbps = 25 Mbps or higher.

Assuming that voice traffic model is available (i.e. offered traffic in Erlangs per eNB), it is possible to dimension the CAC threshold for a given bearer rejection probability using Erlang-B (single GBR service) or multidimensional Erlang-B formulas (multiple GBR services). Other dimensioning techniques have been developed as well, including the effective bandwidth approach (Berger and Whitt, 1998).

[7] IMS voice supports adaptive codec mode change and adaptive frame redundancy to handle packet loss (3GPP, 2013). If these techniques are employed the actual voice bit rate may vary depending on the detected e2e packet loss. The codec adaptation is controlled by the U-plane endpoints (UEs and/or MGW [media gateway]) and it is transparent to eNB and EPC. According to GSM Association (2015) the UE must request to receive a single speech frame per Real-time Transport Protocol (RTP) packet (no frame redundancy).

Example 5.4

Assume that eNB serves 10,000 subscribers (!) and each subscriber generates a busy hour (BH) voice traffic of 25 milliErlangs. The voice codec is narrow-band AMR and there are no other GBR services.

Question: How should the CAC rejection threshold be set if it is desired that at most 0.1% of voice calls are rejected because of transport admission control?

Answer: The total offered voice traffic per eNB is 250 ERL. The Erlang-B formula would suggest that to support 250 ERL traffic with 0.1% blocking there should be at least 291 calls worth of free capacity ("circuits"). Hence the CAC threshold could be set to 291 × 62.4 = 18.2 Mbps.

Although such calculations are possible, usually it would be advisable to let radio admission control handle GBR call rejections and overdimension the transport admission control (or not use it at all).

5.6 Link Bandwidth Dimensioning

Essentially, all models are wrong, but some are useful.

George E. P. Box

One of the most difficult and interesting engineering problems in access transport planning is that of link bandwidth dimensioning. A large amount of scientific papers on bandwidth dimensioning have been written in queuing theory and related fields. Then again, when applying the theory to a mobile access network, where the traffic of the first few miles tends to be dominated by packet flows of a small number of users, it seems that most of the assumptions behind the beautiful theory break down. This, in turn, renders the theoretical formulas inaccurate or at least difficult to use.

The bandwidth dimensioning problem is in principle straightforward if one knows the network topology, egress port buffer sizes, scheduler types, exact mathematical traffic description as well as delay and packet loss target per service. All that needs to be done is to plug this information into a simulator and turn the crank. Of course, in reality the required input information is rarely, if ever, available in this detail. As the input values are themselves often estimates, the dimensioning output cannot be very accurate either, regardless of how sophisticated a calculation method is used. Therefore, for practical bandwidth calculations a simple formula often provides sufficient accuracy and can be much easier to use.

In this section two methods for bandwidth dimensioning are discussed. The first method calculates the average packet-level delay and it is most suitable for voice and signaling applications. The second method calculates the TCP throughput degradation caused by a bandwidth bottleneck and it is most suitable for evaluation of QoS for BE data. Chapter 4 provides a theoretical background of various dimensioning approaches. The methods presented in this section have been selected based on suitability for practical planning purposes, namely the calculations can be implemented in a spreadsheet program without the need for a network simulator. First, however, the traffic parameters needed as an input for dimensioning calculations are discussed.

5.6.1 Obtaining Input Parameters for User Plane Bandwidth Dimensioning

The U-plane traffic is usually stated in terms of the following quantities.

- Average data bit rate per site during packet data BH. One site could have many cells and frequency layers, but for transport dimensioning this is not essential. If there are many data services envisioned then the traffic model is required for each data service.
- Average packet size: this is the size of the end user IP packet, to which S1-U transport overhead has to be added.
- Voice traffic in Erlangs per site during voice BH.

The above parameters may be required separately for the UL and DL, although usually it is enough to dimension the DL if the link physical bandwidth is symmetric. In case multiple non-GBR and GBR services are to be deployed, the traffic parameters should be available per-service.

To take into account traffic growth, the traffic information should be provided for all time epochs of interest.

Depending on the chosen dimensioning method, additional parameters are possibly required. For example, G/G/1 dimensioning formulas (discussed below) require a standard deviation of packet inter-arrival time and packet size.

Typically, the traffic model available at the design phase is somewhat less than complete, and in extreme cases contains only a single parameter: average data volume per subscriber over a certain period of time (month, BH, etc.). For voice, the traffic is almost always given in terms of voice traffic demand in Erlangs per subscriber during BH. The initial step is to convert the available traffic model information into a form required by the dimensioning method to be used. Frequently, this step also requires filling out information missing in the given dimensioning inputs.

5.6.1.1 Average Packet Data Bit Rate Per Site During Busy Hour

The U-plane traffic may be available as gigabytes per month per subscriber or BH traffic demand per subscriber. At this point it is important to distinguish that the traffic figures are provided at the network level or at the site level. Figure 5.17 presents an example of packet traffic distribution over 24 hr at the network level and at site level for three different sites. At the network level the busiest hour carries 6% of total daily traffic, which is a typical value applicable to most networks. At the site level the busiest hour carries 13–14% of the daily traffic, depending on the site. Therefore, if one calculates BH site traffic by simply dividing network-level traffic by the number of sites, the site traffic would be underestimated by over 100% in the example shown. This error would propagate to bandwidth dimensioning.

Since a site's BH depends on its location (due to user mobility), it is possible to exploit this dependency in bandwidth dimensioning by considering the so-called geographic diversity gain, as discussed Chapter 4. Sites that are geographically close to each other typically

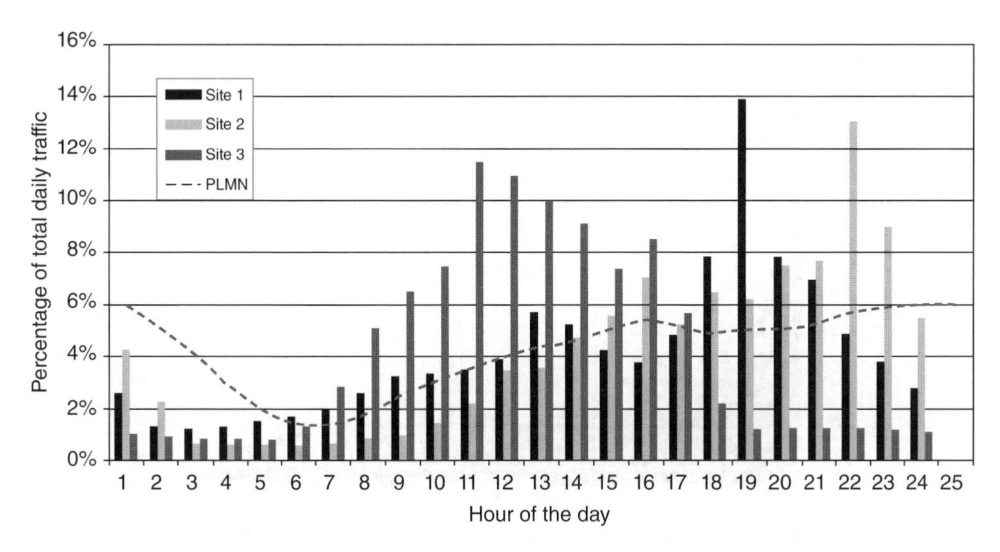

Figure 5.17 An example of distribution of data traffic over 24 hr for three sites and at the network level

experience their BH roughly at the same time, so the geographic diversity gain would be relatively small for an access link serving those sites. On the other hand, for trunk links serving sites of a large geographic area considerable gain could be achieved.

For dimensioning purposes, the average BH traffic of a site is the minimum information required. As mentioned, in some cases the basic traffic model has been obtained from core network statistics or from the billing system, that is the traffic is given at the network level. To convert the network-level BH traffic to site-level the "load distribution factor" (LDF) can be used.[8] LDF is simply the ratio of the BH share of daily traffic at the network level divided by the BH share of daily traffic at the site level. It has to be estimated from network statistics since it depends on subscriber mobility and data usage patterns, which are very network-dependent. LDF is not (very) radio access technology dependent, and hence it can be estimated from statistics of an existing 2G or 3G network, if the LTE network statistics are not available. The LDF of voice and data traffic is usually different and should be estimated separately.

In the example of Figure 5.17, LDF would be 0.14/0.06 = 2.3 for site #1. The site-level BH traffic would then equal the total network BH traffic divided by the number of sites, multiplied by 2.3. Here the LDF = 2.3 is for site #1 only. For simplicity's sake, and in the absence of more accurate information, the average LDF of all sites in the network can be used. Typical values of average LDF are 1.5…4 for packet data traffic (Szukowski et al., 2011) Some individual sites may have a very high LDF due to their special location (e.g. sports stadiums), and such sites may need to be considered on a case-by-case basis, assuming such detail is required.

The BH share of total daily traffic shown in Figure 5.18 illustrates the dependency of LDF on the site traffic volume. There is a clear correlation between the traffic load of a site and the LDF; busy sites tend to have a lower BH share of traffic. This is logical since in order to carry

[8] Sometimes also called the load dispersion factor.

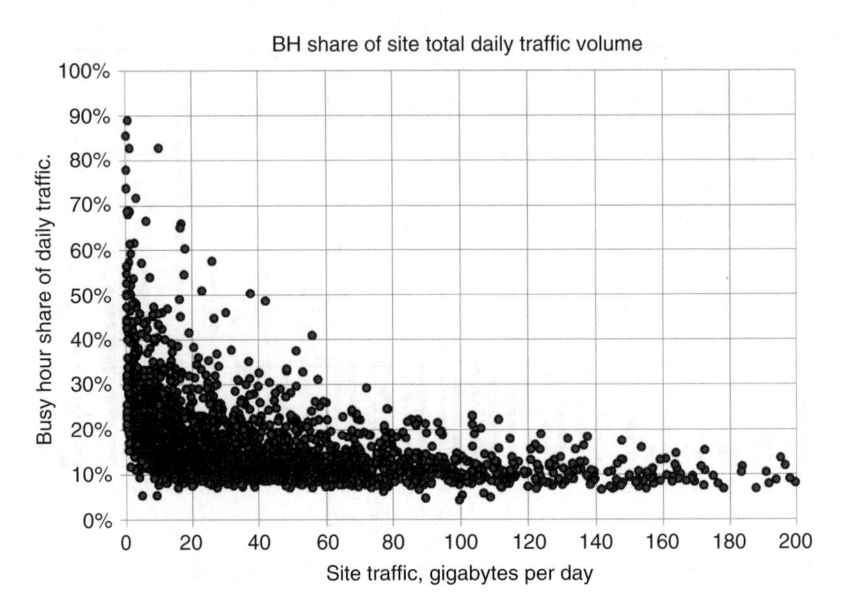

Figure 5.18 Example load distribution factor per site versus site daily traffic volume. Each dot is one site

large data volume within one day's time the site should be highly utilized throughout the day. In turn, low-traffic sites tend to serve only a few UEs which are active only during certain hours of the day, resulting in a very large BH share. If traffic were evenly distributed over the entire day, the BH share would be 1/24 = 4.2%.

For bandwidth dimensioning purposes the most interesting sites are the busy ones. From Figure 5.18, the busiest sites tend to have a BH share of less than 20%, while the average BH share over all the sites is 17%. With a network-level BH share of 7%, this would then result in an average LDF of about $0.17/0.07 \approx 2.4$ for the example network.

Another modifier in traffic calculations is the S1 transport overhead, which depends on the transport protocols used. Table 5.9 shows the transport overheads of protocols commonly used in S1 transport. The S1 packet size is the end user IP packet size sent by the UE (or SGi interface in DL) plus the relevant overhead from Table 5.9. For IPsec the overheads shown in the table should be considered as examples only as the actual overhead depends on the encryption and authentication algorithm used. IP fragmentation increases the transport overhead, which is one reason to use Ethernet jumbo frames whenever possible.

Example 5.5

Traffic model is provided as 5 GB of total UL plus DL data traffic per subscriber per month, measured as total network traffic of one month divided by the number of subscribers. There are 1000 sites and one million forecasted subscribers in the urban area network to be dimensioned. The problem is to calculate the site-level BH traffic.

Depending on the network, the UL traffic is 20–30% of DL traffic at the network level. Assuming that 20% of total traffic is UL, this leaves 4 GB per subscriber per month for the

DL. Statistics usually indicate that there are some differences in data consumption between weekdays, but usually such socio-behavioral aspects can be neglected in backhaul bandwidth dimensioning. Daily DL traffic per subscriber is then 4 GB/30 ≈ 130 MB. From this figure about 6% (≈ 8 MB) is consumed during BH when examined at network level. This would give 1000 × 8 MB = 8 GB traffic volume per BH per site, if all sites in the network had their BH at the same time. Assuming LDF = 2, the BH traffic volume per site is 16 GB. This gives an average BH bit rate of 16000 × 8/3600 ≈ 36 Mbps per site. S1 transport overhead needs to be added to this value. Average DL packet size can be assumed to be 1000 bytes, and therefore with Ethernet-based IPv4 transport without IPsec the headers add 78 bytes per packet with VLAN, producing ≈ 8% S1 transport overhead. This brings the average site transport layer traffic to about 39 Mbps per site. Instantaneous peak traffic can obviously be several times higher.

The calculation above is very simplistic and ignores a lot of other information that could potentially affect it. It is well known that the spatial distribution of network traffic load is very localized. A small number of the busiest sites may easily generate over 50% of the network traffic load. Therefore instead of applying the same traffic model to all sites, a more refined approach would be to check the spatial traffic distribution of existing 2G/3G sites (assuming that LTE will be co-located) and in this way try to obtain more accurate site-level and area-level traffic figures. Similarly, the actual average packet size and other traffic parameters should be estimated from the existing network, whenever possible. A pragmatic alternative is to simply treat all sites as equal in dimensioning (as in the above example), and leave the bandwidth management of busy sites as a post-launch monitoring and optimization task.

If it is necessary to know the bandwidth required by an X2 U-plane, the amount of X2 handovers should be somehow estimated. This can be difficult at best as it depends strongly on the

Table 5.9 Overheads for S1 user plane transport

Overhead type	Bytes
GTP-U header without extension	8
UDP header	8
IPv4 header	20
IPv6 header	40
IPsec ESP header	8
IPsec padding	0–255[a]
IPsec pad length field	1
IPsec next header field	1
IPsec integrity check value, ESP	16
Plain Ethernet header, including FCS	18
Ethernet interframe gap + preamble + SFD	20
VLAN tag	4
Multi-link PPP long header	4
Multi-link PPP short header	2

[a] Depends on ciphering algorithm. For AES block cipher max 16 bytes, for 3DES max 8 bytes.

cell size, cell overlap, UE speed, handover threshold parameters, UE inactivity timer, and so on. Values from real networks show very large variability between networks and enormous differences between sites in a given network. Depending on the network, the number of X2 handover preparations per RRC connected user per hour may vary from five to 50, while at the site level values of several hundred are not uncommon. Not every UE is transferring data at the time of handover, which should also be taken into account.

Example 5.6

Once again, worst-case analysis is used to get some idea of whether X2 U-plane bandwidth should be taken into account in dimensioning. Assume that the UE average DL bit rate over the entire RRC connection is 100 kbps and there are 20 X2 handovers per hour per RRC connected UE.[9] The handover X2 interruption time from X2 handover request to UE context release in the source eNB is usually 100 ms or less if the X2 transport delay is not excessive and radio conditions are normal. Each handover then produces an average of 10 kbps outgoing X2 U-plane load to source eNB and the same incoming load to target eNB. What remains to be done is to calculate the frequency of handovers, which depends on several factors; in this example all these factors are captured in the number of X2 handovers per hour per RRC connected UE (=20, given above). For an average of 100 RRC connected UEs per eNB the amount of X2 handovers in 1 hr would be $100 \times 20 = 2000$, or $2000/3600 \approx 0.6$ handovers per second. Thus, the average outgoing X2 U-plane traffic would be $0.6 \times 10 = 6$ kbps. Since the handovers between cell pairs tends to be roughly symmetric, the source eNB would have the same X2 U-plane traffic in incoming direction, on average. If it was instead assumed that every UE was making 10 Mbps download at the time of handover, the (utterly unrealistic) X2 UP traffic would be a hundred times larger, or 600 Kbps, which is still not very large. Based on the analysis, X2 U-plane traffic can be neglected in almost all practical cases. Advanced radio features in LTE-A may change the analysis assumptions radically, however.

5.6.1.2 Voice Traffic Busy Hour Bit Rate

The total offered BH voice traffic in Erlangs is the traffic of a single subscriber times the number of subscribers served by the site. From the total offered traffic the probability that the number of voice calls at site-level exceeds a threshold value with a certain probability can be computed with the Erlang-B formula. In a circuit-switched world, this would be interpreted as the call blocking probability, that is the probability that there are more offered calls than there are voice circuits available. For IP-based transport there is typically no transport blocking as the available S1 bandwidth should always exceed the voice traffic bandwidth requirement by a hefty margin. In case transport admission control is used for voice service, admission control rejection is, of course, possible.

[9] As an example, if one UE is in RRC-connected state for 30 min during the hour this would count as 0.5 RRC-connected UEs. The UE is active (transferring data) for only a fraction of this time.

Example 5.7

Assuming 25 milliErlangs per subscriber traffic and 1000 subscribers per site, the offered voice traffic would be 25 Erlangs. Using the Erlang-B formula, the probability that there are more than 36 simultaneous voice calls is 1%. In terms of bandwidth, a single narrow-band AMR call with transport overheads (without IPsec) consumes about 62 Kbps, and so the probability of S1 voice bandwidth exceeding 36 × 0.062 = 2.2 Mbps is 1%.

To calculate the voice bandwidth for a trunk link serving multiple sites, the total offered voice traffic served by the link needs to be known. For example, if the link is serving 10 sites with identical offered traffic, using the values of the above example the total offered traffic would be 250 Erlangs. Again, using the Erlang-B formula the probability of exceeding a certain number of simultaneous voice calls carried by the link can be computed. For high offered traffic carried by trunk links the offered voice traffic plus a 10% margin results in a small blocking probability. As a simple rule of thumb, for 200 Erlangs of voice traffic demand the probability that the number of voice calls exceeds 220 is less than 1%. Of course, the Erlang-B formula cannot take into account sudden bursts of calls due to a special event, so an additional safety margin could be applied.

With several GBR services multidimensional Erlang calculation is required. Accurate calculations in this case cannot be made on pencil and paper anymore. The worst-case bandwidth requirement can be obtained by calculating bandwidth for each service separately with the basic Erlang-B and then summing the results for individual GBR services. An activity factor may be used to modify the result.

5.6.2 Obtaining Input Parameters for Control Plane Bandwidth Dimensioning

The volume of S1 C-plane traffic depends on the radio and core network configuration, including UE inactivity timer, tracking area size and S1 paging method, S1/X2 handover frequency, usage of features (e.g. VoLTE, Circuit-Switched Fall-Back [CSFB] with packet service handover [PS HO]), SCTP timers, and so on. X2 C-plane traffic volume can be ignored as it is a fraction of X2 U-plane traffic, which is very small (see previous section).

S1 control traffic depends on the type of UE usage. Smartphones frequently exchange small packets with Internet applications, such as chat messages or email notifications, and while doing so generate a low amount of data traffic but a relatively high signaling load. Stationary data modems on the other hand have low mobility and consume vast amounts of data while causing moderate signaling load. It is thus difficult to relate the amount of signaling traffic to user data volume. Network measurements suggest that the S1 signaling load is much less than 1 Kbps per MME-registered user for a smartphone-dominant network (10-second UE inactivity timer), when S1 paging is excluded. The value includes Ethernet overhead but no IPsec. For many networks, the actual S1 signaling load will be considerably less. As mentioned, however, radio and core network configuration will strongly affect the actual signaling bandwidth.

In case there is a need to accurately estimate control traffic bandwidth, one approach is to multiply the number of signaling procedures per time interval by the number of bytes per procedure, and then normalize the result to bits/s over the time interval of interest. Table 5.10 shows an approximate transport load caused by common S1 signaling procedures. The message size and the number of the messages per signaling procedure depends on various network parameters (authentication, EPS session management [ESM] information request, etc.) so the values should be considered as examples, rather than exact.

In addition to Table 5.10 one needs a traffic model for the number of signaling procedures generated by a subscriber during BH.

Table 5.11 shows an example of a C-plane traffic model. With such information, and the number of MME-registered subscribers (idle and connected UEs) per site, it is possible to estimate signaling traffic load. For example, if there are 1000 subscribers per site, the number of service requests would be 34,000 per site during the BH. Therefore from Table 5.10 the signaling transport load from service requests would be about 34,000 × 0.48 Kbps = 16 Kbps

Table 5.10 Typical cost of selected S1 signaling procedures for DL and UL, including SCTP/IP/Eth/L1 overhead. The UL/DL cost is calculated over 60 minutes

Procedure	DL bytes per procedure	UL bytes per procedure	DL cost of one procedure, bits/second	UL cost of one procedure, bits/second
Attach, including PDN connectivity request	592	1056	1.32	2.35
Detach, UE initiated	212	252	0.48	0.56
PDN connectivity request (additional default bearer)	244	428	0.55	0.96
Tracking area update	220	428	0.49	0.96
CSFB call with redirect	288	472	0.64	1.05
Dedicated bearer activation (QCI1)	380	444	0.85	0.99
Dedicated bearer deactivation	124	236	0.28	0.53
eNB initiated S1 Context release	96	196	0.22	0.44
E-RAB modification	168	244	0.38	0.55
Paging	120	0	0.27	0.00
Service request	212	252	0.48	0.56
PS HO to 3G	404	352	0.90	0.79
S1 HO, source BTS	268	528	0.60	1.18
S1 HO, target BTS	584	484	1.30	1.08
X2 HO, source BTS	256	480	0.57	1.07
X2 HO, target BTS	264	344	0.59	0.77
SRVCC to 3G[a]	488	400	1.09	0.89
SCTP heartbeat / Heartbeat Ack	106	106	0.24	0.24
SCTP stand-alone SACK	72	72	0.16	0.16
One-part downlink SMS over NAS signaling (excluding other S1 procedures)	200	130	0.44	0.29

[a] SRVCC: single radio-voice call continuity

Table 5.11 Example control plane traffic model

Procedure	Procedures per BH per MME-registered subscriber
Attach	0.5
Detach	0.1
Tracking area update	2
X2 HO	8
S1 HO	0.2
PS HO to 3G	1
Paging, CS and PS	15
Service request, MO and MT	34
S1 release (due to UE inactivity)	34
VoLTE call	1
CSFB call with RRC release with redirect	2
SMS	1

DL load and around 19 Kbps UL load per site. The value assumes that the procedures are evenly distributed over the BH, or in other words that there are no bursts of service requests. Summing similarly over all procedures in the traffic model, one obtains around 35 Kbps DL signaling load, before S1 paging or SCTP acknowledgments. The corresponding value for the UL is around 50 Kbps.

For DL, S1 paging will constitute the bulk of signaling load if the paging area is large.[10] Assuming that the MME sends each paging to all eNBs in a tracking area and that the tracking area size is 50 sites with 1000 subscribers in each site, the number of S1 paging messages per hour would be $1000 \times 50 \times 15 = 750e3$ S1 paging messages per hour. At eNB-level this would generate around 200 Kbps of additional DL signaling load. Paging message load depends on the paging method, number of subscribers in the paging area and the number of paging messages per subscriber.

SCTP acknowledgments increase the signaling load slightly. Even with worst-case assumption the additional bandwidth from SCTP is not more than some tens of kilobits per second in this example. SCTP heartbeat messages can be neglected in BH analysis, since they are sent only if there are no data messages sent between the endpoints.

From the above simple calculation one obtains less than 300 Kbps DL and 100 Kbps UL signaling bandwidth per site. Adding a multiplicative safety margin of, say, two, to account for message bursts, the final values would be 600 Kbps DL and 200 Kbps UL.

To summarize, experience from live networks indicates that S1/X2 signaling traffic can usually be neglected as a first approximation in bandwidth dimensioning; in any case signaling should be mapped to a high-priority traffic class to reduce queuing delay. In the capacity management phase, monitoring signaling bandwidth is important to ascertain that QoS is maintained.

[10] Paging is initiated by MME, and the number of eNBs paged depends on the vendor-proprietary paging method. The default method is to page all eNBs in a tracking area or tracking area list. Tracking area size may be anything from about a dozen to several hundred eNBs.

5.6.3 *Link Bandwidth Dimensioning: Single Queue*

In this section formulas for bandwidth dimensioning of a single transport link are given. It is assumed that all traffic shares the same egress FIFO queue, in other words all traffic has the same priority. Generalization to multiple traffic priorities and multiple queues results in fairly complex formulas; approximations for a few special cases are presented in a later section.

The QoS criteria considered are the average packet-level queuing delay (for arbitrary traffic) and average file transfer delay (for elastic traffic). Egress buffer size is assumed infinite throughout the section since no simple formulas are known for finite buffer size and hence computer simulation would be necessary. Since switches and routers available in the market currently tend to have large buffers,[11] with analytical formulas this limitation can be, to a certain extent, taken into account by making the average delay target a small fraction of the buffer size.

Link bandwidth dimensioning formulas are most useful for aggregate link dimensioning. The first mile bandwidth, to be discussed first, is often selected large enough so that no QoS criteria need to be considered.

5.6.3.1 First Mile

The application of a theoretical bandwidth dimensioning formula typically results in a link bandwidth that is less than twice the average traffic demand. As an example for 50% load the average DL throughput of a 20 MHz urban 2x2 MIMO LTE site (three cells) would be 40–60 Mbps. Choosing the first mile bandwidth of 100 Mbps would hence well satisfy the site average traffic demand, with a hefty margin for signaling traffic. On the other hand, with 100 Mbps first mile link bandwidth, a single category 4 UE in excellent radio conditions would not be able to reach its full supported bit rate of 150 Mbps due to limited transport bandwidth. For this reason it has become a common practice among operators to select the first mile bandwidth based on the cell peak rate, rather than the site average traffic demand. After all, if the first mile transport limits the cell peak rate there would not be much sense in investing in peak rate enhancing radio features either (e.g. LTE-A carrier aggregation). Supporting cell peak rate may not be feasible or economical for all sites, however.

Several heuristic rules can be conceived for first mile bandwidth dimensioning, including:

- Choose the first mile bandwidth as site average bandwidth.
- Choose the first mile bandwidth as the cell peak rate.
- Choose the first mile bandwidth as the maximum of cell peak rate and site BH average bit rate (NGMN Alliance, 2011a).
- Choose the first-mile bandwidth as the cell peak rate of one cell plus the average BH traffic of the remaining cells (NGMN Alliance, 2011a).
- Choose the first-mile bandwidth as the cell peak rate times the number of cells.

S1 transport overhead should be added to the result.

Extensive treatment on bandwidth dimensioning based on average and peak rates can be found in NGMN Alliance (2011a). The drawback of such rules of thumb is that they do not

[11] A notable exception to this are microwave outdoor units, which tend to have small buffer sizes.

take into account any QoS criteria. On the other hand, since the resulting bandwidth allocation is usually many times the average traffic demand, almost any QoS criterion of interest is automatically satisfied. Still, mapping delay-sensitive traffic to a queue with strict priority is recommended to avoid unpredictable delay peaks due to occasional high peak rate users.

5.6.3.2 M/G/1 and G/G/1 Formulas for Average Packet Delay

As discussed in Chapter 4, packet delay is not a very meaningful metric for TCP data traffic, for which the effective throughput degradation (excess file transfer delay) caused by the bottleneck link is of most interest, especially for large files that are not sensitive to TCP slow-start. However, for delay-sensitive traffic, such as voice and signaling, it is the packet delay that is relevant and the dimensioning can be based on M/G/1 or G/G/1 formulas.

The average queuing delay in an egress buffer can be calculated with the classical X/Y/c queuing theory formulas. The Kendall notation for a queue is X/Y/c, where X defines the packet inter-arrival time distribution and Y defines the packet size (service time) distribution.[12] The last digit "c" defines the number of "servers," which in the context of a transport node would correspond to the number of egress ports emptying the queue. In this section only the case of c = 1 is considered, in other words each queue is buffering packets of one egress port only. In the most general case (G/G/1) the required input parameters are the means and standard deviations of packet inter-arrival time and packet size. The goal is to calculate the average packet delay in the node, that is the sum of queuing and port serialization delay, for a given egress port bandwidth. More details can be found in the queuing theory literature, a classic reference being Kleinrock (1975). In the remainder of this section, the results are stated without any elaboration on the theoretical background.

Table 5.12 summarizes the classical queue types and example applications.

In the following the focus is on M/G/1 and G/G/1 of which the other formulas are special cases.

Table 5.12 Single-server queue types with example applications

Queue Notation[a]	Packet inter-arrival time distribution	Packet size distribution	Example application
D/D/1	Constant	Constant	IEEE1588-2008, voice
M/D/1	Exponential	Constant	VoIP trunk links, ATM
M/M/1	Exponential	Exponential	Signaling traffic
M/G/1	Exponential	Any	Backbone links, signaling
G/G/1	Any	Any	Any

[a]The letters 'D' and 'M' stand for deterministic and exponential distribution, respectively, while 'G' stands for generic (arbitrary) probability distribution.

[12]Packet size and the output line rate define the time it takes to transmit the packet out from the node egress port (serialization delay). This is called "service time" in the queuing theory literature and is given by the packet size divided by the port line rate.

The M/G/1 average waiting time in the egress queue is given in Equation 5.10.

$$E[d]=\frac{m_s\rho\left(1+c_s^2\right)}{2(1-\rho)}, \tag{5.10}$$

where

m_s mean service time (average packet size divided by egress line rate)
ρ mean egress port utilization (average traffic divided by egress line rate)
c_s coefficient of variation of packet service time (standard deviation divided by mean)

The utilization ρ is defined as the queue input traffic rate divided by the egress port rate, for example if the average ingress traffic is 50 Mbps and the queue egress port rate is 100 Mbps then the utilization would be 50%. The coefficient of variation is defined as the ratio of standard deviation and mean and it characterizes the spread of the distribution about its mean. For example, if the packet size is constant (e.g. VoIP or PTP) the standard deviation of the service time (and packet size) would be zero, and the model collapses to M/D/1. Since in this case $c_s = 0$, the formula can also be used for traffic with constant packet size.

The M/G/1 formula has the built-in assumption that the packet inter-arrival times are exponentially distributed. For bursty packet traffic, such as U-plane traffic, the most general and useful of the models is the G/G/1, which does not make any restrictive assumptions on the distribution of the packet inter-arrival and packet size, even though the mean and standard deviation of the distribution still need to be known. Further, it is assumed that the consecutive packet inter-arrival times and packet sizes are statistically independent. The other queuing models listed in Table 5.12 are special cases of G/G/1. Due to its generality, only approximate delay formulas are available for G/G/1. The mean value of the queuing delay *d* spent in a G/G/1 queue is upper bounded by the Kingman formula (Kleinrock, 1975)—Equation 5.11.

$$E[d]\leq\frac{m_s\rho\left(c_a^2+c_s^2\right)}{2(1-\rho)}, \tag{5.11}$$

where the notation is the same as in the M/G/1 formula with c_a now being the coefficient of variation of packet inter-arrival time. For the exponential case (M/G/1) $c_a = 1$ and the formula can be seen to collapse to M/G/1. In this special case the formula gives the exact result instead of an upper bound. A high value of c_a indicates that there is a considerable probability of the packet inter-arrival time being very short (packets may arrive in bursts) thus increasing the worst-case and average queuing delay and buffer size requirements. Conversely, a low value of c_a would indicate near-constant packet inter-arrival times. For example, PTP timing protocol and speech codecs generate constant size packet packets at constant intervals so both c_s and c_a would be zero if the link carries traffic of a single PTP or speech source only; from the Kingman formula, the queuing delay would be zero in this case. In practice, of course, transport links carry multiple types of traffic from many different sources.

To apply the Kingman formula one needs to know the mean and standard deviation of the packet size distribution, and the standard deviation of packet inter-arrival times. This can be obtained from measurements. In the absence of measurement results some estimate or worst-case assumption of the variance needs to be made. For example, if the frame size is bounded to 64…1522 bytes with a known average value of 1000 bytes, then it is known from general statistical bounding techniques that for an arbitrary packet size distribution the standard deviation must be less than the square root of $(1522-1000)\cdot(1000-64) \approx 700$ bytes. It follows that the worst-case value for the coefficient of variation of the packet size is $c_s \approx 700/1000 = 0.7$. More generally, for packet size restricted to interval $[a\ b]$ with average size of m, the variance of packet size is always smaller than $(b - m) \times (m - a)$; in the absence of better information this simple result can be used to obtain a worst-case estimate for c_s. The upper limit of the Ethernet frame size can be obtained from the IP layer MTU (maximum transfer unit) size by adding Ethernet overheads.

IP layer fragmentation reduces the average packet size but at the same time it increases c_s and c_a so the net result from packet fragmentation is increased queuing delay. Furthermore, some equipment allocates buffer memory per-packet (instead of per-byte) so fragmenting every packet would double the buffer usage, increasing the risk of buffer overflow.

Analysis of packet traces from real networks indicate that packet inter-arrival time of S1 signaling traffic exhibits packet inter-arrival time distribution with coefficient of variation of packet inter-arrivals between one and two ($c_a = 1$ means exponential distribution). This makes M/G/1 and G/G/1 useful models for the calculation of average signaling packet delay. An example of measured packet inter-arrival time for S1 traffic is shown in Figure 5.19 along

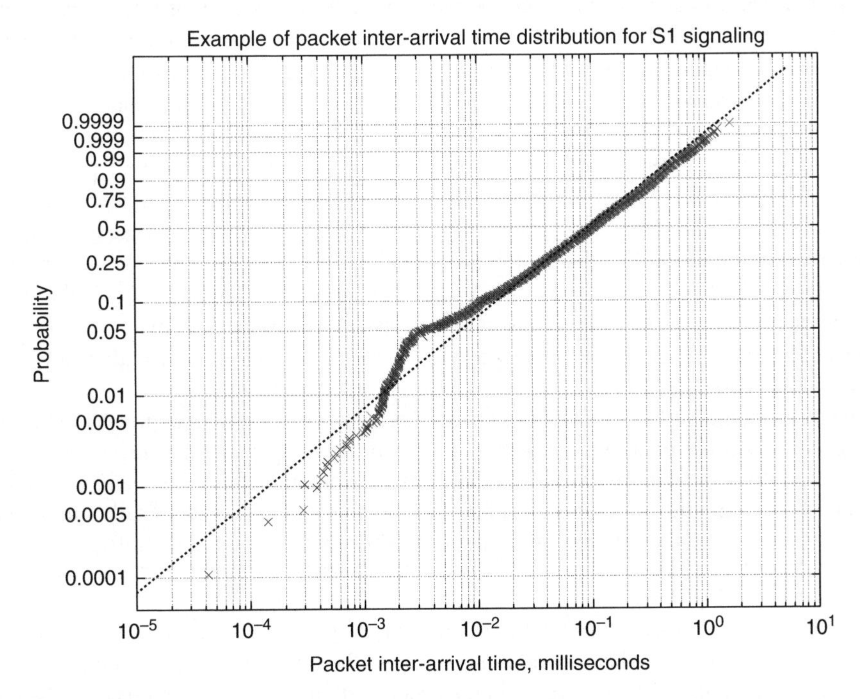

Figure 5.19 Measurement example of S1 signaling packet inter-arrival time distribution with best fit to the exponential distribution

with the best fit to the exponential distribution. The estimated coefficient of variation in the example is $c_a = 1.05$.

While the M/G/1 and G/G/1 are usable for signaling, voice and short data transfers (e.g. messaging or gaming), the applicability of these formulas for bandwidth dimensioning of BE data traffic over access links is debatable. The problem arises from the TCP protocol, which tends to generate packet flows where packet arrivals happen in bursts (i.e. the consecutive packet arrivals are not independent). The burst nature of TCP packet flows violates the assumption underlying the queuing theory formulas (that packets arrive independently). Moreover, TCP is bandwidth-greedy as by design it attempts to increase the data rate up to the speed limited only by the bottleneck link in the e2e path. These effects are especially visible in the first few miles of the access network where the number of active TCP flows over any link typically ranges from a few dozen to a few hundred. Some measurements indicate that for a large number of packet flows (e.g. several thousand) encountered in trunk and backbone links the packet inter-arrival times tend to be exponentially distributed (Cao et al., 2001) and in this case the queuing theory formulas could be applicable also for BE TCP traffic. In other cases the coefficient of variation c_a should be chosen carefully to avoid overly optimistic results.

Example 5.8

Assume a 1 Gbps Ethernet port that is carrying an average of 750 Mbps S1 signaling traffic during BH (this could be, for example, an MME Ethernet port). The average packet size at Ethernet layer from network counters is 200 bytes. Distribution of the packet size is unknown. The problem is to calculate the average queuing delay. For signaling traffic, exponential packet inter-arrival can be assumed at least if the paging areas are not too large (S1 paging generates bursts since many eNBs are paged at the same time). The average link utilization is $\rho = 0.75$. The mean service time is $8 \times 200/1e9 = 1.6$ microseconds. To calculate the standard deviation of service time the packet size distribution would be needed. Taking the worst-case scenario, an upper bound on the deviation of packet size is given by the square root of $(1522 - 200) \times (200 - 64)$, or 424 bytes. This gives the service time coefficient of variation as $424/200 = 2.1$. For simplicity's sake, the Ethernet PHY layer overhead is ignored. With all the required inputs collected, the average queuing delay based on the M/G/1 formula is $1.6 \times 10^{-6}/2 \times 0.75 \times (1 + 2.1^2)/(1 - 0.75) = 13$ microseconds. The total delay is the sum of queuing delay and port service (serialization) delay which for the 200-byte average packet size was already calculated as 1.6 microseconds. All told, the total average node delay is $13 + 1.6 = 14.6$ microseconds.

The result suggests that 1 Gbps links could in principle be operated at quite a high utilization, especially if the underlying assumption of exponential packet inter-arrival time holds. However, if this assumption is violated due to any network anomalies the results can be misleading. Thus, regardless of the results of dimensioning calculations, it is vitally important to monitor packet discard during network operation, especially for signaling traffic.

Since access layer links carry signaling, voice and U-plane traffic in the same pipe, M/G/1 and G/G/1 formulas cannot be applied as such unless the number of simultaneous TCP flows is at least several hundred or preferably in the order of thousands. Even in this case the packet delay is interesting mainly for delay-sensitive high-priority traffic while for the BE TCP data throughput degradation is a more appealing performance metric. Furthermore, the BE data and the delay-sensitive traffic are mapped to different queues, whereas until now the assumption has been that all traffic shares the same queue. For this reason a mix of dimensioning methods will be needed, one for the queue with delay-sensitive traffic and another one for BE data. Before this, however, TCP throughput degradation under limited bandwidth should be addressed, and this is the topic of the following section.

5.6.3.3 M/G/R-PS Formula for Calculation of TCP Throughput Degradation

As seen in the previous section and Chapter 4, the classic approach of calculating delays at the packet level is not suitable for dimensioning BE data for the first few miles, since TCP packet bursts violate the independence assumptions inherent in the queuing theory approach. Moreover, it is difficult to relate the average packet delay to the average throughput degradation experienced by the end user, since, with TCP, a large packet delay does not automatically correspond to a low throughput if the TCP segment loss probability is small.

Partly to address the problems above, the so-called M/G/R-PS approach has gained popularity in the IP traffic dimensioning literature. Instead of the average *packet* delay, the goal of the M/G/R-PS method is to calculate the average *file* transfer delay caused by a transport link having a certain average load.

As an example, assume an FTP server from which 100 clients are downloading a large file simultaneously. In the network somewhere there is a bottleneck link with 100 Mbps bandwidth. Assuming there are no other bottlenecks in the system, the average FTP throughput achieved by each client would be about 1 Mbps since the TCP protocol (ideally) shares the bottleneck link bandwidth equally between flows. In this scenario the QoS parameter from end user point of view parameter is the excess file transfer delay (rather than the packet delay) due to link load from other users. If there were only one client then the download speed would be 100 Mbps. Hence a 100 MB (800 Mb) file would take slightly more than eight seconds to download. One hundred users simultaneously starting the transfer of the same 100 MB file would result in a download time of about 800 seconds, or some 13 minutes. The excess file transfer delay relative to the ideal single-user case would be $f_R \approx 100$, or equivalently the effective throughput would be 1 Mbps.

In the example above it was assumed that the number of users simultaneously downloading a file over the 100 Mbps bottleneck link was fixed at 100. In practice the number of flows sharing the bottleneck link (hence client throughput) varies over time based on the random arrivals of new download requests and random file size. Therefore, the average file transfer delay experienced by a client depends on these statistics. Furthermore, the client may have an access rate limitation of, say, 10 Mbps caused by subscription, so even in the absence of other users his FTP download rate would be limited to 10 Mbps. The dimensioning problem would be to determine the minimum bottleneck link bandwidth that satisfies a given file transfer delay target for the given traffic demand and access rate limitation. The problem can be solved with the M/G/R-PS formula and can be equally well applied to the access link bandwidth dimensioning in cellular networks.

In the M/G/R-PS model inter-arrival time of file requests are assumed exponentially distributed (M) and the requested file sizes can have arbitrary distribution (G) with a known mean file size and variance (both finite). In addition, an additional constraint in the maximum throughput of a single file transfer is posed. For mobile backhaul dimensioning this assumption is quite natural since the access transport link rates are several times larger than the DL throughput of a UE, which is limited by radio conditions or possibly subscription limitations (e.g. APN-AMBR; access point name—aggregate maximum bit rate).

The practical appeal of the M/G/R-PS approach is that it can be applied with very minimal input information: from a traffic model point of view only the average offered traffic is needed. As the second input parameter the maximum achievable TCP throughput of a single TCP flow is required. The limiting mechanism of the TCP flow throughput is not important as such. For the case of LTE access transport dimensioning it is proposed in Li et al. (2011) to adopt the average UE throughput at the radio interface as the TCP throughput bottleneck. The average UE throughput in the radio cell should be available from radio planning. Since the average radio UE throughput will depend on system parameters, for example radio frequency (RF) bandwidth and radio conditions, the transport bandwidth calculation becomes coupled with the radio interface dimensioning. Coming up with a realistic assumption for the single UE radio throughput is typically the trickiest part when using the M/G/R-PS dimensioning.

In the M/G/R-PS the relative increase in file transfer time is calculated in Equation 5.12 (Riedl et al., 2000).

$$f_R = 1 + \frac{E_2(R, R\rho)}{R(1-\rho)} \tag{5.12}$$

where r is the maximum UE throughput, C is the link capacity to be dimensioned, ρ is the link utilization, $R = C/r$ is the relative link bandwidth, and $E_2()$ is the Erlang's second formula defined in Equation 5.13.

$$E_2(k, y) = \frac{\frac{y^k}{k!}\frac{k}{k-y}}{\sum_{i=0}^{k-1}\frac{y^i}{i!} + \frac{y^k}{k!}\frac{k}{k-y}}, \tag{5.13}$$

where k is a positive integer and y is a real number.

The dimensioning problem consists of choosing R and ρ so that the relative delay increase f_R is less than the requirement. Another way of looking at f_R is to interpret it as the throughput reduction caused by the transport link. For example, with $f_R = 2$ the UE air interface download rate would be halved (relative to ideal transport with infinite bandwidth) due to the load of the transport link. In most cases this would be unacceptable; a typical requirement is $f_R = 1.05\ldots1.2$.

For spreadsheet computation, Erlang's second formula can also be expressed as Equation 5.14.

$$E_2(k, y) = \frac{\text{poissonPdf}(k, y)}{\text{poissonPdf}(k, y) + (1-\rho)\cdot\text{poissonCdf}(k-1, y)}, \tag{5.14}$$

where $\rho = y / k$ and poissonPdf(k,y) and poissonCdf(k,y) are the Poisson density and cumulative distribution functions (PDF, CDF) with mean y evaluated at integer k.

It should be noted that in the basic form of the M/G/R-PS formula the variable R is forced to be an integer and hence the dimensioned link bandwidth (R) will be an integer multiple of the maximum single UE maximum TCP throughput.

Example 5.9

Assume that the access transport bottleneck is a microwave link with a bandwidth of 350 Mbps. UE air interface throughput from radio planning is $r = 20$ Mbps. Average BH traffic demand is 100 Mbps per eNB and the link serves three eNBs. What is the throughput degradation caused by the transport link?

In Figure 5.20 the relative delay increase against link utilization is shown for different values of R, that is the link capacity (C) divided by maximum UE throughput (r). In this example $R = 350/20 = 17$ when rounded down to the nearest integer. This means that a maximum of 17 UE TCP flows fit into the link without reducing UE air interface throughput. Throughput degradation results when there are more than 17 UEs sharing the link, and M/G/R-PS can be used to calculate the average degradation. The average utilization of the link is $\rho = 3 \times 100/350 = 0.86$. Looking at the curve R = 16 at an x-axis value of 0.86 it can be seen that f_R should be in the order of 1.2. A more accurate numerical value for R = 17 calculated from the M/G/R-PS formula is $f_R = 1.17$. The interpretation of the result is that the effective UE transfer rate is about 20/1.17 Mbps = 17.1 Mbps where the average throughput degradation of about 3 Mbps is caused by the transport link being congested.

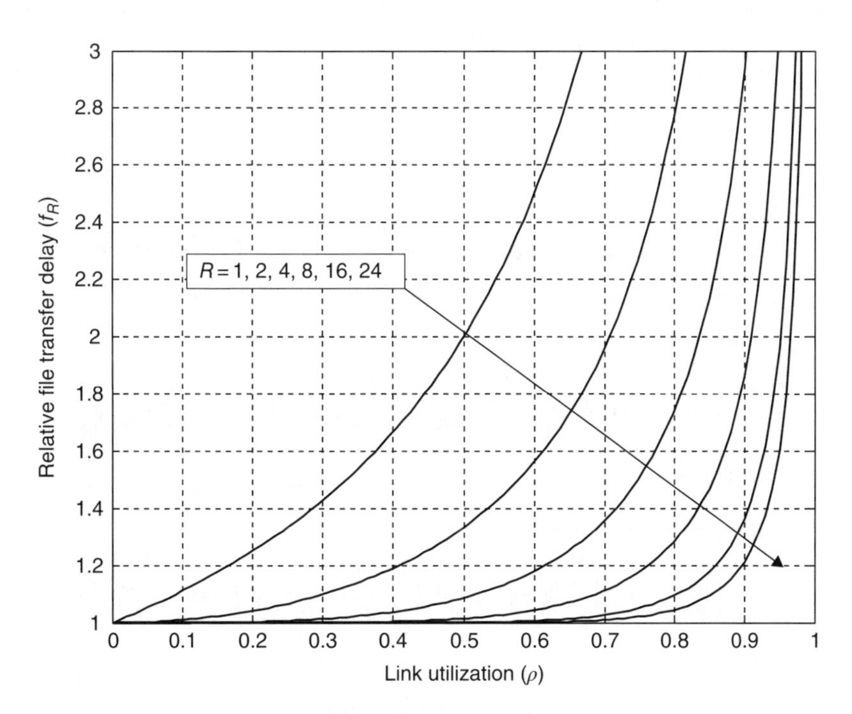

Figure 5.20 Relative download time delay increase as a function of load for different link bandwidths (R = multiple of UE maximum TCP flow rate)

A logical question is how to obtain the total delay increase of the e2e network with multiple links. As the M/G/R-PS method is based on analyzing an isolated link, multiplying relative delays of individual links is not the appropriate way as this could result in a very pessimistic delay value. To obtain an exact e2e delay requires simulation. In practice, after calculating the delay factor f_R for all the links in the e2e network it often happens that one link is the bottleneck (highest f_R). In Riedl et al. (2000) it is suggested to use this maximum f_R as the delay factor of the e2e path.

From Figure 5.21 it can be deduced that, as the number of TCP flows increases, the link utilization can also be increased for a given delay factor. This can be interpreted as a reduction in the peak-to-average ratio of link load, or multiplexing gain. To conclude this section the impact of the number of packet flows on the traffic variability is shown in Figure 5.21. The number of active TCP file transfers (e.g. HTTP) varies from 25 to 200 from top to bottom. For the uppermost picture with 25 active flows the average rate is about 20 Mbps with peaks exceeding 40 Mbps, hence the peak-to-average ratio of the traffic (measured at 100 ms granularity) is about two. When more active flows are present some averaging takes place and the peak-to-average ratio drops to about 1.4 for the case with 200 traffic flows. The example illustrates the multiplexing gain predicted by the M/G/R-PS equation. As the average number of flows with the observation interval increases the peak-to-average ratio reduces and hence the link with high R can be operated at higher load.

5.6.4 Link Bandwidth Dimensioning: Multiple Queues

So far in the bandwidth dimensioning discussion only a single class of traffic and a single egress queue were considered. Returning back to the real world, delay-sensitive traffic should be prioritized in port schedulers. Unfortunately, the bandwidth dimensioning formulas for multiple

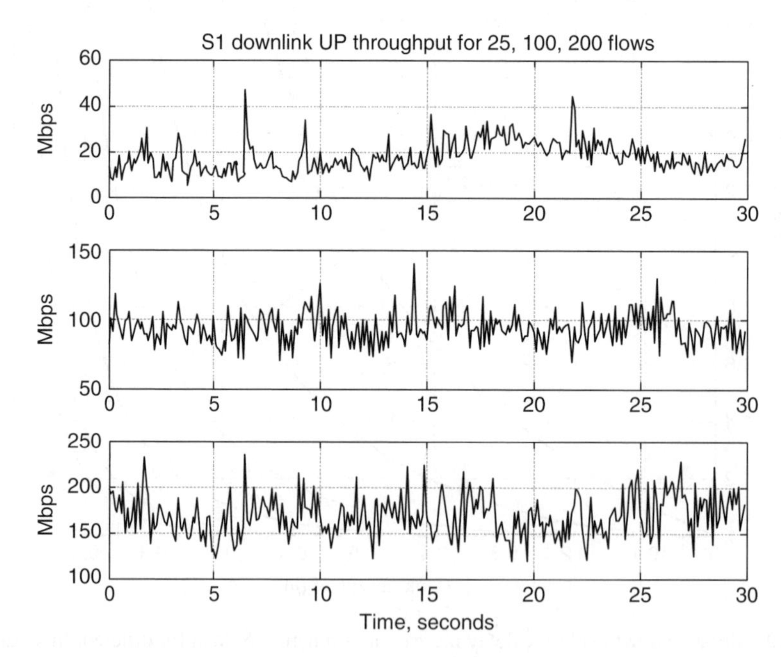

Figure 5.21 Example of S1 DL user plane for 25, 100 and 200 active TCP flows. Throughput is averaged at 100 ms granularity

priorities become either exceedingly complex or analytical formulas are not known at all. If accurate bandwidth dimensioning for all traffic types is required, the only option is to resort to simulations. A few simple special cases can be handled with approximate formulas; some examples are given in the following for strict priority and WFQ schemes. Often it is instructive to check the worst-case packet delay, which can be calculated easily. In many practical cases even the worst-case QoS result is "good enough" and no further calculations are required.

5.6.4.1 Strict Priority Queuing (SPQ) Scheduler

Packet Queuing Delay

In SPQ there are N queues and a higher-priority queue is always emptied first before serving the next highest-priority queue. For the general case of arbitrary packet inter-arrival and packet size distribution (G/G/1) no simple results are available, an approximation can be found in Horváth (2001) and for the case of two queues ($N = 2$) elaborate formulas for exact calculation are provided in Walraevens et al. (2002). If the packet inter-arrival times can be assumed exponentially distributed (M/G/1 model) exact formulas for the SPQ scheduler are known. For non-preemptive queuing the queuing delay of the nth queue is given by ($n = 1,2,\ldots,N$) in Equation 5.15.

$$E[d_n] = \frac{\frac{1}{2}\sum_{k=1}^{N}\lambda_k s_k}{(1-\rho_1-\ldots-\rho_n)(1-\rho_1-\ldots-\rho_{n-1})}, \qquad (5.15)$$

where s_k is the second moment of service time, λ_k is the packet arrival rate, and ρ_k is the link utilization caused by class k traffic, respectively. The special case of two priority queues ($N = 2$) is presented in Equation 5.16.

$$\begin{aligned} E[d_1] &= \frac{\lambda_1 s_1 + \lambda_2 s_2}{2(1-\rho_1)}, \\ E[d_2] &= \frac{\lambda_1 s_1 + \lambda_2 s_2}{2(1-\rho_1-\rho_2)(1-\rho_1)}. \end{aligned} \qquad (5.16)$$

The term in the nominator is the same for all queues and is the so-called residual waiting time, or the average remaining service time of a packet that is already being transmitted out of the egress port; the nominator includes packets of all priorities since in the non-preemptive scheduling considered here an ongoing packet transmission will not be stopped even if a higher-priority packet arrives at the node.

It is worth noting that the queuing delay in M/G/1 is not dependent on the standard deviation (hence coefficient of variation) of the packet inter-arrival times of lower-priority queues. The worst-case waiting time due to a lower-priority packet being transmitted is given by the maximum packet size divided by the egress port line rate.

M/G/R-PS Dimensioning

Mapping bandwidth-greedy BE U-plane traffic to strict priority queue is not advisable as it may starve all lower-priority traffic. If needed, however, approximate M/G/R-PS formulas for the SPQ scheduler can be found in Vranken et al. (2002). Again, the formulas therein are fairly complex to use for practical dimensioning and hence are omitted here. Further, the approximations are not accurate for link loads above 70%.

Several approximation formulas for high-priority streaming and low-priority BE elastic traffic are given in Malhotra and van den Berg (2006). The simplest formula given is based on the quasi-static assumption that the leftover bandwidth C_{eff} available for elastic TCP traffic is the physical link rate minus the streaming traffic bandwidth. This simple approximation is useful if the streaming traffic bandwidth is a small fraction of the link rate. The M/G/R/-PS dimensioning formula can then be used for TCP traffic by using the effective link capacity C_{eff} in place of the actual physical capacity C.

5.6.4.2 Weighted Fair Queuing (WFQ)

Packet Queuing Delay

No formulas for the average packet queuing delay for the WFQ scheduler are available in literature. If the input traffic served by the kth queue is traffic shaped to average rate r_k and maximum burst size b_k, an upper bound on the maximum delay of a WFQ scheduler is given by Stiliadis and Varma (1998); Equation 5.17,

$$d_k \leq \frac{b_k + L_{\max}}{r_k} + \frac{L_{\max}}{C}, \tag{5.17}$$

where d_k is the queuing delay in the kth queue, $L_{\max}$ is the maximum packet size and C is the egress port rate.

The result extends to a series of several WFQ schedulers (Stiliadis and Varma, 1998). The bound is not in terms of average delay, but applies to every packet assuming that the incoming traffic flow has a maximum burst size of b_k. This result captures two important properties of a WFQ scheduler:

- the maximum packet delay for the kth queue is guaranteed to be below a certain upper bound if r_k is less than the WFQ bandwidth allocated to queue k
- the maximum queuing delay can be reduced by shaping, i.e. reducing burst size and average rate relative to link rate.

In contrast, for the SPQ scheduler an upper bound on the packet delay can in general be guaranteed only for the highest-priority queue.

M/G/R-PS Dimensioning

A simple quasi-static approximation can be found in Li et al. (2011), where the idea is to replace the physical link capacity C in the basic M/G/R-PS formula with C_k, which is the average capacity of the kth queue with weight w_k. The value of C_k is given by Li et al. (2011), Equation 5.18.

$$C_k = \max\left(C\frac{w_k}{\sum_{k=1}^{K} w_k}, C - \sum_{j \neq k}^{K} L_j \right), \tag{5.18}$$

where L_j is the bandwidth utilized by the *j*th traffic class. When all traffic classes are fully utilizing their queue bandwidth (link load is 100%) C_k is defined by the queue bandwidth. Otherwise, class *k* traffic can use the excess bandwidth not used by the other queues. Replacing C with C_k in the M/G/R-PS formula is an approximation because it assumes the bandwidth reserved by class *j* traffic is constant, whereas in reality it changes randomly depending on the number of active TCP flows.

5.6.5 Combining Signaling, Voice and Data Traffic

An important use case for wireless access networks is the combination of signaling, voice and elastic BE traffic. Management traffic can be neglected as it occupies considerable bandwidth only in rare circumstances, such as during software upload or interface tracing. The interesting practical design questions for this scenario include:

- Is it necessary to have a dedicated strict priority queue for signaling? Or can signaling be combined in the same queue with voice traffic?
- What is the impact of BE traffic on voice packet delay?
- What if the end user experienced degradation in BE services caused by signaling voice traffic?

To get some feel for the problem, a strict priority scheduler with two queues serving a 100 Mbps link is considered. Voice and signaling are mapped to the high-priority queue and BE data to the low-priority queue. All other traffic types are assumed to have negligible traffic demand. To obtain some concrete numbers, some simple traffic assumptions need to be made.

- Signaling and voice traffic average packet sizes are 150 bytes.
- Data traffic average packet size is 1000 bytes.
- Signaling traffic demand is 5 Mbps.
- Voice traffic demand is 20 Mbps.
- BE data traffic varies from 0 to 75 Mbps.

The average delays can be calculated with M/G/1. Even though speech codec generates packets at constant interval (20 ms for an AMR codec), the packet inter-arrivals of a superposition of multiple VoIP calls can be approximately modeled as random since the individual streams experience some random delays over the e2e path.

The average utilization of the high-priority queue is 25 Mbps. The average packet arrival rate is the traffic volume divided by the average packet size, which for the high-priority queue gives $25e6/(150 \times 8) \approx 21000$ packets per second.

In order to use the M/G/1 formula, the second moment[13] of the service time is needed for both queues since it appears in the nominator of the formula. Thus, the packet size distribution would be needed. Since the distribution of the packet sizes is unknown, the worst-case assumption of the variance of packet size is $(1522 - \text{mean}) \times (\text{mean} - 64)$, where the mean packet size depends on the traffic type and the possible packet lengths are assumed to be between 64 and 1522 bytes. Since the second moment of packet size (and hence service time) is the sum of

[13] The second moment for random variable X is the expected value of X^2.

variance and square of the mean value, this results in a worst-case value of the second moment, and (since second moment appears in the nominator of the delay formula) subsequently also the average queuing delay will be pessimistic.

Figure 5.22 shows the calculated average queuing delays of the two traffic classes for output port rate of 100 Mbps. The link load caused by voice and signaling traffic is fixed at 25 Mbps (25% of the total link load) while low-priority traffic load varies from 0–75 Mbps. It can be seen that even for fully loaded link the high-priority traffic has an average queuing delay of less than 0.1 ms. The low-priority traffic delay is higher and starts increasing sharply when the link load is above 90%. It should be remembered, however, that TCP flows violate the assumptions of M/G/1 formula and hence the calculated delay is indicative only. Moreover, as discussed earlier, the average delay of an individual packet is not a very meaningful metric for measuring TCP throughput.

Port serialization delay should be added to obtain the total packet delay; for a 1522-byte frame and 100 Mbps line rate, this would amount to an additional $8 \times 1522/100e6 \approx 0.12$ ms delay.

Voice codecs have a finite jitter buffer size and hence the packet delay variation they can compensate is also limited. In general, packet delay variation in the e2e LTE system is dominated by the radio interface hybrid automatic repeat request (HARQ) retransmissions, but nonetheless it is of some interest to assess (and minimize) the packet delay variation caused by the transport network. Unfortunately, calculating the delay jitter is very complex and no engineering formulas are available even for the simplest of cases. Bonald et al. (2001) provide some upper bounds and convincing arguments to side with the opinion that the jitter is

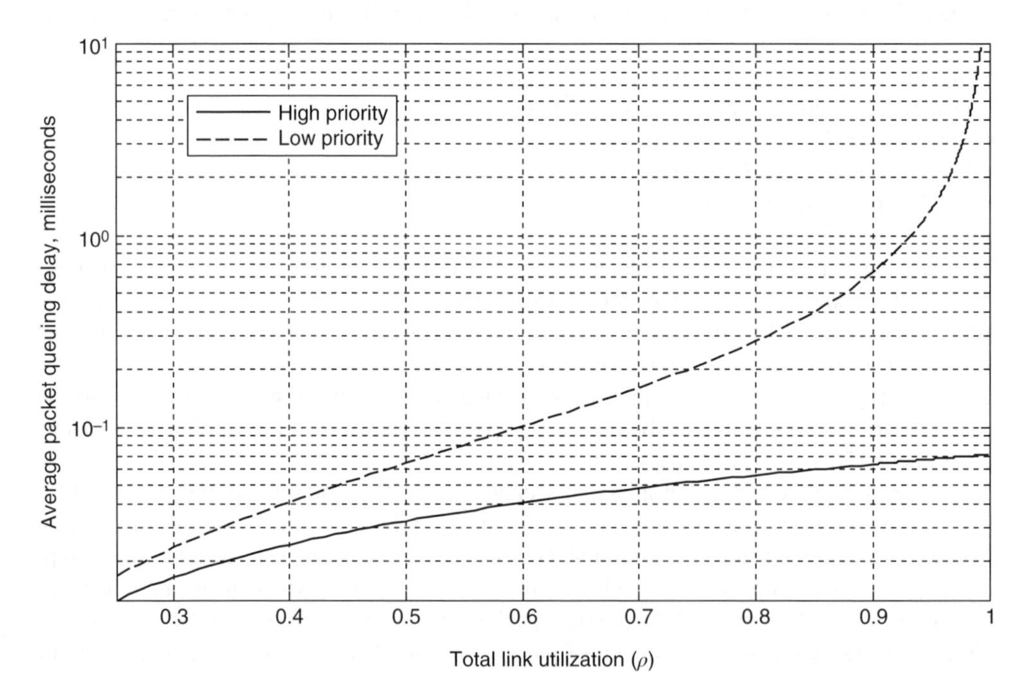

Figure 5.22 Example of queuing delay with two priorities, Link speed is 100 Mbps. Voice and signaling traffic volume is fixed at 25 Mbps and low-priority traffic volume changes from 25–75 Mbps

negligible when voice traffic is mapped to the highest-priority queue and the load of the high-priority queue is a fraction of the total link load. To strengthen this intuition assume that the load of the high-priority queue is low (so that the probability that there is a high-priority packet in the queue at any given time is small) and that the low-priority BE data queue is fully congested, (i.e. the total link load is 100%). In this case the high-priority packet arriving to the queue sees always an ongoing low-priority packet transmission. The resulting maximum waiting time for a 100 Mbps link and 1522-byte packet size would be at most 0.12 ms. If 10 such 100 Mbps links were installed in series, the worst-case queuing delay experienced by the high-priority packet (relative to unloaded link) would be $10 \times 0.12 = 1.2$ ms, in the case that the high-priority packet had to wait in every node.

To estimate the TCP throughput degradation the M/G/R-PS approach can be used. As usual, exact formulas are scarce for the SPQ system. In Malhotra and van den Berg (2006), three approximations are given for the combination of streaming and elastic traffic for the case where streaming traffic has a higher priority than BE data. In the simplest quasi-static approximation physical link capacity is replaced by the bandwidth unused by higher-priority traffic. In the context of the present example, the voice and signaling traffic are assumed to consume a static 25% portion of the link capacity. The approximation is reasonable for links where the high-priority portion of link bandwidth varies slowly over time. In the example the average voice traffic demand is more than 300 Erlangs,[14] and thus from the Erlang-B formula it is known that for a large number of calls the probability that the instantaneous number of voice calls significantly exceeds the average value is fairly small. Therefore, the voice bandwidth can be assumed fairly stable during the BH.

In the example, the average high-priority traffic volume is 25 Mbps. The average link capacity available for elastic BE traffic is 75 Mbps. The variable R in the M/G/R-PS formula is the effective link bandwidth divided by the maximum TCP flow rate limited by the radio interface. The below figure shows the delay increase for $R = 3, 5, 15$ corresponding to 25, 15, 5 Mbps UE air interface average rate (R is rounded down to the nearest integer). Adopting $f_R = 1.2$ as the threshold of acceptable throughput degradation the total link utilization should not exceed a value of 55–85%, depending on the UE air interface rate. For the 25 Mbps air interface rate, the effective bit rate of the UE would be $25/1.2 \approx 21$ Mbps where the 4 Mbps average rate loss is due to the transport link limiting the average TCP throughput.

So far in this section a system with two priority queues has been considered. It could be also possible to map voice traffic to their own queues which would result in a three-queue system. The question arises if the benefit justifies the added complexity. The performance impact would be that the packet delay of the highest-priority traffic (either signaling or voice) would be reduced compared to Figure 5.23, while the second highest priority queue would experience a delay increase. The delay of the lowest-priority BE traffic would not be affected at all, as can be seen from the M/G/1 formula for priority queuing.

As a summary, some guidelines to the questions posed are given:

- It is usually not necessary to have a dedicated queue for voice and another separate dedicated queue for signaling. Both can be mapped to the same queue, especially if the load of the high-priority queue is not high.

[14] The average number of AMR-narrow band voice calls (62 Kbps per call) is about 20 Mbps/0.062 Mbps = 322.

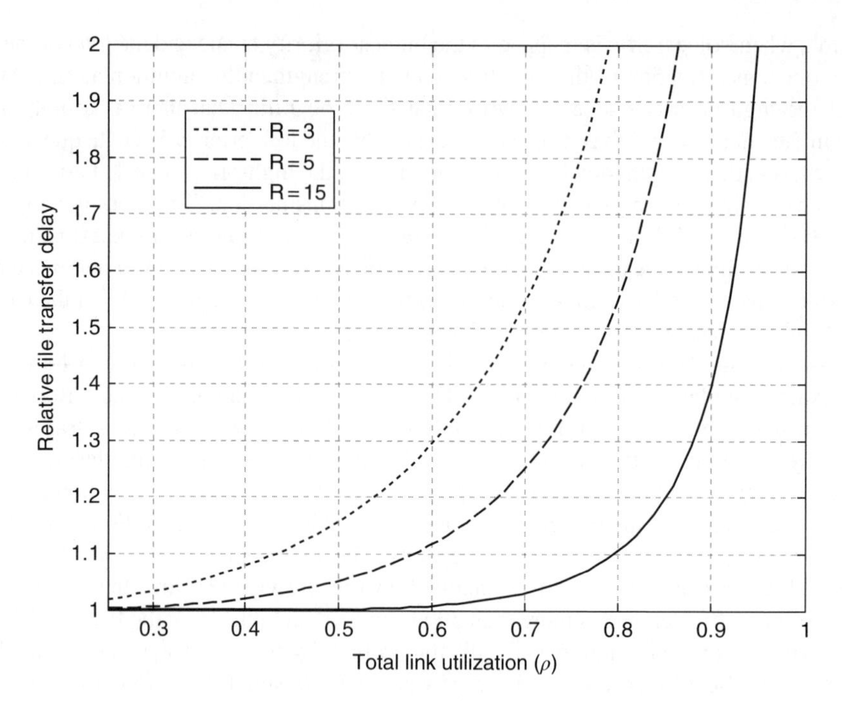

Figure 5.23 TCP file transfer excess delay as a function of total link load for different UE peak rates. The voice and signaling traffic load is a static 25% portion of the link capacity

- When signaling and voice are mapped to the strict priority queue, even a high link load due to BE data has only a minimal effect on the average queuing delay of the high-priority packets.
- The impact of high-priority traffic on the low-priority BE data throughput depends mainly on the effective bandwidth available for the low-priority TCP traffic.

5.6.6 Comparison of Bandwidth Dimensioning Formulas

Table 5.13 summarizes the dimensioning approaches discussed in this section. The G/G/1 and M/G/1 calculate the average packet delay, making them suitable for delay-sensitive traffic such as signaling. M/G/R-PS in turn produces the excess file transfer delay (throughput degradation) as its output.

None of the formulas in the table has a simple extension to multiple priority classes or networks of queues. For accurate results with multiple traffic classes, simulations are required. Due to the lack of usable analytical results, a pragmatic approach is to simply map delay-sensitive voice and signaling traffic to a strict priority queue, since in the majority of cases this can be trusted to result in low enough delay and packet delay variation even without detailed delay calculations. Due to the many approximations and uncertainties in the dimensioning process, a capacity-monitoring process based on counters or active measurements should be in place to be able to track and tune the network QoS parameters.

Table 5.13 Summary of dimensioning approaches

Method	QoS criterion	Type of traffic	Assumptions	Notes
First mile	none	any	none	
G/G/1	average queuing delay	any	Distribution of packet inter-arrival time has finite mean and variance, packet inter-arrivals are statistically independent	Requires coefficient of variation of the packet inter-arrival time and packet size as input
M/G/1	average queuing delay	any	Packet inter-arrival time is exponentially distributed, packet inter-arrivals are statistically independent	Gives optimistic delay values for bursty traffic Delay formula for strict priority queuing exists
M/G/R-PS	Relative file transfer delay increase	Elastic (TCP)	Flow arrivals are Poisson distributed	UE air interface throughput defines the maximum TCP flow throughput R

5.7 Dimensioning Other Traffic Types

5.7.1 Management Traffic

Management traffic can be usually neglected in bandwidth dimensioning. Some exceptional cases that may need to be considered include remote software download to base station and cell tracing, which consume DL and UL bandwidth, respectively.

The management traffic is usually transported over TCP. TCP protocol will adapt to the available bandwidth in the e2e path and, given enough data to transmit, will try to fill the transport pipe up to a rate limited only by the bottleneck link in the path. For this reason, a useful design approach is to prioritize management traffic over BE data, but give it lower priority than network signaling and control traffic. With WFQ-based schedulers the management traffic bandwidth can be limited per link. This enables responsiveness of remote network element management even over congested links, but will not allow TCP to hog excessive bandwidth from network control traffic. Some network vendors may also support limiting management traffic bandwidth at the application layer. Maintaining reliable remote management connectivity to the eNB also under abnormal congestion scenarios (e.g. congested backup link) is important since otherwise a site visit may be required.

5.7.2 Synchronization Traffic

Unicast IEEE1588 (2008) for frequency synchronization traffic produces 44-byte Sync messages n times per second in DL, where n is a power of two, for example 16 or 32 messages per second. With UDP and IP headers (28 bytes) the message size is 72 bytes before L1/L2 overhead. With 16 messages per second and Ethernet L1/L2 transport overhead (42 bytes) the bit rate per unicast stream is 14.6 Kbps in the DL. The UL bit rate is negligible if frequency synchronization is based on time-stamped Sync messages.

In Unicast IEEE1588 (2008) for phase synchronization DL there are two messages: Sync message and the Delay_Resp message.[15] The payload size of the Delay_Resp message is 54 bytes before UDP/IP/Ethernet overhead. Delay_Resp messages are sent as a response to Delay_Req sent by eNB in the UL and hence the frequency of Delay_Resp messages depends on how often eNB sends Delay_Req. The UL rate of sending Delay_Req messages depends on the master clock which informs the slave of the Delay_Req message interval in the Delay_Resp message. The minimum interval is the same as the Sync message interval and the maximum interval is one Delay_ Req every 32 Sync messages. Assuming 128 messages per second for all messages the bit rate over UDP/IP/Ethernet is 244 Kbps DL and 117 Kbps UL.

SyncE traffic load is negligible since the extraction of timing is based on the PHY Ethernet signal carrying normal packets. The synchronization status message (SSM) rate over the Ethernet slow channel is limited to at most 10 packets per second by the IEEE802.3 standard, and the maximum recommended frame payload size is 128 bytes, although larger payload sizes are permitted. Assuming the maximum frame size of 128 bytes the bandwidth consumption of SSM messaging should not be more than ≈ 13 Kbps, and usually much less.[16]

5.7.3 Other Traffic Types

Depending on the network, other traffic types present may include, for example, webcam or site-monitoring traffic (e.g. security and temperature measurement devices).

In some cases, bandwidth for network control traffic (e.g. VRRP, OSPF, BFD, IKE) may need to be estimated. Note that critical network control traffic is mapped to the queue 1 (SPQ) in Table 5.7. Especially in cases of network anomalies and problems, more control messages are transmitted than typically. Of course, in this situation it may be that anyway there is no possibility to correctly forward U-plane or any other traffic, until the network situation is corrected by the control messages. In that scenario, dimensioning of the network control traffic is not relevant. If deemed necessary, the bandwidth consumed by network control traffic will have to be estimated on a case-by-case basis.

Bandwidth reserved by active transport monitoring, such as UDP Echo or Two-Way Active Measurement Protocol (TWAMP) consume some bandwidth. To get some feel of the bandwidth required by the active transport measurement the bandwidth of TWAMP traffic can be calculated. Assuming that four different streams with different DSCPs per eNB are probed and that the packet rate and packet size for each class is one packet per second and 1000 bytes, respectively, the TWAMP bandwidth would be 32 Kbps in UL and DL. This is the bandwidth per eNB and can be considered negligible in almost all real-world use cases.

5.8 Base Station Site Solutions

This section briefly describes the site where eNBs are installed and the relevant backhaul equipment and their functionality at that site.

[15] Follow_Up message is rarely used.

[16] In SyncE messaging channel specification the information-carrying part of the payload is only 24 bytes, but in addition to this there is padding to minimum Ethernet frame size of 64 bytes and an optional data field ITU-T (2008).

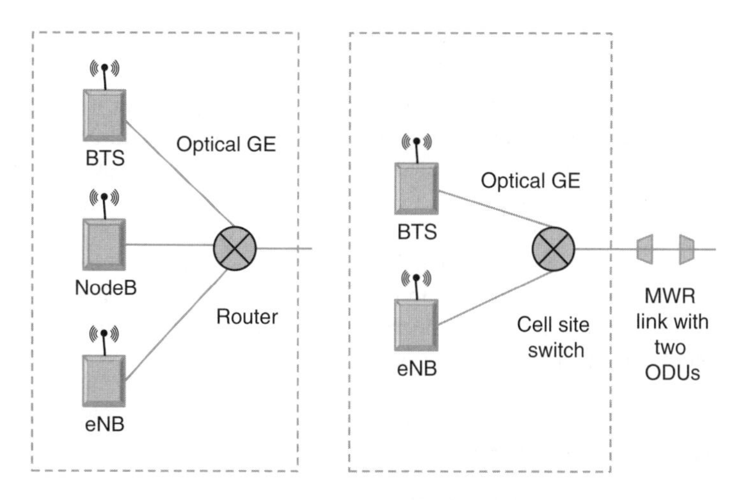

Figure 5.24 Base station site solutions examples

Base station site solution is considered here from a transport planning perspective leaving out the planning of the additional site support equipment, battery backup systems and alarm systems needed at the base station site. A base station site may consist of multiple base station equipment for different technologies such as GSM, W-CDMA, time division duplex LTE (TD-LTE) and full duplex LTE (FD-LTE). In case there is multiple base station equipment at the base station site, at the bare minimum there is a need for an Ethernet switch at the site. In some cases there is no need to add additional equipment as the Ethernet switch can be also integrated in the base station equipment or the indoor unit of the microwave equipment can act as the site switch. Recently, MPLS has become more common at the site level as well as bringing routing to the base station site. A couple of simple examples are given in Figure 5.24.

In addition to switching or routing the traffic toward the transport, the site equipment could be used for testing the backhaul connectivity and quality in case site equipment supports the TWAMP, as described in Section 5.13, or some other proprietary testing functionality. This could be used to complement such testing capabilities available in eNB.

Even organizational boundaries can be a good enough reason to have a clear point of responsibility between organizations operating radio access and transport networks.

IPsec integrated into the eNB can be considered the preferred solution, but IPsec support in the site router would enable the creation of a single tunnel to the core network site and this would make the IPsec implementation and operation simpler provided that traffic is secured between eNB and site router, or the site is physically protected.

The following section will describe in more detail the whole security solution from eNB toward the core network.

5.9 Security Solutions

From a security point of view, the difference between LTE and W-CDMA is that in W-CDMA the traffic is encrypted at the end user device and terminated in the controller, whereas in LTE the encryption terminates at the eNB. The use of untrusted backhaul requires traffic encryption and authentication in order to secure the communication. Even if many of the LTE networks may have been initially rolled out without securing the backhaul, the operator awareness of the existing threats increases and security solutions will be deployed.

The small cells deployments can be more vulnerable compared to macro cells as they usually use various access methods with no security mechanism. In addition, given the typical use case for small cells, they may be located in public areas, where the potential attacker can easily tap into the backhaul network.

In this section the security is handled from the point of view of the network planner that plans the eNB security, transport and related IPsec parameters for the eNBs and SEGs. As an example, the comparison of encryption/ciphering algorithms and their differences is left out of this section assuming that the mobile operator has some overall security policy for helping the planner select the right protocols for use.

5.9.1 Network Element Hardening

One of the foundations for a secure network is to have secure network elements. This requirement applies not just to the eNB or the EPC but to any node which is connected to the network.

Network element security is a wide topic that covers multiple aspects of the node functionality. Some of them need to be considered by the manufacturer at the early design and implementation phase, while others need to be taken into account by the operator during the deployment and maintenance phases.

One of the activities to be done is the system hardening. It comprises a number of design, patching and configuration actions that are meant to minimize the probability of success of any attack against the network element.

The following list includes some of the hardening actions that should be performed as part of the deployment and maintenance of the network element:

- Software should be kept up to date, including the latest security patches and bug fixes.
- Default passwords are changed.
- Passwords are regularly changed and only strong passwords are used.
- Login banners are configured so that they do not provide any details about the system.
- Unnecessary physical interfaces are disabled.
- Unnecessary services and ports are disabled.
- Firewall is enabled in all the network interfaces.

Additional hardening actions might be recommended by the vendor.

5.9.2 Network Security High-Level Architecture

LTE backhaul network security can be classified into the following major functions:

- protection of the traffic between the different network elements
- authentication of the communication peers.

Protection of the traffic within the mobile addresses the threats of eavesdropping and tampering, and it can be and should be accomplished at multiple levels. While end users data is a crucial asset to protect, a reliable network operation also requires that management traffic, mobile network signaling and network control traffic are also protected.

Often the protection at each level requires dedicated solutions. For instance, routing traffic can be protected by mechanisms embedded in the routing protocol, which calls for separate planning and configuration activities.

Some other times the protection mechanisms are of more general scope, allowing multiple protocols and applications to be protected by a single security framework. Such is the case of IPsec and TLS.

In LTE backhaul, 3GPP standards cover the protection of the traffic between the eNB and core network (3GPP 33.210), and the selected mechanism is IPsec.

IPsec is a generic security architecture which can be used to protect any kind of IP traffic in different configurations and network deployments. It can be either integrated into the network elements or it can be deployed at the edge of the trusted network. IPsec can be used on its own as the only security solution for the LTE traffic, or it can be combined with Transport Layer Security protocol (TLS), the latter used, for example, for network management traffic. With the network management channel secured separately, it may be carried either within the IPsec tunnel or outside of it, without compromising security.

Equally important as the traffic protection is the authentication of the communication peers when the secure connections are established. Without authentication, rogue network elements could be used to gain access and launch further attacks or retrieve confidential information. Authentication can also take place at multiple layers. For instance, a network switch might authenticate the peer using 802.1x before opening the network port. Most commonly, authentication takes place as part of a VPN establishment, as a preliminary step before the tunnels are established.

In LTE backhaul with no authentication, rogue eNBs could be used to gain access to the core network and launch further attacks. Likewise, the eNBs might get connected to an attacker impersonating the network. The authentication in LTE backhaul is based on public key cryptography and X.509 certificates (3GPP 33.410) since it provides the required scalability for the typical size of the networks. The use of certificates requires the so-called public key infrastructure (PKI), which encompasses the network elements and procedures required for the issuing and maintenance of the certificates. Further aspects of the PKI are discussed in Section 5.9.4.

The authentication procedure is embedded in the TLS and IPsec protocols, integrating the connection setup and the authentication, and avoiding separate steps. However the certificates still need to be installed.

The network elements involved in the LTE backhaul security are:

- eNB
- SEG
- CA.

The eNB usually integrates the security functions, which ensure that all the traffic sent and received is already protected within the node. This option also provides a common management interface and processes with the rest of the eNB functionality. An alternative for eNB native security would be the use of a base station site SEG. However, this solution is not secure in some environments (public locations). It is also not cost effective, due to the additional equipment and maintenance needs.

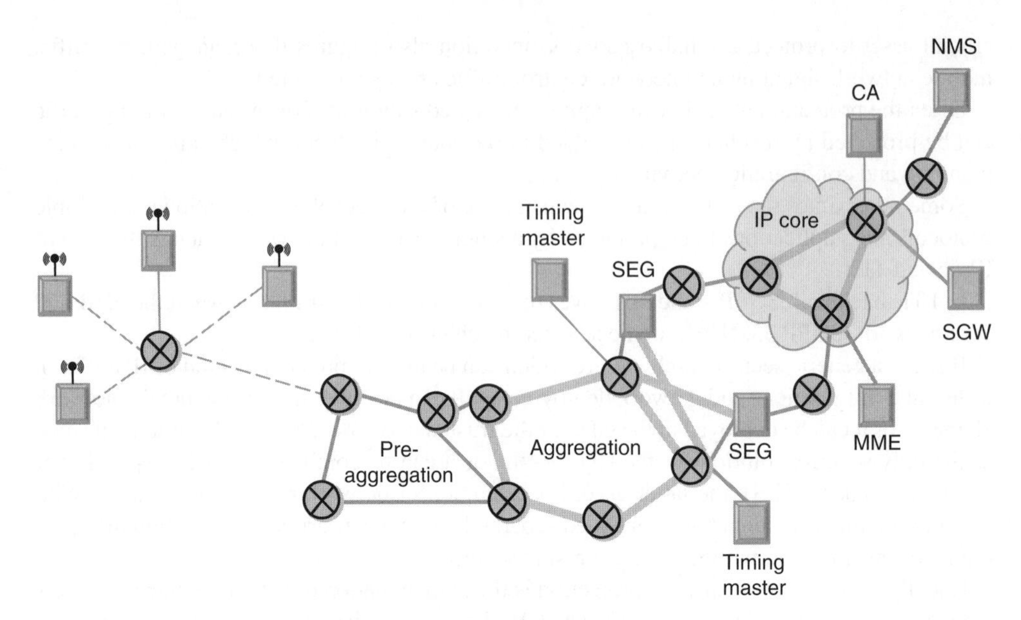

Figure 5.25 Security solution with geo-redundancy

The security functions at the core network end are carried out by the SEGs. They are placed between the radio aggregation network and the core network to protect traffic from the eNB in UL direction and toward the subscribers in the DL direction. They have the following functions:

- Authenticate the eNBs when they connect to the network.
- Provide encryption and integrity protection of the traffic between eNB and SEG.

SEGs need to be protected against failure. They can operate either in active/passive or active/active mode and these are further described in Section 5.9.3. When multiple clusters are needed, those should be located also on different locations to provide geo-redundancy and protection against a site failure, like problems with power supply of the site.

The CA is the central element of the PKI, and the root of trust for the authentication process. Section 5.9.5 describes the main aspect to consider for the CA.

Figure 5.25 is simplified to present a few eNBs and the geo-redundant security solution in the LTE mobile network.

5.9.3 Security Gateway High Availability

Security gateways (SEGs) are high-capacity equipment which can handle the traffic of over a thousand eNBs. A failure of a single device might cause, depending on the network planning, the outage of a large service area, with the associated loss of revenue for the operator. It is therefore important to guarantee that there is always a point of access for the eNB to the core network.

High availability can be implemented in a number of ways, with different recovery times and with different degrees of involvement of the eNB. A few possible mechanisms are listed below:

- SEG stateful failover
- SEG stateless failover
- backup tunnel
- traffic rerouting
- load sharing.

Many SEGs support redundant high-availability multi-chassis configuration that allows combining multiple physical equipment into one logical element. In other cases, it may be advisable to use multiple devices so that they can be located in different locations providing geo-redundancy.

With stateful failover, the SEGs communicate with each other and continuously synchronize their state (state of the tunnels, session keys, etc.). Then, in the event of failure of the active SEG, the tunnel endpoint IP address is moved to the passive SEG without the need to re-establish the tunnel. This failover is transparent to eNB, which sees only one remote IPsec tunnel endpoint IP address. Only one tunnel is established to the SEGs, which is terminated at the active one.

Figure 5.26 shows a typical configuration for the stateful failover.

Stateless failover is similar to the stateful failover, but there is no synchronization between the SEGs. This is a typical situation when the SEGs are in different locations, as the synchronization requires both SEGs to be in the same subnet. The SEGs still share the same tunnel endpoint IP address and have the same IPsec configuration, but they do not share the state of the tunnels. The practical effect of the stateless failover is that the tunnels need to be re-established, either by the eNB or by the active SEG, which increases the duration of the outage.

The traffic can be also protected by selecting a different route. In this case two tunnels are established simultaneously from the eNB to two different SEGs and their availability is monitored (using, for instance, BFD). When the eNB and the network detect that the route through one of the tunnels is not available, the traffic is rerouted through other tunnel.

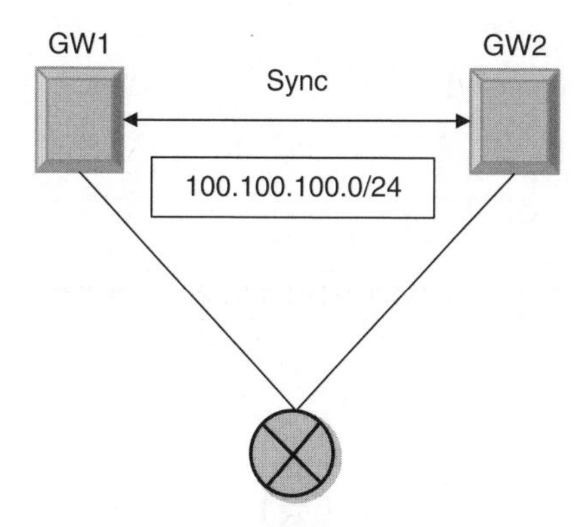

Figure 5.26 Stateful failover protection

A variant of the stateless failover protection is the use of a backup tunnel. In this case, there is also only one tunnel established between the eNB and the SEG. When the eNB detects that the SEG is not available (using, for instance, dead peer detection [DPD]), a new tunnel is established to the passive SEG. Each SEG has a different tunnel endpoint address, but otherwise they will normally have the same security configuration.

In all the approaches above, only one SEG is active (carries traffic) at a given point in time while the passive SEG is idle, at most with some tunnels established but without carrying traffic.

The SEG protection can be also implemented using a load-sharing configuration, as in Figure 5.27, where all equipment is active and sharing the load in normal conditions. In case there is failure in one of the nodes, the remaining node provides full redundancy.

When considering implementing a load-sharing configuration for SEG protection, the following benefits can be identified:

- Lower system level utilization on each node.
- Higher throughput and lower latency.
- Traffic peaks which would lead to oversubscription in stateful/stateless configurations can be covered.
- Only half of the number of tunnels will be affected in case of failover.

Some disadvantages of load sharing compared to other mechanism are that configuration and routing are more complex. Also maintenance, like software or hardware updates, is more complex as both the nodes are active and carrying traffic. Additionally, the load-sharing configuration by itself does not ensure the availability of all the tunnels. In order to do that, it should be combined with some of the other mechanisms described above.

In all cases, the SEGs need to be dimensioned similarly as in the event of failure all traffic needs to be carried by a single node.

SEGs can also be protected against complete site failure and by locating them at different sites and providing geo-redundancy. The same mechanism depicted above can be used in this scenario, with the exception of the stateful failover, for the reason already stated.

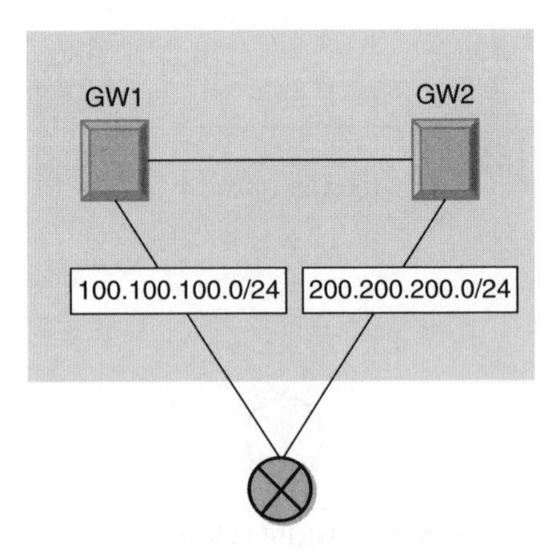

Figure 5.27 Load-sharing configuration

Figure 5.28 illustrates a geo-redundant configuration together with the backup tunnel mechanism.

SEG redundancy might be also deployed not as a standalone measure but together with the core network redundancy, as shown in Figure 5.29. With redundant core networks, two or more MMEs and SEGs are available sharing the load of the eNBs. They are typically located at different sites and each site is equipped with a SEG. In this case, the failure of one SEG is seen by the eNB as a failure of the whole core network site, and the core network recovery mechanisms are applied.

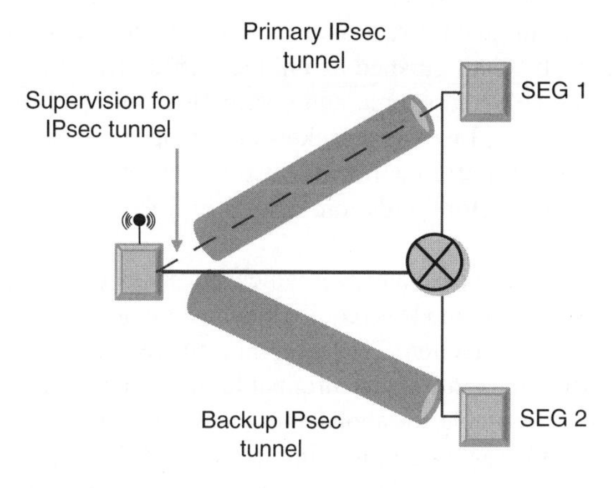

Figure 5.28 Geo-redundancy for security gateways

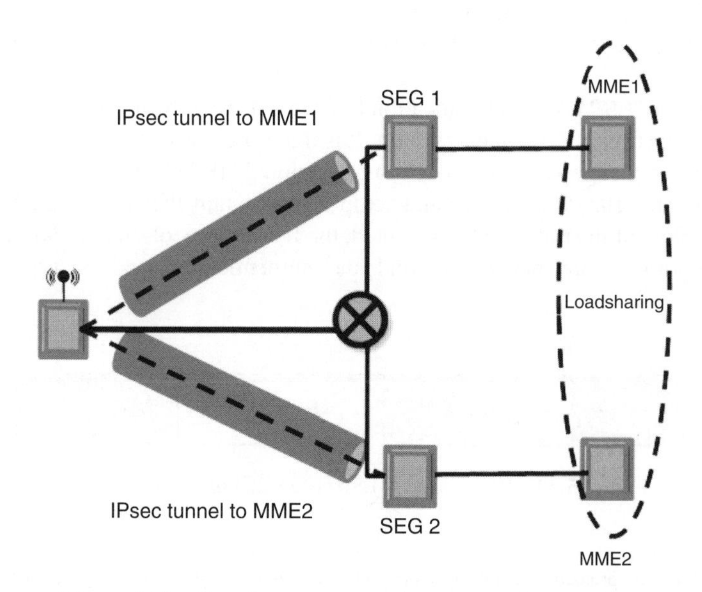

Figure 5.29 Security gateways and core network redundancy

The time to restore the service is a combination of time to detect failure and the time needed to restore the tunnels.

For stateful failover mechanism, the failure detection occurs between the SEGs and should occur within one second. For the other protection mechanisms, the failure detection strongly depends on the detection mechanism used as well as the timers involved. The delay is typically in the range of one second to several minutes.

5.9.4 IPsec Parameter Planning

IPsec-secured connections are defined in terms of IPsec security associations (SAs). Each IPsec SA is defined as a single unidirectional flow of data between two peers or endpoints covering traffic that can be distinguished by unique traffic selector, that is source and destination addresses, protocols and source and destination ports. All traffic flowing over a single SA is treated the same. Incoming packets are mapped to a particular SA within the SA's database by the three defining fields: destination IP address, security parameter index and security protocol. Both ends must agree on the SA for the secure connection to work.

Since the security solution is based on SEGs which encapsulate traffic generated by the core network devices, an IPsec tunnel mode is required. Tunnel mode is also the standardized mode in 3GPP for LTE backhaul protection. The difference between tunnel and transport mode is that, in tunnel mode, IPsec encapsulates the original IP header and the payload and adds a new IP header, which makes it suitable for applications where the IP packet is generated by one node and IPsec is applied by another (bump in the wire [BITW]). Tunnel mode can also be used for integrated implementations. With transport mode, on the other hand, only the IP payload is encapsulated, which requires integration of IPsec and IP stacks. This mode is suitable for integrated implementations but not for SEGs.

IPsec provides two protocols to protect the traffic. The IPsec security protocol used for LTE is the Encapsulating Security Payload (ESP).[17] Figure 5.30 shows the IP frame after encapsulation.

An alternative to ESP is the Authentication Header (AH). As the name indicates, it provides an authentication service, but not encryption. It is therefore not suitable for LTE.

Internet Key Exchange version 1 (IKEv1) or version 2 (IKEv2) is used to set up the IPsec SAs and negotiate related parameter and setup and maintain the IPsec tunnels. During the setup, the selectors of the traffic to be protected, the IPsec protocols, the hash function and the encryption algorithm are negotiated, and the authentication and encryption keys are generated.

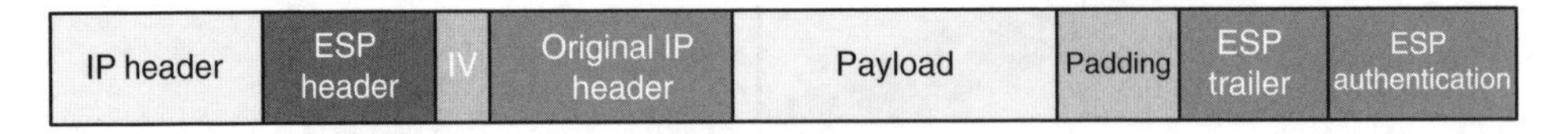

Figure 5.30 IP frame after ESP encapsulation in tunnel mode

[17] Use of the ESP security protocol is assumed as it provides data confidentiality, data integrity, data source authentication and anti-replay capability.

The IKE protocol consists of two phases:

- A security association (IKE SA) is established in order to protect the messages during the phase two exchanges.
- An IPsec SA(s) is established.

IKE SAs are protected in a similar way to the IPsec SAs, and similar negotiation takes place during the setup (except for the traffic to be protected). An important step of the IKE SA setup is the authentication of the peer. In this phase, each node will provide the authentication information which enables the peer to verify the authenticity of the certificates. Optionally, the certificate (containing the public key) can also be delivered in the same step.

In 3GPP 33.210 it is required that the SEGs support both IKE protocol versions, but the use of IKEv2 has security and performance advantages such as QoS handling, protocol simplicity and robustness against DoS attacks over IKEv1 and therefore it is recommended. However, not all implementations support IKEv2 and in those cases IKEv1 is provided for interoperability.

5.9.4.1 Re-Keying

IKE and IPsec SAs have a maximum lifetime each, which defines for how long the SAs can be used. In this way the encryption and authentication session keys are renewed often enough so that the likelihood of a brute force attack against the keys is reduced. When the SA lifetime expires, the SA re-keying is triggered.

IKE SA and IPsec lifetimes are operator's security policy parameters, and they need to be configured in the eNBs and the SEGs. It is usually enough to configure the required lifetime in either the eNB or the SEG, while the other node will have a long enough lifetime so that it does not initiate the re-keying.

Depending on the IKE version and re-keying method, the IKE SA re-keying will cause the IPsec SA to be re-keyed as well. Therefore the IKE SA lifetime should be longer than the IPsec SAs lifetime to avoid excessive IPsec SA re-keying and the subsequent impact on the service availability.

When using IKEv2, there are two possible methods for re-keying:

- When using IKE_SA_INIT and IKE_AUTH exchanges, the IPsec SAs under the IKE SA being re-keyed are deleted and recreated. This causes a brief interruption to the traffic, and some operators regard this interruption as unacceptable. On the other hand, this re-rekeying method allows the peers to re-authenticate, which is especially relevant when supporting certificate revocation lists (CRLs).
- The alternative is to use the CREATE_CHILD_SA exchange, which allows a smooth hand-over of the IPsec SAs from the old IKE SA to the new IKE SA. The drawback is that the peer is not authenticated again, so there is no way to check if the peer certificate is still valid.

Some implementations might offer the possibility to choose the re-keying method, which should be selected depending on the operator requirements for service availability and authentication.

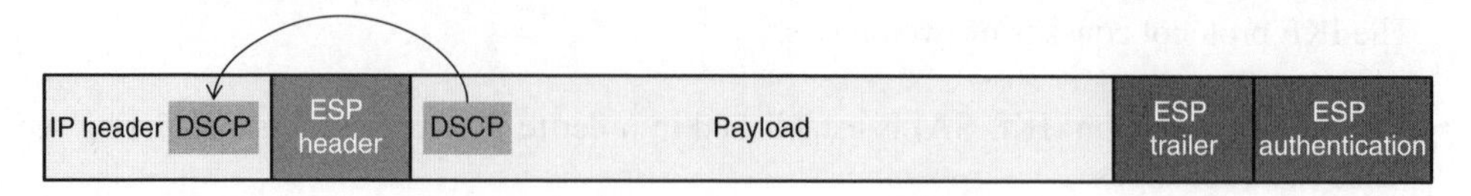

Figure 5.31 DSCP copy with ESP tunnel mode

5.9.4.2 Quality of Service (QoS)

QoS needs to be preserved for the tunneled traffic, also when the traffic is protected. In tunnel mode, when encryption is enabled, the DSCP field carried in the inner IP header cannot be inspected by the backhaul network equipment. To avoid this limitation, the DSCP is copied from the inner IP header to the outer IP header and thus QoS can be applied by the network equipment based on this outer header DSCP, as in Figure 5.31. This also introduces a security risk since some traffic types have a very distinctive DSCP and the packets could be identified just by inspecting the outer DSCP field. Also the outer DSCP field is not authenticated, thus it could be modified by the attacker and affect the QoS mechanisms.

In addition to the tunneled traffic, QoS needs to be applied to the IKE protocol messages. This can be simply done by defining the DSCP to be applied to the IP frames carrying IKE messages.

IPsec uses an anti-replay window for each IPsec SA to provide a partial sequence integrity check which protects the system against the replay of packets by an attacker. The anti-replay window mechanism is based on checking the sequence number of the received packets. If the packet has been already received or if it occurs in the past outside the rear end of the window the packet is discarded. The packet is accepted if it falls within the window and it has not been received already. If the packet is received ahead of the window, the packet is accepted and the window slides forward.

A side effect of using the anti-replay window is that it requires the packets to arrive roughly in order. If the packets suffer a significant level of reordering in the backhaul network, the lower-priority packets (which arrive later) will be dropped by the receiver since they miss the window rear limit. The tolerable level of reordering depends on the size of the anti-replay window. This situation might happen when the packets traverse schedulers with multiple queues for different QoS classes and there is congestion.

In order to minimize packet losses due to the anti-replay window, the network should be dimensioned according to the expected traffic demand, as per dimensioning practices presented earlier. Additionally, the biggest available window size should be selected at the receiver end. If none of the previous measures is sufficient and heavy congestion is experienced, it might be necessary to disable the anti-replay window to avoid excessive packet drops.

When IKEv2 is used, it is also possible to establish multiple IPsec SAs for the same traffic descriptors, but with a unique DSCP for each IPsec SA. Each IPsec SA has its own anti-replay window. As packets are not usually reordered for a given DSCP, the packets will not be discarded by the receiver even in the event of heavy congestion. However, this mechanism is not currently available in most implementations.

5.9.4.3 IPsec Parameters

There are a number of parameters that need to be defined at the eNB and the SEG. Some of them need to exactly match or have some common range as they are part of the SA negotiation, while others are relevant for the node where they are configured. Those parameters

Table 5.14 Planning parameters for IPsec

Parameters	Example value	Need to match
Traffic selector: Source IP address	N/A	Yes*
Traffic selector: Source port number	N/A	Yes*
Traffic selector: Protocol	N/A	Yes*
Traffic selector: Destination IP address	N/A	Yes*
Traffic selector: Destination port number	N/A	Yes*
IPsec action (PROTECT, BYPASS, DISCARD)	N/A	Yes
IKE ID	enb1@mynetwork.com segw1@mynetwork.com	No
IKE DSCP value	34	No
DPD timeouts	N/A	No
IKE Diffie–Hellman group	DH group 2	Yes
IKE encryption algorithm	AES-128-CBC	Yes
IKE integrity algorithm	HMAC-SHA-1-96	Yes
IKE SA maximum lifetime	1 day	IKEv1: Yes, IKEv2: No
IKE PFS enabling	Yes	Yes
IKE protocol variant	IKEv2	Yes
IKE SA rekeying procedure selector (IKEv2 only)		No
Local IPsec tunnel endpoint IP address	N/A	Yes
Remote IPsec tunnel endpoint IP address	N/A	Yes
Anti-replay enabling	Yes	No
Size of the anti-replay window	256	No
ESP encryption algorithm	AES-128-CBC	Yes
ESP authentication algorithm	HMAC-SHA-1-96	Yes
PFS Diffie–Hellman group	DH group 2	Yes
IPsec SA lifetime	1 hour	No

*The traffic selectors do not need to strictly match. However, they need to overlap so that they both apply to the traffic to be protected.

depend of course on the vendors and their security implementations, but Table 5.14 presents a subset of parameters needed for setting up an IPsec connection.

5.9.4.4 Tunnel Architecture

IPsec offers the capability to define which flows should be protected and how they should be protected. Given this capability, the network planner needs to consider the following questions:

- Does all the traffic need or can be protected?
- Does all the traffic need to follow the same network path?
- Does the operator have different security policies for different traffic types?

In the simplest case, each eNB is configured only with a security policy which creates one IPsec SA for the outbound traffic and another for IPsec SA for the inbound traffic (note that the IPsec SAs are unidirectional, but they are always established in pairs). This can be easily configured by defining a single "PROTECT all" type of policy.

In some occasions, the performance of some applications might be negatively affected by IPsec processing delay, such as eNB synchronization. In this case the corresponding traffic type needs to be sent in clear text, which would require the configuration of a specific BYPASS policy.

A more complicated situation appears when the operator requires that different traffic types use different networks. One example is the separation of production traffic (U-plane and C-plane) and management traffic. To protect both traffic types, separate PROTECT policies are required. Protecting O&M traffic separately using other high-level methods (SSH, TLS) prevents losing remote management to the eNB in the event of an IPsec tunnel being down.

The traffic can be split further so that, for instance, C-plane traffic uses a highly available network, while U-plane traffic uses a somewhat less-available but higher-capacity network. These situations need to be handled by defining the corresponding policies.

5.9.4.5 Traffic Selectors

The tunnel architectures depicted above should be configured by using the traffic selectors. The IPsec implementations allow the selection of the packets to be protected based on the source and destination addresses, protocol type and source and destination ports. While not all the implementations support all the options, they can at least define the source and destination addresses.

Using all five parameters allows a fine granularity, but at the expense of configuration complexity and maintenance cost. In most cases, it is sufficient to use the source and destination addresses and, if additional granularity is needed, the protocol type.

When configuring the traffic selectors in the eNB and in the SEG, it should be noted that it is not necessary to define the full granularity in both sides. It is sufficient to define, for instance, a wide range in both nodes so that it is ensured it applies to the traffic to be protected. For IKEv2 it is also possible to define a wide range in the SEG side, and a narrow range in the eNB side. During the IPsec SA setup, both nodes would negotiate and agree on using the narrower range. Consider the following configuration:

eNB – Policy 1:
Protect, source from 192.168.1.0 to 192.168.1.255, destination any

SEG – Policy 1:
Protect, source any, destination from 192.0.0.0 to 192.255.255.255

The agreed traffic selectors would be:

eNB – Policy 1:
Protect, source from 192.168.1.0 to 192.168.1.255, destination any

SEG – Policy 1:
Protect, source any, destination from 192.168.1.0 to 192.168.1.255

This approach has the additional benefit that a single policy can be used in one of the nodes (in the example above in the SEG) to match the policies in multiple peers, therefore reducing the design and maintenance effort.

5.9.5 Public Key Infrastructure (PKI)

The PKI addresses all items that are needed to enroll and maintain certificates throughout their lifecycle. The main component is the CA, which is the part of a PKI that signs the X.509 certificates containing the public keys of the clients with its own private key. The CA is needed for the lifecycle management of certificates; it is needed when issuing the eNB and SEG certificates in the first place, for renewing the certificates when they expire and for revoking the certificates when the old ones are not trusted anymore.

Having thousands of eNBs in the network, the PKI solution should support high-availability configuration to prevent potential problems in certificate renewal and corresponding eNB outages. The CA server should have high-availability hardware platform capable of responding to requests with minimal downtime. Also the IP transport network should support high-availability access to the CA server.

5.9.5.1 Trust Model

The trust model defines which network elements are to be trusted and how the trust between those network elements is created. Different trust models can be designed to suit the operator's needs, ranging from a simple authentication of the eNBs and the SEGs to interconnections to other operators (for RAN sharing applications).

It should also be considered who is issuing the network element certificates (the operator, the vendor, a trust supplier), how many certificates are installed in the network elements, when to use them and which ones need to be used for a given security protocol.

In this section, the one-level hierarchical trust model of Figure 5.32 is used as an example.

In this example it is considered that the operator has an own CA which is the root of trust (i.e. any node in the network will trust a peer which presents a valid certificate signed by this CA). Hence, there are two kinds of certificates in this model:

- Root CA certificate: there is only one certificate, and it contains the public key used to verify the validity of the entity certificates.
- Entity certificate: there is one certificate for each network element (eNB and SEGs), and it contains the public key associated to the network element.

The root CA certificate is installed in each of the network elements and is required to validate the peer's certificate. The entity certificate is installed in each of the entities.

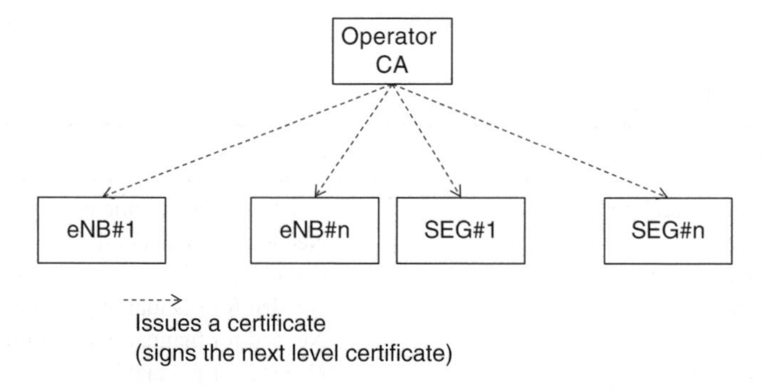

Figure 5.32 One-level hierarchical trust model

5.9.5.2 Certificate Lifecycle Management

Certificates need to be provisioned in the network elements before they can be used by any security application. They can be pre-installed at the factory by the equipment vendor, they can be installed during the commissioning of the node or they can be installed later on when the node connects to the network for the first time. Certificates also have a lifetime so they also need to be updated regularly.

The exact procedure to provision the certificates will depend on the equipment capabilities as well as on the operator's operational needs. It is possible to use a combination of the mechanisms above and, for instance, pre-provision the network element with a certificate, and update the certificate on the first contact to the network.

The most straightforward way to install the certificates is manually, via the management system. This method might be convenient when doing the commissioning, for one-time installation or for a few network elements in a central location such as the SEGs. However, it is not a scalable solution for a big number of devices with a regular update of the certificates, such as the eNBs.

A more useful way to install and update the certificates is doing it automatically by the network element on first contact to the network or when the certificate is approaching the end of the lifetime. The certificate retrieval and enrolment operations can be one with a protocol such as the CMP. Using CMP, the network element first retrieves the root CA from the certificate repository in the CA. Then it generates the key pair (public key and private key) used for authentication, and sends to the CA a "certificate signing request," including the public key, the identity and other relevant information. The CA signs the request and sends the certificate back to the network element. When the certificates approach the expiration date, the network element automatically repeats the process, ensuring the certificates are always valid.

An additional important management operation is the revocation. It provides a mechanism for the operator to make sure that certificates which have been compromised (i.e. the associated private key has been exposed) are not used any more for authentication. When a certificate is revoked and presented during IKE SA setup, the connection should be rejected by the authenticating party.

There are several methods to implement certificate revocation but the one required by the 3GPP for LTE backhaul is the CRL. The CRL is based on a regular download of the list of the serial numbers of the revoked certificates. The list is signed by the CA and it can be retrieved from the CRL distribution point on a regular basis, or when pushed by the CA. The local copy of the CRL is checked by the network element every time a certificate needs to be verified.

Table 5.15 Planning parameters for PKI

Parameters	Comments
Root CA address and port	Needed for automatic management
Subject name	Needed as node identity
CRL distribution point	Could be retrieved from the certificates
CA certificate	Needed for manual installation
Private key	Needed for manual installation
Entity certificate	Needed for manual installation

5.9.5.3 PKI Parameters

The number and kind of PKI parameters that need to be configured in the network element strongly depend on implementation (Table 5.15). However, there are a few of them that can be normally found.

5.9.6 Self-Organizing Networks (SONs) and Security

3GPP has defined a way to start the chain of trust already during the production process, and then hand it over to the operator's CA. In this way it is possible for the eNB to be authenticated by the network without the need for manually installing the certificates during commissioning.

The procedure is started by installing a "vendor entity certificate" in the eNB at the factory. The vendor root of trust (vendor CA certificate) is also delivered and installed in the operator root CA. In this way the operator will trust the eNBs when they connect to the network for the first time.

Upon connection of the eNB to the network, the eNB retrieves the basic configuration such as addresses and the root CA address. In some implementations this configuration can also be done manually. Once the eNB knows the root CA address, it proceeds to create a key pair and enrolls the new entity certificate to the root CA using the vendor's entity certificate for authentication. Any subsequent authentication operation is done using the operator's entity certificate and the vendor's certificate does not need to be used anymore.

5.10 IP Planning

In this section IP subnetting and routing are discussed. The focus is on the telecom layer aspects of access planning. Specifically, this means that planning of MPLS transport (L2 VPN or L3 VPN) is beyond the scope of this chapter (there is a considerable amount of literature already available on this topic). The access planning scope considered here is illustrated in Figure 5.33. The L2/L3 service would typically be either L2 VPN or L3 VPN, transported over MPLS.

In this section the terms "L2" and "Ethernet" are used interchangeably as are "L3" and "IP." The term "application layer" means telecom protocols transported over L4 (UDP, SCTP, TCP), including S1AP, X2AP and U-plane GTP.

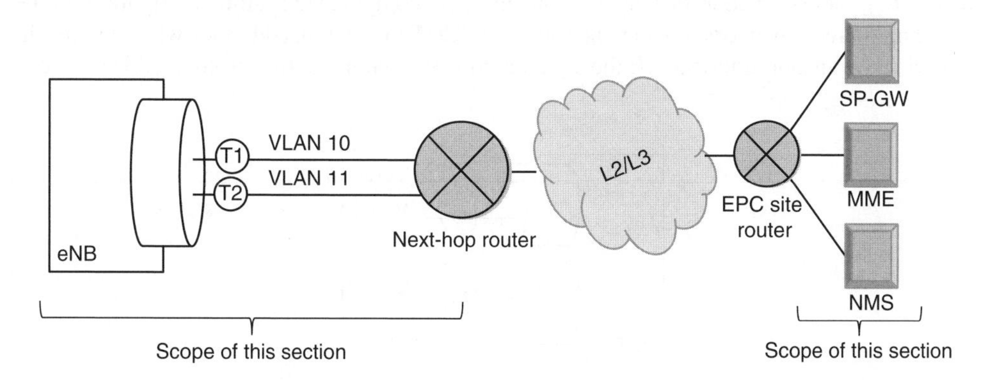

Figure 5.33 Telecom layer IP planning scope

Two backhaul deployment scenarios are introduced in the beginning of this chapter and a more thorough presentation on various backhaul deployment scenarios can be found in NGMN Alliance (2011b).

5.10.1 IP Addressing Alternatives for eNB

In this section some generic IP addressing models for eNB are presented. These models serve as a basis for discussion on subnetting and routing design later in this section. The difference between the models is in the termination point of various traffic types, such as U-plane traffic (U), C-plane traffic (C) and M-plane traffic (M).

5.10.1.1 Model #1: Traffic Terminated at Network Interface IP Addresses

In this model, the eNB may support one or more physical Ethernet interfaces. Each physical Ethernet interface may, in turn, host one or more network interface IP addresses that are bound to an untagged sub-interface and/or one or more tagged VLANs. Application layer traffic is terminated at the eNB network interface IP address. An example of this addressing model with three application addresses, three VLANs and two Ethernet interfaces is shown in Figure 5.34.

This addressing model can be considered the baseline addressing model when IPsec is not used. Since the application addresses are bound to transport addresses, static IP route configuration is not needed at the next-hop router to reach the application addresses. An exception would be the case where eNB has an internal router that forwards packets to other base stations or site support equipment.

5.10.1.2 Model #2: Traffic Terminated at Internal Host IP Addresses

This addressing model is shown in Figure 5.35. In this model the network interface IP address (denoted T1, T2, T3) is different from the application layer traffic termination point, which is now a base station-internal host address. The concept is similar to loopback addresses used in routers which terminate router management traffic at a router-internal host address.

The drawback of this configuration is the more complicated design (with respect to Model #1), increased address consumption and more complicated routing configuration. The next-hop router now needs to know the route to U/C/M application addresses which are reachable via the transport interfaces. If the eNB supports a dynamic routing protocol, this route can

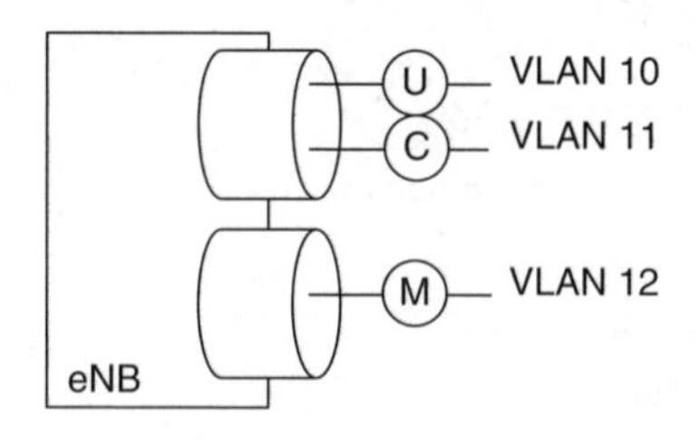

Figure 5.34 User/control/management traffic sharing network interface IP address

be automatically advertised to the external transport network; otherwise, static routes need to be configured at the next-hop device; this may result in considerable configuration overhead in large networks. Proxy ARP (RFC1027), if supported by the eNB, may be used to mask the eNB-internal IP addresses, so that the next-hop router does not need to know the routes to the U/C/M host addresses, though from a deployment effort point of view this merely shifts the configuration effort from the next-hop router to the eNB.

The model is useful in deployments that will be migrating to IPsec later. If suitably planned, there is no need for eNB IP address reconfiguration during IPsec deployment. Another use case could be a scenario where the transport network provider administrates the allocation of transport layer IP addresses, but it is desirable to have application IP addressing in a separate subnet, for example because transport network addressing is from a routable public IP range while the eNB application addresses should be allocated from a private IP range.

Depending on eNB implementation, the outgoing VLAN would typically be selected based on IP routing, that is the packet destination address and eNB routing table. Alternative implementations are possible, though. For example, the outgoing VLAN could also be selected based on a more complicated routing policy, such as source-address based routing or a configurable mapping table that links a protocol/port (application) to an outgoing VLAN interface.

5.10.1.3 Model #3: IPsec

The IPsec tunnel mode may be utilized to protect all or some application layer traffic, depending on the operator security policy. Figure 5.36 presents an IP addressing model for IPsec. Although strictly speaking not necessary, the typical IPsec site-to-site VPN implementation uses the addressing model where the applications are bound to dedicated host (/32) addresses.

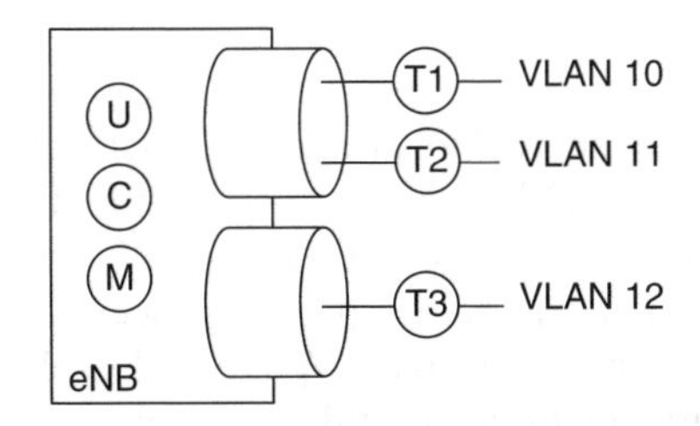

Figure 5.35 User/control/management traffic bound to stand-alone IP addresses

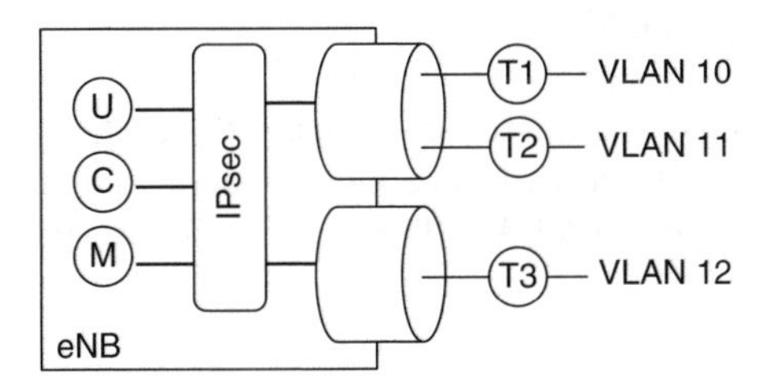

Figure 5.36 IPsec addressing model

In the UL direction, IPsec applies a traffic selector to each packet and, based on the user-defined security policy, IPsec bypasses the packet unmodified to a transport interface, protects the packet with IPsec or discards the packet. In IPsec tunnel mode, the protected inner packet is prefixed by a plaintext outer IP header; the destination IP address in the outer IP header is the remote IPsec SEG address.

In this model the next-hop router does not need to know the DL routes to the application layer host addresses. For non-protected traffic, these routes still need to be configured, unless of course in the special case that the application is bound to a network interface IP address.

5.10.2 VLAN Planning

5.10.2.1 How Many VLANs are Needed Per One eNB?

A typical question during the planning phase is the number of tagged VLANs that should be allocated per eNB.[18] Obviously tagged VLANs do not need to be used at all, if it is possible to cope without them. The drawback of having many VLANs is not obvious if considering only a single or a few eNBs typical in initial technology trials, since in this case the configuration and maintenance effort is not an issue. However, the situation is quite different if operating a network of thousands of eNBs. Real-world network experience has shown that the difference between having four VLANs per eNB versus a single VLAN per eNB can result in an almost cataclysmic increase in management and troubleshooting of the network, and consequently overall network quality. This is one of the main differences between classic backbone IP planning and access IP planning; in a backbone or aggregation layer, the number of nodes that need to be administered is in the order of dozens, whereas in a mobile access network the number of nodes is easily two or three magnitudes higher. It is usually worth the effort to include the "ease of configuration and maintenance planning" criterion.

Oftentimes the answer to the number of VLANs is naturally given by the existing transport network design. For example:

- If the eNB is to be directly connected to a next-hop router using VRF then the VRF design directly determines the number of VLANs required at eNB. This is because, by default, the VRF selection (routing table used) in a router is typically based on the incoming VLAN, although other VRF selection criteria are possible depending on router features. As an example, if the router implements separate U-plane and C-plane VRFs, a separate VLAN must be used for these traffic types in the eNB, and so on.
- If the eNB is to be directly connected to a Metro Ethernet-compliant leased transport provider, the provider service offering in terms of pricing and QoS features will be the key in deciding the number of VLANs.

With the possible exception of the leased line use case, segregating traffic to different VLANs because of QoS alone is rarely needed since most modern equipment supports multi-queue scheduling and shaping within one VLAN. Therefore, QoS can typically be controlled even in a single-VLAN deployment.

[18] "Tagged VLAN" here means that the Ethernet frame has at least one VLAN header of four bytes.

Given complete freedom from external transport network constraints, it is a useful rule to keep the number of VLANs to a minimum in order to simplify configuration, network maintenance and troubleshooting. In practice, this means one or two VLANs per eNB. An often-followed best practice in the IP planning community is to allocate a dedicated VLAN for management traffic. While not by any means mandatory, this approach does have some benefits in terms of potentially allowing remote management connectivity to the eNB in case reconfiguration of U/C IP addressing and routing is required.

Often the access router directly connected to the eNB is deployed as a redundant pair in active/standby configuration, and the router switchover is controlled by VRRP/HSRP type protocols. In such cases, the number of supported VRRP/HSRP groups should be taken into account since in basic configuration a separate VRRP/HSRP process is controlling the redundancy of a single VLAN.[19] If there are several VLANs per eNB, the number of supported VRRP/HSRP processes may limit the number of eNBs that can be connected to the router. One remedy for this is to have more than one eNB share the same VLAN, which conveniently brings us to the question of how many eNBs should there be per VLAN.

5.10.2.2 How Many eNBs Should There be Per VLAN?

There are nearly as many opinions on the title question as there are access transport planners. Two alternative viewpoints are the physical port capacity of the next-hop router and the size of the L2 broadcast domain.

Often the answer is given by the L2 port capacity of the next-hop router. As an example, if it is assumed that the average DL throughput of a macro eNB with three 20 MHz cells is 150 Mbps, it is not advisable to have more than three eNBs per physical 1 Gbps port as this would result in an above 50% average port load. If the L2 port is 10 Gbps an integer multiple of 1 Gbps, the number of eNBs would scale accordingly. In either case the number of eNBs per VLAN is upper bounded by the L2 port capacity. Calculating the aggregate traffic of several eNBs is further discussed elsewhere in this chapter.

From a broadcast domain sizing point of view, under normal operating conditions, the number of hosts in VLAN can be quite large, assuming there are no applications generating abnormal amounts of broadcast traffic. A rule of thumb from office LANs is to allocate a 24-bit subnet per L2 segment. On the other hand, an often-overlooked virtue of an access transport design is that of ease of management and troubleshooting, and from this perspective allocating a few or only one eNB per VLAN is the favored approach. While single eNB per VLAN wastes some IP addresses, there are clear advantages in terms of connectivity troubleshooting as well as performance monitoring (e.g. eNB traffic can be monitored from VLAN traffic counters). Also robustness against broadcast storms caused by configuration errors is increased. On the other hand, subnetting should be done more carefully to allow efficient route summarization for access PE routers; in other words, all eNBs connected to a given access router should share the same supernet.

[19] Router implementation may support controlling several VRRP/HSRP groups (VLANs) with a single process.

5.10.3 IP Addressing

An addressing plan should satisfy at least the following conditions.

- It should be simple and follow some easily memorized, logical rules.
- It should be scalable up to a predicted number of eNBs to be deployed in the network with sufficient margin for future expansions.
- It should allow simple routing configuration and efficient route summarization.

There are two basic templates for allocating IP addresses: location first and application second or application first and location second. In the location-first approach a contiguous subnet is allocated to each one of operator core sites and regional hubs, after which the subnet for a given location is divided between different applications. This scheme results in efficient route summarization. On the other hand, a static routing configuration at path endpoints may be more complex since to send a given traffic type out of a specific VLAN will require one static route per location. Therefore, this approach is not recommended. With the application-first approach, each eNB VLAN is carrying traffic of certain applications (depending on VLAN split), and the static route configuration is independent of the number of core site locations. This results in a simpler static route configuration at eNBs than in the location-first approach.

Using the location-first approach with IPsec, several traffic selectors per application are required since the application subnet can be non-contiguous. Therefore, from an eNB perspective, the approach where each application is allocated a contiguous subnet is usually more appropriate also with IPsec as it results in a simpler traffic selector configuration.

After the VLAN planning step has been completed, the planner should know the number of VLANs per eNB and the number of eNBs per VLAN. From this information it is easy to calculate the minimum required subnet size. What is more difficult to predict is the future requirement of IP addresses. Experience has shown that it is practically always beneficial to overestimate the future network growth rather than allocate "just enough for current requirement." In case addresses are depleted at some point, eNB IP address or routing reconfiguration may be required, in the worst case resulting in maintenance outages.

Besides eNB IP addresses, the external subnets that need to be allocated include:

- S1-MME IP addresses for S1 C-plane connectivity
- S1-U IP addresses for U-plane connectivity
- synchronization over packet servers, e.g. IEEE1588 (2008) grandmaster
- O&M system
- DNS, NTP, tracing servers
- IPsec SEGs
- CA server for IPsec.

If each of the above applications is allocated a contiguous subnet, the routing parameter configuration becomes simpler at eNBs (if multiple VLANs are used). For IPsec this translates to simple traffic selector configuration.

For one eNB per VLAN the smallest IP subnet allocated to a VLAN is usually /29 (six hosts), to allow one eNB network interface IP address and three VRRP/HSRP addresses. In principle, allocating /30 provides less wasteful addressing and one could use it if it is absolutely certain there is no need for additional hosts (e.g. co-sited eNBs) in the subnet in the future.

5.10.3.1 Subnetting Example #1: Two VLANs Per eNB, Traffic Terminated at Network Interface IP Addresses

Figure 5.37 presents a simple design with two VLANs and two IP subnets per eNB. The following is assumed for the network:

- All eNBs in the network follow the same scheme.
- Operator has 5000 base station sites.
- Application traffic is terminated at network interface IP addresses.
- Two VLANs per eNB. One VLAN for User, control and synchronization traffic and another VLAN for management traffic.
- Each VLAN has one eNB in it.
- Allocated subnet per eNB VLAN is /29 to allow for VRRP/HSRP, where applicable. Hence one eNB consumes two times /29 subnets.
- S1-MME subnet is 10.1.1.0/24, from which all current and future S1-MME termination points are to be allocated.
- S1-U subnet is 10.1.2.0/24, from which all current and future S1-U termination points are to be allocated.
- OSS system subnet is 10.10.0.0/16, including NTP, DNS and tracing servers.
- Packet synchronization servers are allocated from 10.16.1.0/24.
- eNB outgoing VLAN selection is based on destination address based routing.
- eNB C-plane and U-plane addresses are the same for S1 and X2.

The external addresses should be further subdivided (by location) to operator data centers, or physical core sites. For example, assuming the operator has two EPC core sites, the S1-MME/S1-U subnets could be further subdivided to two /25 ranges, one /25 range per core site.

Assuming one eNB per base station site, the total U/C/S and M addresses required by eNBs is 2 × 5000 × /29. Rounding this up to the nearest power of two gives the minimum requirement as 16384 × /29 = 2 ^ 14 × 2 ^ 3 addresses = 2 ^ 17 addresses, or one /16 subnet for each U/C/S and M traffic, altogether one /15 subnet.

Considering the operator may subsequently deploy at least one more frequency layer per site, this addressing requirement could be easily doubled. It is also possible to allocate addresses from the same /29 subnet for the co-located eNBs, assuming the eNB or the base station site has an L2 switch.

With the given external address subnetting, each eNB can be configured with one static route to 10.10.0.0/16 via VLAN11 (management traffic) and all other UL traffic is sent to the default gateway via VLAN10. UL QoS within the U/C/S VLAN is handled by the eNB VLAN scheduler. All X2 U/C traffic is forwarded to the default gateway, hence the solution does not require maintenance of X2 routes while keeping QoS implementation simple.

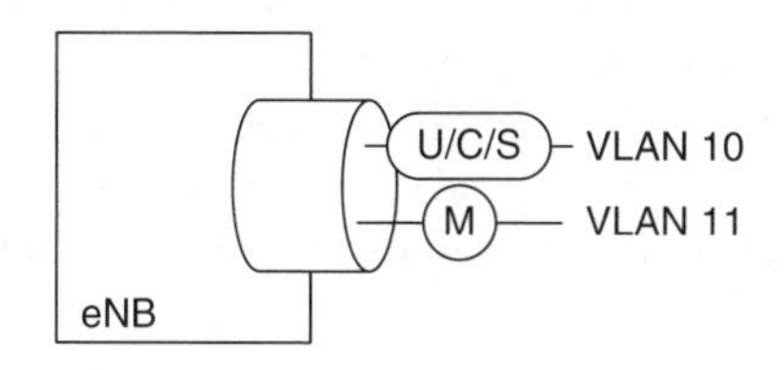

Figure 5.37 A basic eNB transport design. "S" means packet synchronization traffic

5.10.3.2 Subnetting Example #2: IPv6 for Transport

By the IPv6 standard recommendation, each IPv6 host is allocated a /64 prefix which is four billion times the number of addresses in the entire IPv4 address space. Even so, there are still addresses to spare in the 128-bit IPv6 address range. With the same assumptions as in the above IPv4 example, one could allocate the 2 ^ 17 addresses for the eNBs. With IPv6, however, there is little need to spare addresses and, more importantly, the subnet sizes should be aligned with the hexadecimal letters in blocks of four bits (2 ^ 4). The next largest allocation for U/C/S and M plane addresses is hence a block of 2 ^ 20 addresses, corresponding to one /44 prefix for eNBs. For IPv6 this could still be considered a tight allocation, and /40 prefix could also be considered (to take into account future expansion, such as small cells). A large organization such as a telecom operator would probably have a site prefix of /32.

The eNB addresses are then divided first based on U/C/M application and then based on location.[20] One possible scheme would be:

$$2001:0db8:00xy:yyyy::/64$$

where 2001:0db8::/32 is the operator IPv6 range allocated by a regional Internet registry. The letters "x" and "y" are used to denote application and a physical eNB respectively. For example, using x = "a" for LTE U/C/S and x = "b" for LTE management traffic the address ranges would be

$$2001:0db8:00a\mathit{y}:\mathit{yyyy}::/44 \text{ for U / C / S traffic, and}$$
$$2001:0db8:00b\mathit{y}:\mathit{yyyy}::/44 \text{ for management traffic.}$$

With this allocation there is room for 2 ^ 20 eNBs, each having a /64 prefix. The reason for choosing 20 bits for eNB identification is that this coincides with the 20-bit length of the macro eNB ID in the 3GPP standard. Thus, the 20 bits marked with "y" can be directly taken as the eNB ID converted to hexadecimal. This allows for easy conversion between IP address and the eNB ID without any need for lookup tables. Such simple logical mapping rules can make network planning and operation a great deal easier. Such a wasteful IP allocation rule would be impossible with IPv4.

The aforementioned scheme enables easy expansion. Small cells could be considered an "application" with x between "0" and "7"; this would further allow devising various schemes on how to divide this prefix range by geographic area or small cell customer type. Obviously, different radio access technologies could also be given their own prefix (e.g. x = "c" for 3G). If IPsec is used, encrypted and non-encrypted applications should be assigned a different prefix.

Since in the example the IPv6 global unicast range is used and hence routable by public Internet routers (unlike the IPv4 private 10.0.0.0/8 range for example) the operator firewalls will need to be configured to block access to the eNB range from public Internet. This is again made easier when the prefix design is simple. In the above example, the operator would block access to the entire /44 range (or /40) used by eNBs.

The last 64 bits of an eNB address can be assigned manually or using stateless address auto-configuration. The details would depend very much on how the configuration management (including discovery of neighbor addresses for handovers) was implemented.

[20] In case IPsec is used, another option is to split applications into encrypted and unencrypted ones.

5.10.4 *Dynamic Versus Static Routing*

Dynamic routing is typically employed to automatically distribute routes between L3 nodes in the backhaul. This does not mean it is required to have dynamic routing protocol running in eNB, as in many cases it does not bring much added benefit. Furthermore, depending on the vendor, eNB may not support dynamic routing.

A possible use case for dynamic routing is the model shown in Figure 5.38, where the eNB has separate application addresses for S1 U-/C-plane, X2 U-/C-plane and management traffic. S1/X2/M traffic are all using a dedicated VLAN, hence routes to these addresses need to be known by the next-hop router. If the eNB supports a dynamic routing protocol, these routes can be advertised to the next-hop router, and from there on to the rest of the backhaul. Moreover, the eNB can obtain X2 routes to other eNBs dynamically from the next-hop router, which is beneficial if the X2 address range in the network is non-contiguous and will otherwise require static route configuration.

Besides a few specific use cases such as the one shown in Figure 5.38, it is debatable whether deploying dynamic routing at the eNB is justified over simple static route configuration at the eNB and at the next-hop router. Ending up with a network requiring first-mile dynamic routing can sometimes be circumvented by opting for a different approach altogether, at least if the design is not overly constrained by legacy transport network design.

5.10.5 *Examples*

In the examples we assume that the external subnets are as given in the example in Section 5.10.3.1. In summary:

- S1-MME subnet is 10.1.1.0/24.
- S1-U subnet is 10.1.2.0/24.
- Management server subnet is 10.10.0.0/16.

5.10.5.1 Without IPsec

Example #1: Simplest Case: Single Transport VLAN

To begin with the simplest—and because of this a very useful—case, in Figure 5.39 an eNB consuming one IP address and one /29 subnet is shown. In principle, a /30 would be enough but this seems slightly restrictive in terms of future-proofing. In eNB only the default route via 10.100.1.1 needs to be configured. This makes the configuration extremely simple and useful for mass eNB deployment.

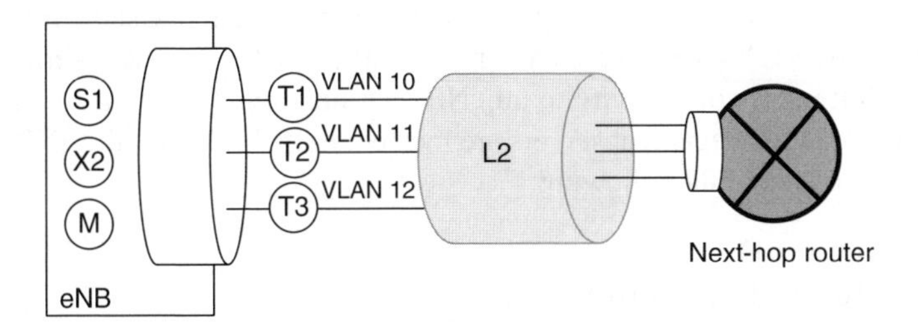

Figure 5.38 eNB connected to the next-hop router

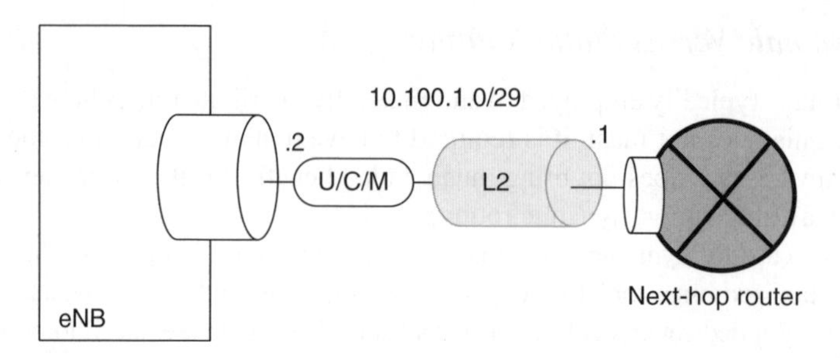

Figure 5.39 Simplest case: one transport VLAN. Single IP address per eNB

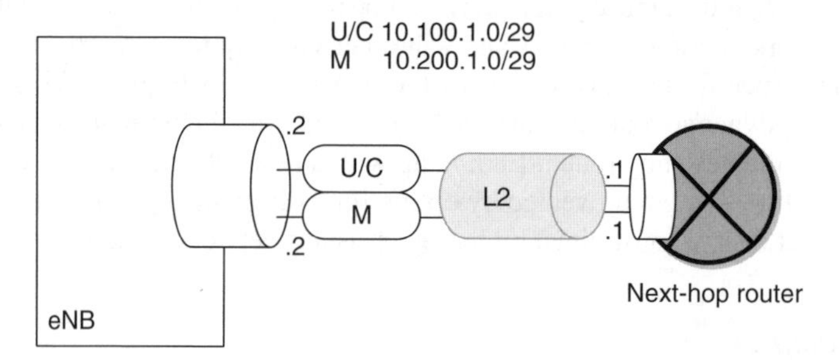

Figure 5.40 Second-simplest case: two transport VLANs. Two IP addresses per eNB

The L2 segment connecting the eNB could be simply a single Cat6 cable (if a next-hop router is at the same base station site) or it could be a carrier Ethernet realized by means of a multi-hop microwave radio network or a leased line network.

This simple solution may not be sufficient if the L2 is leased transport and the service provider only supports QoS at the VLAN level. Another aspect is that remote reconfiguration of the eNB IP parameters may result in management connection loss since management traffic itself is transported in the same VLAN.

Example #2: Second-Simplest Case: Two Transport VLANs

The model shown in Figure 5.40 fixes the problem that modification of, for example, routing or VLAN parameters of U/C VLAN would result in remote management connection loss, which in the worst case could result in a site visit of a remote corner of the operator network. Although not shown, the management VLAN could also be under its own physical interface to further increase resiliency.

This configuration requires, at the very least, a default route and one static route configured in the eNB. The U/C VLAN should typically be used as the default route to simplify X2 ANR, which requires full mesh connectivity to all eNBs within a regional area. Assuming that the OSS subnet, where all management servers are located, is 10.10.0.0/16, the eNB routing configuration would consist of the following:

- default route via 10.100.1.1
- static route to 10.10.0.0/16 via 10.200.1.1.

In case the management servers are not all from the same contiguous subnet, more static routes may be required.

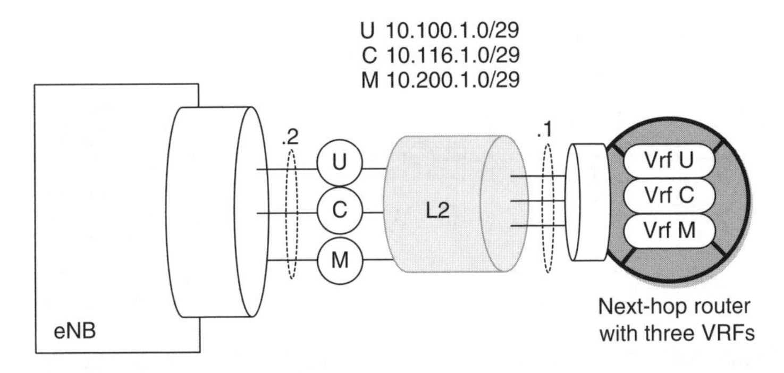

Figure 5.41 All traffic segregated to own VLAN, three IP addresses per eNB

In the next-hop router no route configuration is required since the U/C/M applications are directly connected.

Example #3: Traffic Segregation Based on VRF Design

Figure 5.41 presents an example where the eNB is connected to a next-hop router with VRFs for U-, C- and M-plane traffic. For this reason, three VLANs are deployed, each corresponding to a VRF. Based on the incoming IP interface (VLAN), the next-hop router knows which routing table should be used to route the packet with. Alternative ways for selecting VRF are possible, depending on the router features.

Note: in this example it is assumed that the next-hop device is a provider edge router implementing L3 VPN.

The configuration requires a minimum of two static routes and a default route, if dynamic routing is not used in the last hop. More routes would be required if U- and C-plane subnets for S1 and X2 are not contiguous. Assuming that all eNB application addresses are allocated from 10.x.0.0/16 subnet and external addresses are as assumed in the beginning of this section, we could define the static routes as:

- default route via 10.100.1.1 (X2 U-plane)
- static route to 10.1.2.0/24 via 10.100.1.1 (S1 U-plane)
- static route to 10.116.0.0/16 via 10.116.1.1 (X2 C-plane)
- static route to 10.1.1.0/24 via 10.116.1.1 (S1 C-plane)
- static route to 10.10.0.0/16 via 10.200.1.1 (M-plane).

In this case an outgoing VLAN selection based on the service (traffic type) would be more efficient configuration-wise than a routing-based one.

In the next-hop router, no static routes need to be configured as all application addresses are directly connected.

5.10.5.2 With IPsec

With IPsec, protected applications are terminated in base station-internal host addresses while traffic bypassing IPsec could also be terminated in the transport interface. The eNB implements SEG functionality. In the example shown in Figure 5.42, U/C traffic is protected by IPsec while management traffic bypasses IPsec, as it is assumed that it is protected by

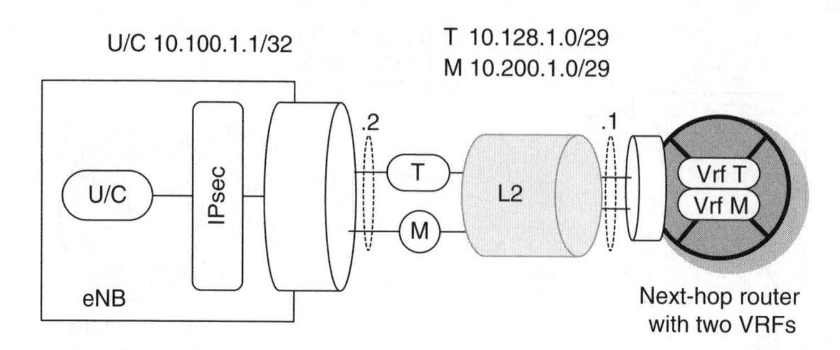

Figure 5.42 IPsec example two VLANs and U/C traffic protected by IPsec

higher-layer protocols such as SSH or TLS. The two transport VLANs correspond to two VRFs in the next-hop router. Encrypted traffic is transported in the T VLAN and destined to the remote SEG. Bypassed traffic is transported in the M VLAN. The traffic in these VLANs is correspondingly routed using transport and management VRF instances in the next-hop router.

It can be noticed that segregation of traffic to transport (carrying IPsec VPNs) and management to certain extent simplifies also VRF design in the access network, as there is no need to extend U/C VRFs to the edge of the access network. On the other hand, in this design the X2 delay may become prohibitive if the remote SEG is located far from the base station site.

Protecting management traffic at a higher layer and letting it bypass IPsec protection brings some operational benefits in case there are problems with IPsec connectivity or a need to reconfigure IPsec parameters using a remote O&M connection. This is especially useful since IPsec implementations from gateway and eNB equipment vendors have been notoriously incompatible.

The configuration requires a minimum of one static route and one default route. Assuming that all remote SEG transport addresses are allocated from the transport subnet 10.128.0.0/16, the static routes could be defined as:

- default route via 10.200.0.0/16
- static route to 10.128.0.0/16 via 10.128.1.1 (transport plane).

No routes need to be defined in the next-hop router.

5.11 Synchronization Planning

As mentioned in Chapter 2, the types of synchronization methods of interest for a mobile network are phase/time synchronization and frequency synchronization. TDD-LTE, positioning techniques, multimedia broadcast multicast service (MBMS) and certain advanced LTE features (coordinated multi-point [CoMP], enhanced inter-cell interference cancellation, [eICIC]) require that all eNBs are phase synchronized in the sense that they must start radio frame transmission at exactly the same instant. This can be achieved via satellite-based methods (e.g. GPS or Global Navigation Satellite System [GLONASS]) or IEEE1588 (2008) based packet synchronization. Phase synchronization also provides a reference signal for carrier frequency synchronization, although vice versa is not true.

FDD-LTE networks without special radio features requiring phase synchronization may be deployed with frequency-only synchronization. In this case the synchronization signal is needed only for keeping the carrier frequency within required limits. In addition to satellite-based methods and IEEE1588 (2008), SyncE may also be used for frequency synchronization. NTPv4 could also be able to provide sufficiently accurate reference in some scenarios, albeit due to a lack of standardized hardware support it cannot compete with IEEE1588 (2008) or SyncE in terms of synchronization accuracy. NTP-based packet synchronization will not be considered further in this section.

Time synchronization means that the accurate value of the current absolute time is provided to the eNB. Both satellite-based and IEEE1588 (2008) are capable of providing this at accuracy in the order of microseconds or better. SyncE is based on the PHY Ethernet signal and does not convey any absolute time information.

Detailed expositions of packet-based synchronization for mobile networks can be found in Ferrant et al. (2013) and Metsälä and Salmelin (2012), and the basic protocol information found therein will not be repeated in this section. Instead, the target is to provide a digest of information relevant to the planning of synchronization for the IP-based mobile access network.

5.11.1 Global Navigation Satellite System (GNSS)

Satellite navigation systems include GPS and GLONASS. The list of planning requirements for GNSS is fairly short. Obviously, and most importantly, the location of the GNSS receiver should allow an unobstructed view of the satellites of the used GNSS. If the distance between the GNSS receiver and the eNB is large, the cable length may need to be changed by a manual configuration of the delay, or automatic delay measurement if supported by the equipment. This is to ensure that the exactly same radio frame timing is used in all eNBs. For example, a 100-meter cable introduces a propagation delay in the order of 0.5 microseconds, depending on the cable electrical properties. As the phase timing requirements for LTE radio frame transmission are about 1 microsecond, it is advisable to configure cable delays to avoid any unnecessary radio performance impairments.

5.11.2 Synchronous Ethernet (SyncE)

The Ethernet interfaces based on IEEE802.3 are non-synchronous. SyncE defined in ITU-T G.8261/8262/8264 increments the IEEE802.3 baseline capability with more accurate Ethernet PHY clock requirements as well as the capability to transmit and receive SSMs similar to those in SDH/SONET. A SyncE-capable Ethernet port recovers the frequency from the received PHY line code zero crossings taking place roughly in nanoseconds.[21] Therefore, SyncE is not a packet-based but rather "bit-based" synchronization method. The frequency reference derived from the synchronous mode input port is passed to the transport node

[21] In comparison, the inter-arrival time of PTP packets in IEEE1588-2008 is in the order of milliseconds, which explains the complex algorithms required to achieve acceptable performance with IEEE1588.

oscillator, which uses it as a reference to maintain frequency accuracy within the ±4.6ppm requirement defined in ITU-T (2013a) Rec. G.8261.[22]

In addition to PHY clock recovery, a SyncE-capable interface can send and receive SSMs that inform peer ports of the synchronization quality level of the Ethernet PHY signal.

Depending on the configuration, a SyncE port can operate in either synchronous or non-synchronous mode. Ports facing each other must be configured in the same mode. A port in non-synchronous mode can be connected to a port that is not SyncE-capable, facilitating mixing of equipment with different capabilities in the network. Some equipment vendors implement SyncE support only for optical Ethernet interfaces. Another point worth noting is that for copper-based gigabit Ethernet interfaces the auto-negotiation procedure determines the master and slave clock for the link. It should be ensured that the master node is selected as the SyncE source node,[23] consistent with the synchronization plan (ITU-T [2008] Rec. G.8264).

From an access planning point of view, SyncE has two important differences to IEEE1588 (2008) frequency synchronization: the stability of the recovered frequency does not depend on the link traffic load and both master and slave nodes and ports must support SyncE at the hardware level. Consequently, SyncE has to be implemented in all intermediate nodes on the SyncE timing path, which may require hardware upgrades in the network. Obviously, the timing reference need not be transported from a core site all the way to the eNB. A more suitable alternative is to arrange a timing reference to a regional site by some other method, for example IEEE1588, SDH or GNSS. The final leg from the regional site to the base station site could use SyncE.

As in SDH/SONET, the sending SyncE interface transmits periodically clock quality information in SSMs. ITU-T (2008) Rec. G.8264 defines an ESMC for SyncE links, in which the SSM are exchanged. The ESMC messages are carried by Ethernet Slow Protocol (i.e. the messages are sent to multicast Ethernet destination address 01-80-C2-00-00-02 and use EtherType 88-09). “Slow” in this context means that IEEE802.3 restricts the maximum sending rate to 10 frames per second per protocol. Slow protocol frames in general (and hence ESMC frames) are not switched from one Ethernet port to another, to avoid accidental leaking of incorrect synchronization information. In some special cases, care must be taken to ensure this. For example, L2 VPN based on RFC4448 Ethernet raw encapsulation may, depending on implementation, forward ESMC frames over non-synchronized PHYs.

By suitable configuration of the port synchronization priorities and outgoing SSM messages, timing loops can be prevented (Ferrant et al., 2013). The node should send SSM value “Do Not Use” out from the port that the node uses as its SyncE timing input. As in a large network timing loops are difficult to detect and troubleshoot, simple and consistent planning rules should be created to facilitate robust network deployment.

An example of ring topology is shown in Figure 5.43. The leftmost node is configured to receive timing only from one port, and send SSM information to both ring-facing ports. The ports of three base station nodes have been configured with timing input priorities, denoted in

[22]The basic IEEE802.3 clock accuracy requirement of ±100 ppm is inadequate for radio interface frequency synchronization.

[23]If the GE (gigabit Ethernet) port is forced to be a clock slave then the remote port should be the clock master. In the event of a failure situation (e.g. misconfiguration), if both ports are forced, slave or master the link will not come up. Unless the equipment is capable of intelligently preventing misconfigurations, some planning may be required to avoid such failures.

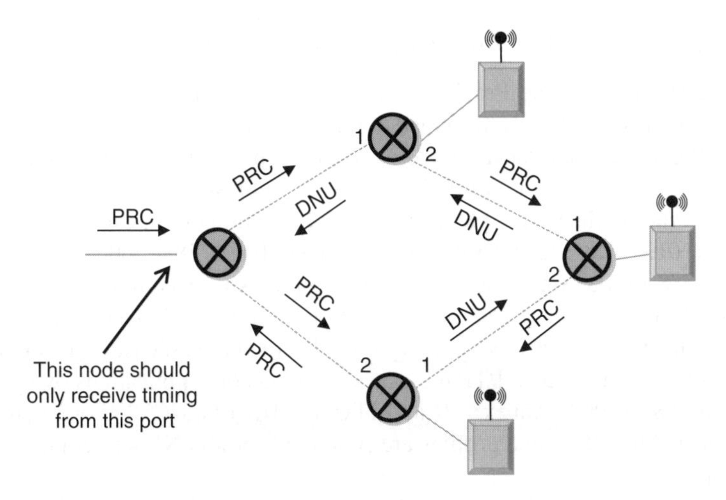

Figure 5.43 SyncE timing loop prevention in a ring

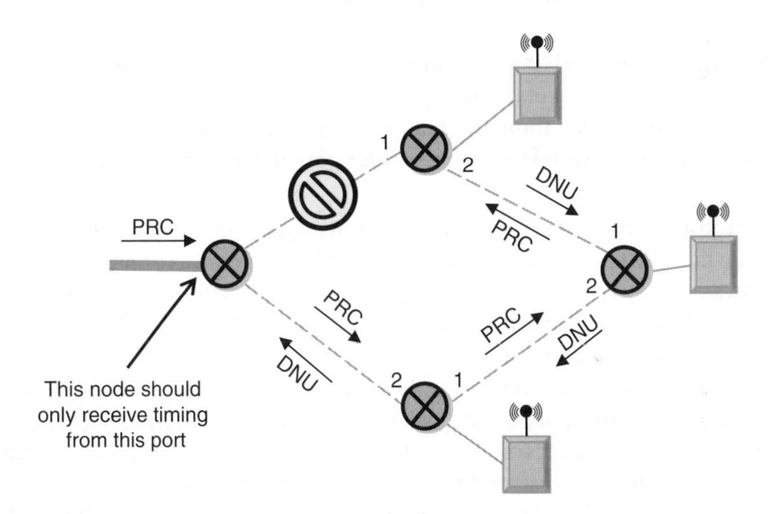

Figure 5.44 Link failure causes timing signal to traverse the ring counter-clockwise

the figure with “1” and “2.” A DNU message is automatically sent out from the port being used as a synchronization input. In normal operation, seen in Figure 5.43, timing traverses the ring clock-wise.

Considering the accidental configuration error where the leftmost root node accepted timing input from all ports, a danger of timing loop would occur in failure situations. For example, if the primary reference clock (PRC) input feeding the ring failed for some reason the node could select the ring-facing input port with incoming PRC SSM, hence creating a timing loop.

In the event of a link failure direction, the nodes will select the second priority SyncE port and the direction of the timing path will be reversed, as shown in Figure 5.44. Also the port from where DNU message is sent out is updated.

5.11.3 IEEE1588 (2008) Frequency Synchronization

IEEE1588 (2008) defines the Precision Timing Protocol version 2 (PTPv2). The earlier protocol version 1 was initially designed for industrial automation. The protocol version 2 defines PTP over various transport protocols but in practice PTP over unicast UDP/IP is the operation mode favored in mobile backhaul. The purpose of this section is to focus on the aspects relevant to the access planner. More information on the protocol can be found in Ferrant et al. (2013), Metsälä and Salmelin (2012) and IEEE1588 (2008).

The IEEE1588 (2008) standard has numerous options to accommodate a wide range of applications. To ease the implementation, different PTP profiles have been defined. The profiles define the value ranges for various protocol parameters, such as transmission intervals of event messages (time-critical PTP messages). For telecom applications two profiles are of interest: IEEE1588 (2008) Annex A.9 and ITU-T (2014) G.8265.1. For interoperability it should be verified that the same profiles are supported for all eNB and grandmaster vendors in the network.

The standard has its special terminology for various node types. A highly condensed description of the node types relevant to the present discussion is:

- Grandmaster: the node receiving its timing reference directly from primary reference signal
- Master: port generating packet timing signal
- Slave: port receiving packet timing signal from master or grandmaster
- Boundary clock: a node with at least one master port and one slave port.

The main purpose of BCs is to regenerate the timing signal, in order to remove packet delay variation from the PTP packet stream, for example in long microwave link chains. BCs require special hardware implementation, which is optionally available in most access layer switches. In the simplest deployment scenario, BCs are not used at all and eNBs receive the packet timing directly from the grandmaster.

Also so-called transparent clocks have been defined in the IEEE1588 (2008) standard. A switch implementing the transparent clock registers the time difference between the packet entering the input port and leaving the output port, and writes the packet residence time within the node into the PTP packet header. Based on this a slave can compensate queuing and processing delays in the e2e chain. Hardware-level support is needed to implement the transparent clock.

The detailed method for timing signal recovery at the slave node has not been specified and is therefore vendor-dependent. Invariably, regardless of the algorithm, it is affected by the path delay variation of PTP event messages. A typical timing extraction scheme would use only the fastest percentage of the event messages, and discard the rest.

In a simple "DL implementation" the master sends Sync event packets to the slave, say, 16 times a second, and the slave extracts the reference timing from the incoming Sync packet stream.[24] In the "UL implementation" the slave sends Delay_Req event packets to the master which then responds with Delay_Resp that contains the time stamp of receiving

[24] The master can include the transmit time stamp either directly in the Sync message or in a separate Follow_Up message which is not delay-critical and sent after the Sync message. Follow_Up message is mainly needed if the master does not support PTP event message time stamping in hardware.

the Delay_Req message. The slave is able to adjust its reference timing based on the UL delay calculated from the time stamps received in the Delay_Resp. As the UL link load is usually lower, the UL method, if supported by the eNB, provides some performance advantage at the cost of an increased processing load at the master. For access planning it is important to check the PTP implementation specifics to be able to design QoS parameters and monitor the network performance. In practice the chosen PTP method is usually determined by the slave implementation.

Regardless of the implementation, the absolute packet delay between master and slave is not relevant. It is rather the packet delay variation introduced by the transport network that degrades the performance. This implies that in any transport node along the timing path each PTP event message should be forwarded with constant processing delay and without queuing. Depending on whether a UL or a DL method is used, the direction of time-critical messages is different. Since the forwarding delay should be constant, proper equipment selection and configuration are also important. There should not be any transport equipment with software-based packet inspection or forwarding at any protocol layer as this is prone to introduce CPU-load-dependent packet delay variation. For example, it is not advisable to use firewalls or software-based, policy-based routing between the master and the slave. From a delay variation perspective, IPsec could be used if all cryptographic processing was done on hardware; however, prior testing of the equipment before mass deployment is recommended. Since IPsec mangles the packet contents, BCs and transparent clocks cannot be used in the protected path.

The key question in access planning is how many link hops can there be between a master and a slave before the PTP packet stream quality deteriorates to an unacceptable level. Long chains may necessitate relocating the grand master node closer to the slave nodes or introducing BCs to the network in order to de-noise the timing packet stream.

Microwave links changing modulation adaptively can cause delay jumps in the timing packet stream. Too frequent delay jumps may prevent the slave from converging to steady state. Therefore, the frequency of modulation changes should be minimized by tuning link adaptation thresholds if possible. In long microwave link chains the frequency of the delay jumps may be too high depending on the link budget and propagation conditions, and disabling adaptive modulation may need to be considered. Assuming stable link rates higher than 150 Mbps, a crude rule of thumb is that no more than 10 microwave links should be chained before introducing a BC (Metsälä and Salmelin, 2012). The robustness of the slave implementation and the link speed play a role here. For low-speed links and sub-standard slave implementation, 10 microwave links in a chain will probably result in unacceptable packet delay variation. Exact planning rules that fit all scenarios do not exist since a multitude of factors influence the packet delay variation. In practice, active monitoring of slave performance versus packet delay variation at the slave input should be carried out in production networks.

Slave performance degradation can also be caused by long-term link congestion where almost all timing packets are delayed for extended periods of time (e.g. hours). The impact of congestion can be reduced—but not eliminated—by giving timing packets strict priority in all schedulers along the path. This reduces the packet delay variation seen by the slave during congestion, enabling it to maintain steady-state performance for a longer time. The load of high-speed links is allowed to be higher than low-speed links. Given that the load of the strict priority queue is low so that there are no other high-priority packets in the queue, a PTP event packet arriving to a fully congested (full egress buffer) node will in the worst case have to wait at most 12 microseconds until a 1500-byte low-priority packet is sent out of a 1 Gbps egress port (for MTU of 1500 bytes). For a 100 Mbps port the corresponding queuing time would be 120 microseconds. Comparing to the ITU-T requirements, discussed below, the conclusion is that a single fully loaded 100 Mbps link can consume most of the allowed 150 microseconds delay variation. It can be seen that a fully utilized 1 Gbps link produces much less packet delay variation to the strict priority PTP packet stream. Conversely, if the strict priority traffic load is high, a PTP packet arriving at the node is most of the time forced to wait until all other strict priority packets in the queue ahead of it are sent out. Thus, keeping the highest strict priority queue utilization at a low level is important.

As mentioned, a typical PTP slave implementation uses only the fastest PTP packets (time stamped event messages). For this reason it is not the average PTP packet delay that is of importance but the percentile delay. Figure 5.45 shows a measurement example of the delay of high-priority PTP packets and low-priority best effort (BE) packets over a chain of four 100 Mbps Ethernet links with strict priority schedulers. The link load of the egress ports in the chain increases at each hop from additional BE traffic accumulated from other ingress ports. The load of the last hop in the chain is shown in the horizontal axis. Only PTP traffic is utilizing the higher-priority queue, this is called "EF" (expedited forwarding) in Figure 5.45. The volume of PTP traffic is a negligible fraction of the total traffic.

The result shows the average delay for PTP and BE packets as well as the delay of 1% of the fastest PTP packets. The interesting observation is that the PTP delay for fastest packets is quite stable for loads up to 90%. From queuing theory it is known that the probability that a packet entering a queue has to wait is equal to the link load.[25] This roughly explains how the fastest few

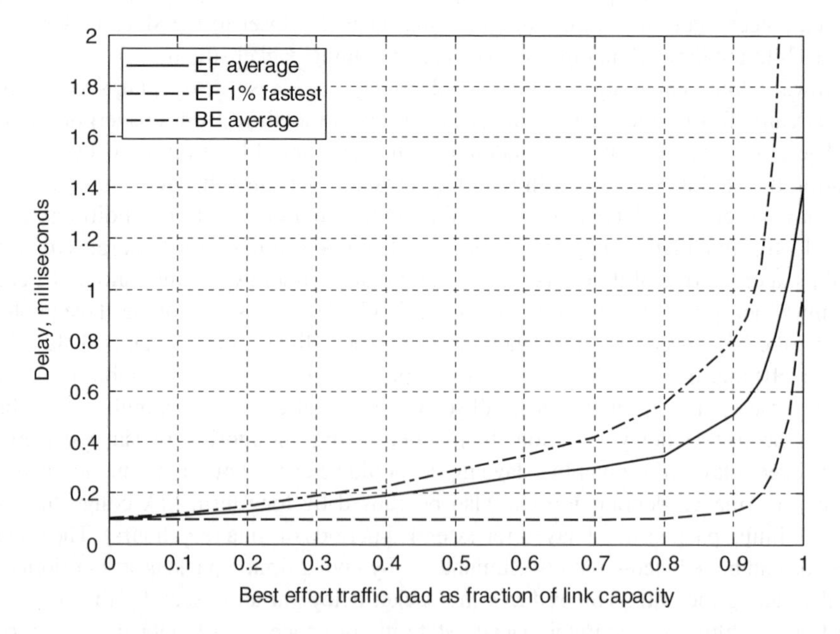

Figure 5.45 Delay of expedited forwarding and best effort traffic over a chain of four 100 Mbps Ethernet links. The x-axis shows the best effort traffic load of the last link in the chain

[25] Strictly speaking, this is true only for exponentially distributed packet inter-arrival times.

percent of PTP packets do not have to wait in the queue until the link load of the last hop is approaching 100%. When BE packets fill up the transmit buffer completely (100% load), every incoming PTP packet will have to wait at the last-hop egress buffer until the transmission of the BE packet ahead of it has been finished. There is a relatively high probability that a PTP packet has to queue already before reaching the last hop, which adds to the overall path delay.

Since the PTP slave implementation usually makes use of only the fastest few percent of the packets, the result implies that 100 Mbps links can be operated at quite high utilization if the PTP packets are mapped to the highest strict priority queue and the highest priority traffic load is a small fraction of the link capacity. Higher capacity links could in principle operate at even higher loads. On the other hand, at link loads of above 90% even a small increase in traffic results in large increase in packet delay and therefore performance monitoring should be carried out more frequently. Furthermore, time to react to congestion (e.g. upgrade the link capacity) should be short. Taking into account these operational and capacity management process requirements the savings from operating mobile backhaul links at above 90% loads is seldom worth the additional risk and effort.

ITU-T has published a network performance requirement for PTP-based frequency synchronization (IEEE1588, 2008) for the hypothetical reference model (HRM) shown in Figure 5.46. The model consists of three optical 10 Gbps links and seven 1 Gbps optical links between the master and the slave clock. The packet nodes in between do not support a boundary or transparent clock. The ITU-T recommendation (IEEE1588, 2008) is the first attempt to standardize equipment performance requirement in terms of the packet delay variation at the input of the slave clock, shown as reference point “C” in Figure 5.46.

An HRM-1 compliant implementation should be able to deliver the timing reference at the slave output where the fastest 1% of received packets at the slave input (point C) are within 150 microseconds of the fastest packet. The requirement should be satisfied for any time window of 200 seconds’ length. If the network satisfies this requirement, the error of the timing reference generated by the slave should be less than 16 parts-per-billion (ppb).

Live network experience has shown that PTP-based frequency synchronization in general works well even without boundary or transparent clocks. This may be partly because LTE radio interface PHY is by design required to tolerate large Doppler shifts (for high-speed-train scenarios), and even relatively large carrier frequency offsets between eNBs do not cause noticeable handover problems. Further, provided that the eNB oscillator is reasonably robust, short-lived quality impairments at the slave clock output (e.g. due to delay jumps or congestion)

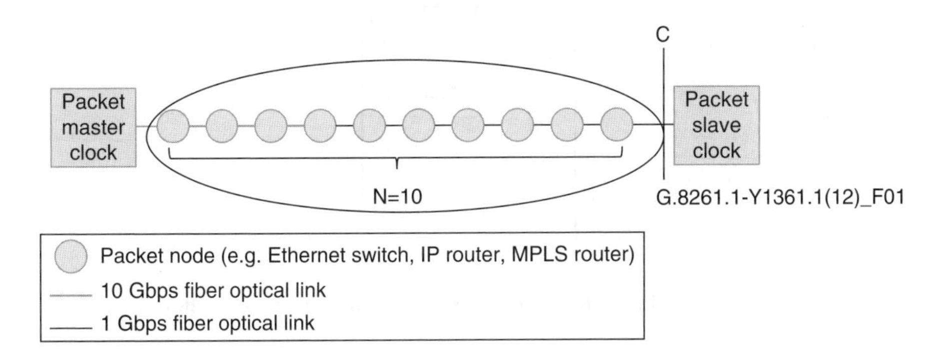

Figure 5.46 HRM-1: Network using only 1 Gbps and 10 Gbps connections (IEEE1588, 2008)

do not result in large errors in carrier frequency. However, as mentioned above, a notable use case for deploying BCs is to break up long chains of microwave links.

While frequency synchronization without PTP support in intermediate nodes is possible in most cases, the same cannot be said for PTP phase synchronization, which tolerates "packet delay variation" to the order of 1 microsecond at the input of the slave, that is a requirement two orders of magnitude tougher than the corresponding requirement for frequency-only synchronization. This is the topic of the next section.

5.11.4 IEEE1588 (2008) Phase Synchronization

The IEEE1588 (2008) phase synchronization increases the complexity of network design relative to the frequency synchronization scheme. The goal of the phase synchronization is to convey to the slave clock the exact phase of the master clock so that both master and slave start their clock cycles to within 1 microsecond of each other.[26] The problem is fairly easily solved via GPS/GNSS-based synchronization as the satellite clock in the sky provides a common time reference to all slaves. Distributing the phase alignment over a packet-based telecommunication network provides some challenges due to several error sources, including queuing delays in the order of milliseconds, packet forwarding delays in the order of microseconds and propagation media and equipment-dependent path asymmetries in the order of nanoseconds.

PTP-based phase synchronization comes in two flavors: full PTP support in every intermediate node of the path and partial timing support where only some (or none) of the intermediate nodes support BC.

Figure 5.47 shows the simplest case of the PTP message flow for phase synchronization. The target is to calculate the slave clock error relative to the master. The phase of the master clock could be calculated as $t_1 = t_2 - d$ if the slave phase offset Δ relative to the master was known. However, since both Δ and d are unknown, two equations are needed to solve them.

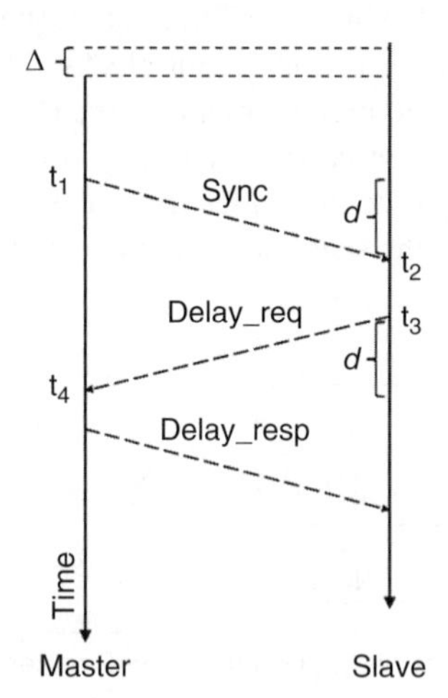

Figure 5.47 IEEE1588 (2008) phase synchronization message flow

[26] For most applications the requirement for radio frame timing error is about 1–3 microseconds.

The first equation is obtained from the Sync message which the master time stamps with value t_1. The slave clock receives the Sync message when its own clock shows time t_2. The slave clock cannot calculate the path delay simply as $t_2 - t_1$ since its own time error Δ relative to the master is unknown. A second equation is formed by sending the Delay_Req message to the master at time instant t_3. After receiving the Delay_Req message at time t_4 the master sends the time stamp t_4 back to the slave in the Delay_Resp message. Assuming that the path delay d is the same in both directions, the slave has now enough information to solve the unknowns, the path delay d and the clock phase offset Δ. The path delay d can be written in two ways, as in Equation 5.19.

$$\begin{aligned} d &= t_2 + \Delta - t_1 \\ d &= t_4 - t_3 - \Delta. \end{aligned} \tag{5.19}$$

Summing up the two equations and dividing the result gives the two results in Equation 5.20.

$$\begin{aligned} d &= \frac{t_2 + \Delta - t_1 + t_4 - t_3 - \Delta}{2}, \\ &= \frac{t_2 - t_1 + t_4 - t_3}{2} \end{aligned} \tag{5.20}$$

where the unknown phase difference conveniently cancels out. Finally, the unknown Δ can be solved from either of the original equations for d.

In the message exchange the delay critical messages (called “event messages” in the standard) are Sync in DL and Delay_Req in UL. For access planning, an immediate consequence is that both UL and DL delays are critical in phase synchronization. Another practical deployment problem is that the UL and DL paths are required to have the same path delay, which was not the case with frequency-only synchronization. With asymmetric path delays the slave will have an unknown phase offset. The delay asymmetry may result in difficult operational problems and lengthy field troubleshooting if not taken into account in the planning and acceptance testing phase.

Measured delay asymmetry of routers and switches ranges from about 10 nanoseconds to a worst-case value of several hundred nanoseconds, depending on hardware implementation (Pietilainen and Virta, 2013). To make matters more complicated, measurements indicate that there can be differences in forward and reverse delays between different port pairs of the same physical chassis, and furthermore these delays may change when the equipment is restarted. A design rule of thumb is to allocate 100 nanoseconds to delay asymmetry per switch, which, based on measurements, works for most access layer switches in the market at the time of writing.

A microwave radio link may have delay asymmetry of up to several hundred nanoseconds. Thus, a worst-case assumption for microwave radios, in the absence of any testing results, is that already two chained microwave hops result in an accumulated asymmetry that exceeds the allowance of 1 microsecond. Also the link rate should not be less than 150 Mbps. Either adaptive modulation should be disabled completely or modulation switching should be such that both directions use the same link rate at all times, and the amount of rate switching per path should be minimized.

Because of the strict delay symmetry requirements xDSL (digital subscriber line) modems cannot be used with phase synchronization. IPsec for PTP traffic cannot be recommended either. Asymmetric routing, meaning that traffic in UL and DL directions uses different paths through the network, may cause issues as well. Note that it cannot in general be assumed in IP networks (nor LTE backhaul) that the very same path is used both in the UL and the DL, so especially if delays through the paths differ the issue needs to be solved.

The packet network performance requirement for phase synchronization is defined in ITU-T (2013b) Rec. G.8271.1 for a chain of 10 BCs between the master and the slave. This reference network model is also called HRM-1 but it should not be confused with the HRM-1 for frequency synchronization presented in the previous section.[27] The network performance requirement for HRM-1 is that the absolute packet timing error caused by the network should not exceed 1.1 microseconds at reference point C in Figure 5.46; this can be loosely interpreted as a 100-nanosecond timing error per node. It can be seen that the network delay variation requirement is about two orders of magnitude stricter than the frequency synchronization requirements.

At the time of writing, ITU-T has not defined any network performance limits for partial PTP support where some of the packet nodes in the timing path do not support BC. Again, the key planning question is the number of hops that can be chained before packet delay variation deteriorates to an unacceptable level and a BC is needed. The result will depend on link utilizations, total delay asymmetry of the path and on the radio frame timing accuracy requirement, which may be different for different applications.

Figure 5.48 shows an example measurement result where 12 commercially available access layer switches are chained at 1 Gbps port speed. The label on top of the curves shows the

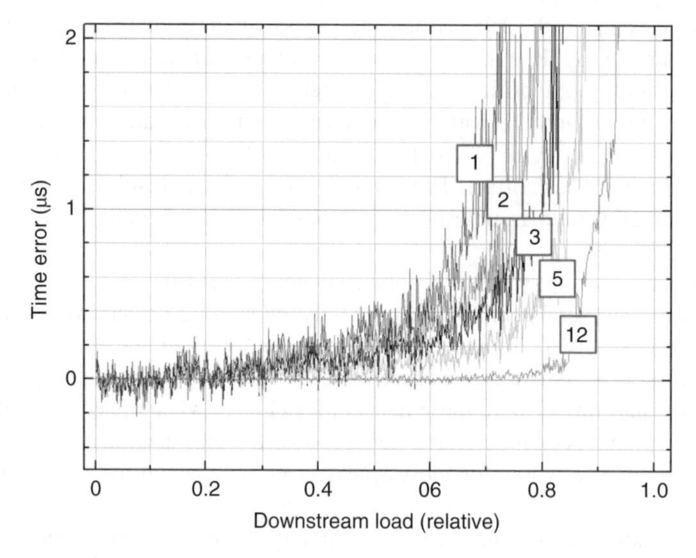

Figure 5.48 Timing error for different number of BCs in a chain of 12 switches. Impact of path delay asymmetry was removed in processing. All hops have the same load and the UL load is 25% of the DL load. *Source:* Pietilainen and Virta, 2013

[27] ITU-T (2013b) G.8271.1 defines the number of BCs cascaded in the HRM-1 as 10. Where the PTP slave is embedded in the eNB equipment the complete chain consists of one grandmaster node, 10 BCs nodes and the slave node.

number of BCs in the chain. For example, for the leftmost curve only the node in the middle has a BC while all the other nodes are PTP-unaware. The rightmost curve in turn has BC implemented in every node. Path delay asymmetry has been removed from the figure which is evident by noting that for the unloaded case the timing error is zero; with path delay asymmetry there would be a non-zero offset in the timing error for 0% load. PTP packets have been mapped to the highest strict priority queue while all other traffic has a lower priority.

Taking 1 microsecond as the maximum allowed timing error, it can be seen that with every node having BC the DL load can be almost 90%. The maximum load reduces to about 70% with only a single BC in the middle of the chain. The result also indicates that, in the ideal delay-asymmetric case, it is possible to deploy up to six packet nodes in a 1-Gbps link chain without BC support, assuming that the link loads are less than 70%. In practice, path delay asymmetry reduces the maximum number of PTP-unaware nodes in the chain for a given load.

An interesting planning question is the number of link hops that can be chained without BC, so that the timing error does not grow too large. Figure 5.49 shows a measurement example where the number of 1-Gbps link hops increases from one to seven. None of the switches in the chain has BC. The path delay asymmetries have not been removed and in this measurement have the same sign. Looking at the zoomed picture on the right-hand side, it can be seen that as the number of link hops increases the systematic timing error from delay asymmetry accumulates to 500–600 nanoseconds, even without any background load. Increasing the load to about 60% results in an unacceptable timing error of 1 microsecond (all link hops have the same load). On the other hand, for the same timing error a single 1-Gbps link hop load can be up to 90%. The result indicates that even in the absence of any network load it is not possible to chain too many link hops, due to equipment delay asymmetries. Depending on the implementation, a BC can remove some or most of the asymmetry introduced by a physical node (compared to the reference case that an identical node was deployed without BC), but it cannot, for example, help with asymmetry caused by the link propagation media.

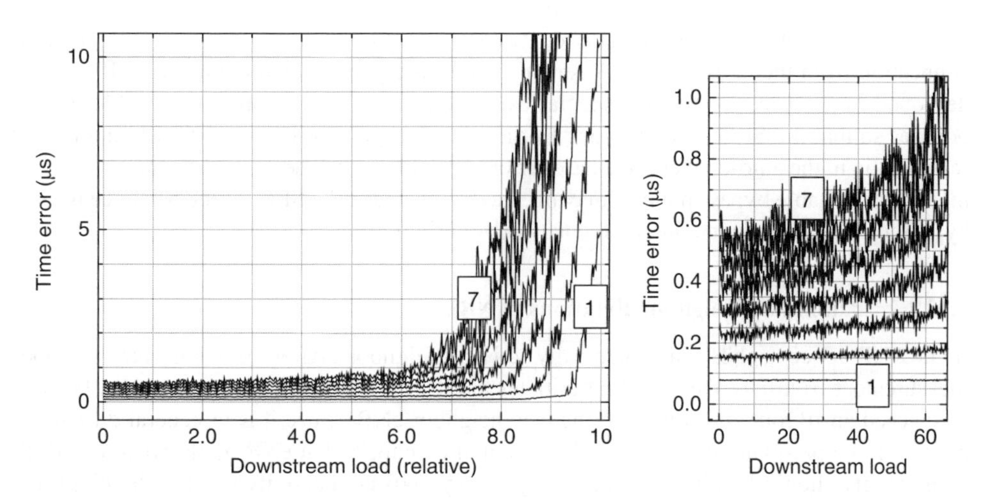

Figure 5.49 Timing error for different number of chained 1-Gbps link hops without BCs. All hops have the same load and the UL load is 25% of the DL load. The labels on top of the curves indicate the number hops in the chain

Finally, when deploying phase synchronization over packet, it should be kept in mind that the availability requirements are also more stringent than with frequency-only synchronization. Unlike loss of frequency synchronization, which in the worst cases results in a degraded service (failed handovers), loss of eNB radio frame timing can result in a service outage, since the oscillator holdover time is typically some hours only. These aspects are especially important for TD-LTE, where loss of accurate radio frame timing may cause site outage, or network radio interference if the radio transmission is not shut down early enough. Convergence time to steady state after packet timing recovery could also be considered part of the mean time to restore in availability calculations.

5.12 Self-Organizing Networks (SON) and Management System Connectivity

5.12.1 Planning for SON

Transport network needs to support SON auto-connection and automatic neighbor management (if used). Auto-configuration option should also be taken into account by access planners.

5.12.1.1 Auto-Connection Planning Aspects

During auto-connection phase, eNB broadcasts Dynamic Host Configuration Protocol (DHCP) DISCOVER message to all VLAN IDs. Access routers need to be configured to relay DHCP messages to a centralized DHCP server, or configured with DHCP service themselves. Usually DHCP relaying would be configured for one "auto-connection VLAN" only; this VLAN can be different from the final VLAN used by the eNB after configuration data have been automatically downloaded from OSS. The DHCP server IP address pools must be configured according to the auto-connection IP plan; the DHCP tags must also be configured. Note that the DHCP server must support vendor-specific tags.

During SON auto-connection (prior to IPsec tunnel setup), there has to be a non-secured IP connection from the eNB to the CA server for retrieving the operator certificate. If IPsec is deployed in the access network, special care should be taken to provide suitable bypass policies so that the SEG will not block Certificate Management Protocol (CMP) and DHCP relay traffic in the auto-connection phase, prior to establishing IPsec secured connection. This means that suitable bypass policies may need to be created at the SEG, or the SEG needs to be bypassed altogether, depending on the network configuration.

5.12.1.2 Automatic Neighbor Relations (ANR)

In the ANR concept, eNB can dynamically add new neighbors. After receiving a new neighbor measurement from UE, the eNB first initiates an SCTP connection to the new eNB. For this to succeed an IP route must exist to the new neighbor eNB. Since it is in general difficult to predict in the pre-planning phase all the potential neighbors that eNB might wish to add, IP connectivity should be planned so that any-to-any eNB connectivity is possible within the network area of interest. This requires careful routing design in the Layer 3 network.

If IPsec is deployed together with ANR, it should be noted that the required configuration transfer message as defined by 3GPP up to Release 9 did not support IPsec endpoint, and this was added with in 3GPP Release 10. This should be taken into account when planning neighbor connectivity; in practice this means that only the X2 star configuration is the most reasonable configuration with IPsec for most networks with ANR; in some special cases X2 mesh may be used.

5.12.2 Data Communications Network (DCN) Planning for Transport Network and the Base Stations

A common best practice inherited from the IP planning community is that the base station management (M) plane is in a separate VLAN from user, control and synchronization traffic. All base stations at a site can use the same VLAN for M-plane, as that might bring some OAM (operations administration and maintenance) simplifications. For example, one benefit from M-plane VLAN separation is that management actions on other VLANs do not necessarily interrupt the remote O&M connection to the base station.

If untagged plain Ethernet is used for M-plane for all base stations of the site, it should be noted that the base station internal L2 switch can be set to a mode that supports switching tagged VLAN frames only (from user-configurable VLAN ID list), or in a mode that also switches untagged plain Ethernet frames.

Transport DCN planning is strongly dependent on the transport equipment. A common best practice is that transport equipment uses different management VLAN from base stations. The management VLAN may be carried in out-of-band (proprietary) low-bandwidth wayside channel or in-band with other Ethernet traffic. In-band management has typically higher bandwidth than out-of-band, and should preferably have its own DSCP/PCP value, having higher QoS treatment than BE but lower than real-time traffic.

Remote connectivity is essential to the operation of the mobile network reducing site visits and operation costs and its planning requires specific consideration.

5.13 LTE Backhaul Optimization

This section describes different concepts for performance monitoring and how these can be applied to the mobile network and to backhaul optimization. The section concentrates on what should be monitored and why; whereas how a network is monitored is described more in detail in Chapter 7.

5.13.1 Introduction to LTE Backhaul Optimization

LTE backhaul monitoring enables LTE backhaul optimization: the corrective actions needed to optimize the capacities or configuration for the traffic to flow so that it complies with the QoS requirements set by the end user services and end use expectations. LTE backhaul optimization can be based on the LTE user traffic monitoring from the radio network or the backhaul network monitoring, or both, depending on the monitoring capabilities of the equipment.

Essentially, it is about capacity management and how to ensure that the user traffic is not limited by the backhaul network, preventing packet loss, too high delay or jitter by having the equipment correctly configured. Avoiding and detecting fragmentation of the mobile network traffic is also one of the main objectives in LTE backhaul monitoring and optimization.

The user traffic also places demands on the LTE backhaul network. If a network has been initially designed for data access, it probably requires extensive assessment and some optimization when, for example, VoLTE is being introduced.

5.13.2 Proactive Methods

LTE backhaul network monitoring is necessary to be able to detect and prevent any problems in the backhaul network that can degrade user services. The main objective of proactive optimization is to prevent, for example, capacity bottlenecks in the backhaul network. A general objective when building a mobile network has been that the transport should not limit the mobile network traffic, but this is becoming more and more challenging with the increasing end-user equipment capabilities and peak rates. The majority of traffic in the current LTE networks tolerates sporadic delay and delay variation, but ultimately capacity management is a strategic decision made by the operator in its competitive environment. If there is a lot of competition in the market and users are changing operator eagerly, capacity management plays a more important role than in less competitive environments. It is also an important marketing message to be benchmarked as the number-one operator in a country and this requires consistent monitoring and optimization of the mobile network.

5.13.2.1 Busy Hour Definition

Traffic varies during the day and there are differences between weekdays and weekends. Traditionally, it has been necessary to monitor the highest traffic hours to prevent congestion at the so-called busy hour (BH). There are many ways of determining the BH in telecom networks (CCITT, 1988) and for different traffic types the BH can be at different times of the day. In LTE networks the majority of the traffic is data traffic and therefore it is somewhat easier to define the BH—and network operation people know quite well when is the highest load in the network even if the BH can be different in different areas of the network: city centers, office areas and residential areas just to name few examples. Also so-called special events require specific planning as for a short period of time there can be many times more users located in distinct areas in the network who cannot be served by the normal capacity planning.

For backhaul network, the traffic consists, of course, of other radio access technologies than LTE and other traffic types than mobile network traffic. Operators have been building multiservice networks that are carrying many other services as well, like residential broadband access.

What makes monitoring LTE mobile networks, from a radio network perspective, more difficult than monitoring, for example, W-CDMA networks is that there is no controller in the radio network. Where as in, for example, W-CDMA a controller can be used for monitoring traffic sent to each NodeB, in LTE the U-plane traffic sent by the S-GW or the C-plane traffic sent by the MME might not support traffic monitoring per eNB. This means that from the mobile network equipment the traffic can be monitored from the ingress direction at the eNB and can only report traffic that has reached the eNB. In practice this means that for the purposes of S1 capacity management and traffic measurements, QoS performance control as well as to support centralized troubleshooting, it is well worth the effort to plan how the transport network capacity and QoS optimization and fault management is to be facilitated during network operation in a centralized way.

One simple solution is to assign each eNB its own VLANs and monitor per-eNB traffic in the centralized router based on VLAN ID. In addition to basic router performance monitoring, various enhanced solutions are available.[28] These special methods can differentiate between individual packet flows based on IP header information (the so-called IP 5-tuple), but may not be capable of monitoring high-traffic loads due to CPU performance constraints; individual product details should be checked case by case against monitoring requirements.

Traffic in the LTE network is bursty in nature; therefore, instantaneous traffic peaks can be very high compared to average traffic. Monitoring maximum traffic per interface, link or connection only can be misleading and provide too high values and does not give a good idea of the overall traffic in the network. This means: monitoring maximum traffic using either interface only, link only or connection only.

Average traffic is monitored often by measuring the bytes or octets during the measurement period and calculated to throughput by dividing the data volume by the measurement period to obtain, for example, megabits per second. Therefore, monitoring average traffic can, on the other hand, imply that the traffic is low as the traffic is typically averaged over longer periods and you can have very high peaks for some minutes, for example on train stations that have trains leaving once an hour. Figure 5.50 shows the same traffic over a five-day period with a one-hour measurement period versus 15-minute and 1-minute measurement periods.

As one can see from the uppermost graph in Figure 5.50, using a shorter measurement period for the example data produces more variation in the traffic. The traffic peaks are higher and the low points are lower when the measurement period that is used for averaging the traffic is shorter. The highest traffic data point is illustrated separately with a value to show the difference in the measured traffic. Using a shorter averaging period produces more accurate measurement from the traffic, but at the same time produces much more measurement data. Using a longer averaging period requires estimating peak-to-average ratio and introducing an upgrade threshold in order to prevent congestion.

The measurement result also depends on the number of eNBs aggregated on the link as the traffic peaks happen at different times and this aggregation is seen as a lower variation of traffic. The example data are aggregated traffic of multiple traffic sources; therefore, the difference between those is not that big as if it would be the traffic for a single traffic source.

When monitoring average traffic per eNB (first mile), access or aggregation link, one has to set a threshold for upgrading the link. To be able to determine the right threshold for upgrade when monitoring average traffic, requires an understanding of the peak traffic, and the following section gives an example.

5.13.2.2 Peak-to-Average Ratio

Measuring peak traffic and comparing this to the average traffic at BH allows one to create a threshold for monitoring average traffic. For example, if during the BH eNB has peak traffic of 140 Mbps and average traffic of the BH is 50 Mbps, the peak-to-average ratio is 2.8. When

[28]Performance monitoring solutions include. for example. Cisco NetFlow. NetFlow-enabled routers export traffic statistics which are then collected by a NetFlow collector. The collector does the actual traffic analysis and presentation. The term "NetFlow" has become a de-facto industry standard and is supported by platforms other like Juniper J-Flow, Huawei NetStream, Alcatel-Lucent Cflow and Ericsson Rflow.

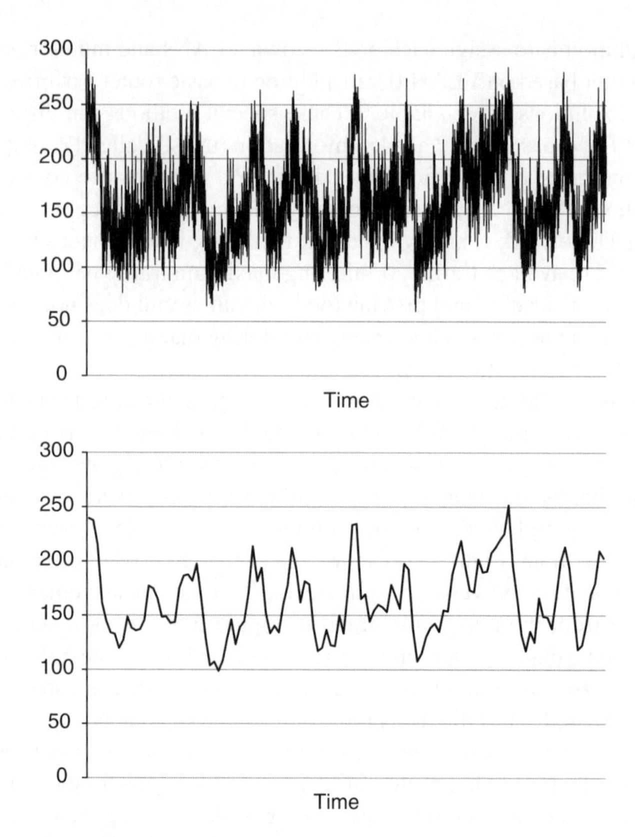

Figure 5.50 Example traffic averaged over a one-hour (lower graph), and 1-minute period (upper graph)

setting a threshold for average traffic, this would imply 36% utilization as the threshold for upgrade and for a 1-Gbps interface; it would imply that once the average traffic has reached 360 Mbps the peak traffic is being limited by the interface capacity.

The peak traffic can be, for example, 1 or 5 minutes average, which is fairly commonly available in different equipment used for backhaul or routing in the networks. Note that the peak-to-average ratio is very different, if we are monitoring the first mile to the eNB or we are monitoring a trunk in the network that carries the traffic for tens or hundreds of eNBs. When aggregating traffic of the eNBs, the traffic peaks and BHs of those eNBs can be different, which of course means that one gets multiplexing gains in concentrating the traffic, and the peak-to-average ratio for aggregation links is lower and the utilization threshold can be higher.

Peak-to-average ratio is introduced in Section 5.6.3.3, where the example shown in Figure 5.21 illustrates how the multiplexing gain has an effect on the peak-to-average ratio. With 25 active flows the average rate is about 20 Mbps with peaks exceeding 40 Mbps, hence the peak-to-average ratio of the traffic (measured at 100-ms granularity) is about two. When more active flows are present some averaging takes place and the peak-to-average ratio drops to about 1.4 with 200 traffic flows.

5.13.3 Reactive Methods

Reactive monitoring is needed to detect unavailability or the degradation of the quality of the backhaul links, site equipment or the eNB as fast as possible. The quality of the backhaul network or eNB can be affected by:

- configuration errors in the equipment
- software and hardware failures in the equipment
- weather conditions
- unexpected or sudden traffic growth in certain areas.

The reasons are many and the problem should be isolated or the root cause of the problems should be identified as fast as possible to minimize the effect to the end users. If reactive monitoring fails, the failures are found only when end users start to report for accessibility problems, low throughput or call drops.

The most severe problems are naturally reported as alarms, when network equipment reports that equipment is out of order. For example, MME reports lost connectivity to the eNB when the first mile link is cut. If the router supports Ethernet OAM, centralized fault management can also be facilitated.

More difficult is to identify what to monitor in the backhaul network to detect problems that are causing degradation of user services, like packet loss, packet error or fragmentation. The network may consist of many different vendors' equipment in different parts of the network; there could be multiple management systems. Sometimes it is easiest to monitor the throughput or packet loss from the eNB and only start investigating the backhaul network when there is an indication that the backhaul network is not performing. One common problem is leased lines and their performance.

Reactive monitoring is focused on:

- packet loss
- packet errors (cyclic redundancy check [CRC])
- packet fragmentation problems, like failed reassembly of packets
- throughput.

Such measurements are available in most equipment, the implementations of course are different; therefore, it is also important to know what the measurements actually report.

Packet loss is a major contributor to problems causing low throughput for user services that are using TCP, but it is difficult to troubleshoot depending on the complexity of the backhaul network. LTE user traffic can vary a lot and the buffers in different equipment vary in size; therefore, it is essential that it would be easy to monitor in the event of packet loss in any part of the network.

Buffer size is an important parameter in backhaul network optimization and it has an effect on the packet loss, delay and delay variation of the user services: the bigger the buffer, the lower the packet loss as the buffers can absorb the traffic bursts of the different user services. For user services having smaller packet size and lower delay or delay variation requirements, smaller buffers would be better as the delay or delay variation would be smaller. A study by Sequeira et al. (2013) indicates that the bursty nature of some applications can affect the voice quality and having a high-link utilization of above 70% prevents achieving good voice quality.

Shaping is used to prevent packet loss when there is a high bandwidth connectivity to a network element but a low bandwidth connectivity from the same network element. Also, leased line operators police the traffic according to traffic contract and excessive traffic is dropped, as described earlier. As mentioned in Section 5.5.4, shaping increases jitter, but optimizing access network parameters can make a significant improvement to the user throughputs if sources of packets loss are identified and configuration optimized.

Fragmentation can also be a major cause of user service degradation. Fragmentation happens when the IP packet to be transported exceeds the MTU size of the egress interface and the packet needs to be fragmented in order to be sent forward. At the receiving end, all the fragments need to arrive before the packet can be processed. Compared to transporting the same traffic over the Internet, due to GTP-U tunneling, the packets grow when transported over the mobile network and fragmentation cannot be completely avoided.

Efficient reassembly is, of course, important, but limiting fragmentation as much as possible would be better to ensure high throughputs, lower delay and delay variation. Path MTU discovery, if supported by the nodes, would discover the smallest MTU size along the path to the destination and avoid any further fragmentation in the network until the destination. The interface MTU can be also manually configured to avoid fragmentation, but not all equipment supports interface MTU configuration and setting this value correctly would require information on the capabilities of all the equipment in the network.

Especially when using IPsec, overhead increases and fragmentation are more likely and the use of a Jumbo frame is recommended to prevent this. As also described in Metsälä and Salmelin (2012), there are alternative ways of performing fragmentation:

- Pre-fragmentation is performed when the IP packet is expected to exceed the interface MTU after encapsulation and the IPsec stack performs fragmentation before encryption. The benefit of this is that only the final destination needs to perform the reassembly.
- Post-fragmentation is performed when there is no information on the MTU to avoid fragmentation and when using IPv6.

5.13.4 Active vs. Passive Methods

Another way of looking at the LTE backhaul network monitoring is the way the monitoring is done. Everything that has been described until now is considered passive monitoring that is done by retrieving statistical information from the network elements. As already mentioned, the capabilities and functionalities are of course dependent on the measurements implemented by different vendors. In addition to passive monitoring, it is possible to introduce e2e active monitoring of the backhaul.

5.13.4.1 Active Monitoring

When test traffic is generated and sent over the backhaul network to detect, for example, delay or packet loss, this is regarded as active monitoring. It requires equipment that sends test traffic with different QoS characteristics—DSCP, VLAN p-bit—and packet size to equipment that reflects it back to the sender equipment for analysis. RFC5357 (2008) specifies a standard defined for the functionality called TWAMP (Two Way Active Measurement Protocol). The principle of the operation is shown in Figure 5.51.

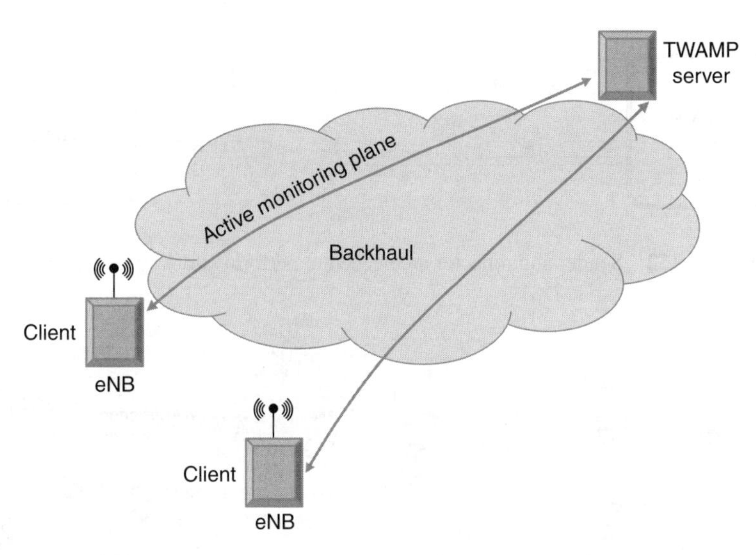

Figure 5.51 Example configuration where a TWAMP server sends the traffic that is reflected by the client back to the server

Before TWAMP, TCP/IP protocol itself has provided tools like ping, Traceroute and User Datagram Protocol (UDP) Echo. However, these tools can be used for simple troubleshooting but not for overall performance measurements. As a result, there was a need for a standards-based, effective performance monitoring method.

TWAMP defines two protocols: a control protocol for setting up performance-measurement sessions and another for the transmission and reception of those performance measurements. The control protocol enables endpoints to negotiate and start a performance monitoring session. There is also a simplified version of TWAMP where the control protocol is not used and the endpoints are configured for sending the test traffic as well as measuring the test traffic. This method is called TWAMP Light.

The functionality needed for sending or mirroring the traffic back to the sender can be either external to the mobile network equipment or integrated into, for example, the eNB providing real e2e performance monitoring of the backhaul. Using a different type of test traffic that reflects the mobile network traffic types like VoIP, streaming or Web browsing, it is possible to detect, for example, congestion in the backhaul network through delay and packet loss.

TWAMP provides measurement for:

- RTT that can be used to estimate one-way delay by dividing the RTT into two, assuming the delay is symmetric
- packet delay variation
- packet loss.

Figure 5.52 illustrates the active measurement in more detail.

The packets have time stamps when they are sent by the sender (T_1), received by the reflector (T_2), sent by the reflector (T_3) and received by the sender (T_4). The reflector's time stamps are needed to exclude its internal packet processing delay in the reflector. RTT is the time between sending and receiving the packet at the sender minus the processing time at the reflector. The RTT calculation in Equation 5.21 is measured based on four time stamps.

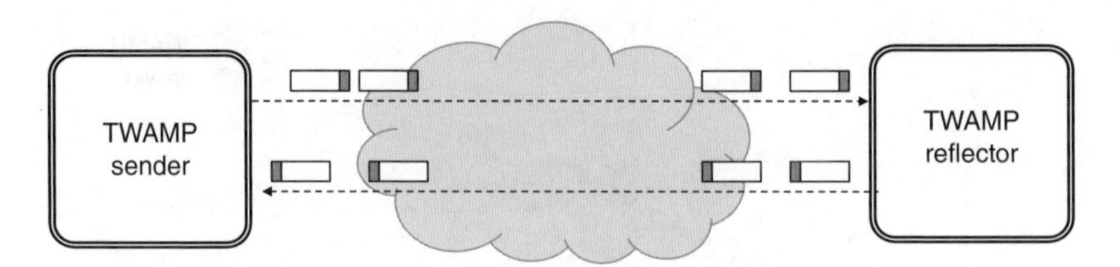

Figure 5.52 Sender and reflector are needed to perform two-way measurement

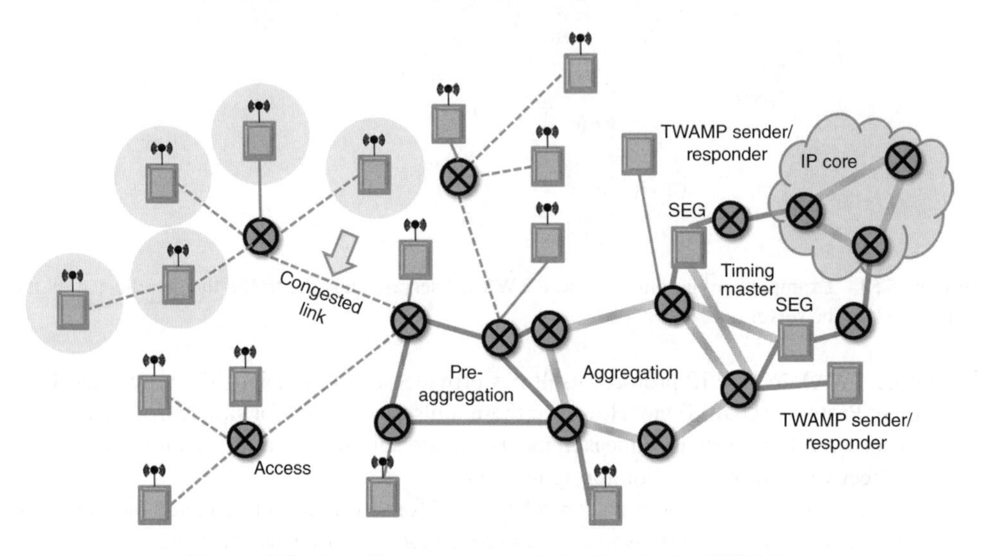

Figure 5.53 Identifying congestion in backhaul using TWAMP

$$RTT = (T_4 - T_3) + (T_2 - T_1) \tag{5.21}$$

Delay or one-way delay can then be estimated by dividing the RTT in two. Monitoring delay or delay variation is important, for example for enabling LTE network for VoLTE. It has been identified that high delay peaks have a higher impact on voice quality compared to average jitter.

The packet sequence number is used to calculate the packet loss ratio. The packet loss ratio as defined in Equation 5.22 defines the proportion of the packets lost during the test.

$$PLR = 100\% \times \frac{Lost\ TWAMP\ packets}{sent\ TWAMP\ packets} \tag{5.22}$$

The above testing is non-intrusive as the test traffic sent over the transport path is negligible compared to the actual traffic. It gets the same treatment as the actual traffic in the network but has none or only a minor impact on it.

The results of active monitoring can reveal problems in the first mile, but if measurement results are analyzed at the network level it is possible to identify problems in access or aggregation links as well. In Figure 5.53 a few eNBs have been highlighted (on the top left of the figure).

If similar results like increase delay or packet loss are identified in a set of eNBs, analyzing the transport links can identify, for example, common congested links in the access network.

There are other active monitoring methods as well that can measure delay, jitter, packet loss and throughput. For monitoring SLA conformance of the external transport network, there are intrusive test methods that can be used to verify the performance of the connection before service is activated or during the maintenance window. Measurement methodology and measurement quantities for SLA measurement should be defined carefully. Two useful standards in this regard are:

- IETF (1999) RFC2544, *Benchmarking Methodology for Network Interconnect Devices*
- ITU-T (2011) Y.1564, *Ethernet Service Activation Test Methodology*

ITU-T Y.1564 is a newer standard compared to RFC2544, which has been used for benchmarking for years. RFC2544 was initially designed for testing network devices in a lab but since it was able to test throughput, frame loss and delay, it became common in the real network and is supported by many vendors. However, RFC2544 lacks the possibility to test jitter and is built for testing the IP layer, while ITU-T Y.1564 is designed for testing Ethernet services by running multiple tests in sequence and collecting statistics in parallel. ITU-T Y.1564 defines test streams with service attributes aligned to MEF 10.2 definitions. A test sequence could consist of different frame sizes and throughputs and each of the test results includes the sent and received load, throughput, frame loss, latency and jitter.

Besides SLA monitoring, the measurement methods and quantities described in the above standards are useful references also more generally (e.g. as a basis for detailed customer discussions) and as formal definitions of transport network quality metrics, their measurement methods and reporting in technical reports.

In case timing over packet or SyncE is used, the provider capability to support these technologies should be agreed and monitored afterward.

Non-intrusive active monitoring can be done continuously to proactively identify problems in the mobile backhaul network. Alternatively, active and passive monitoring can be combined, passive monitoring can be used to find problems in the network and active monitoring is used to troubleshoot the problem. In case such functionality is embedded in the eNB, it is fairly easy to set up a connection to the eNB to check the e2e QoS of the backhaul, which would otherwise be difficult, due to various different equipment, leased network connections and limitations in the measurement ability of the equipment involved in the backhaul network or how the network has been setup.

5.13.4.2 TWAMP Session Configuration Example

As described at the beginning of this section, it is good to measure with different QoS characteristics, like DSCP value and packet size, when using active monitoring. The following QoS setting defined in Table 5.16 is assumed for user, control S- and M-plane.

In the defined QoS configuration the highest priority U-plane traffic is configured the same as the S- and C-plane traffic; therefore, it makes sense to configure one TWAMP session having the QoS characteristics of QCI1 and using, for example, a packet length of 100 bytes. Additional TWAMP sessions can be configured for different QCI classes, but additionally at

Table 5.16 Example for QoS setting for user, control, synchronization and management planes

Plane[a]	QCI/Protocol	DSCP	PHB	PCP
Control plane	S1, X2	46	EF	6
User plane	QCI1	46	EF	6
	QCI2	26	AF31	3
	QCI3	46	EF	6
	QCI4	28	AF32	3
	QCI5	34	AF41	4
	QCI6	18	AF21	2
	QCI7	20	AF22	2
	QCI8	10	AF11	1
	QCI9	0	BE	0
Synchronization plane	ToP	46	EF	6
Management plane	CM, FM, PM	34	AF41	2

[a] Network control plane traffic, like ICMP, BFD and IKE were left out of the example.

least one session more is needed representing for the lowest-priority U-plane traffic. This session can have the QoS characteristics of QCI8 and a packet length of 1000–1500 bytes representing data traffic like Web browsing or file download/upload.

References

3GPP (2013) *TS 26.114: IP Multimedia Subsystem (IMS); Multimedia Telephony; Media handling and interaction v10.8.0*, December 2013.

Adams R. (2013) Active queue management: A survey. *IEEE Communications Surveys & Tutorials* 15(3).

Berger A. and Whitt W. (1998) Extending the effective-bandwidth concept to networks with priority classes. *IEEE Communications Magazine* 36(9).

Bonald T., Proutière A. and Roberts J. W. (2001) Statistical performance guarantees for streaming flows using expedited forwarding. *Proceedings of the IEEE INFOCOM*, http://perso.telecom-paristech.fr/~bonald/Publications_files/BPR-infocom01.pdf, accessed 1st June 2015.

Cao J., Cleveland W. S., Lin D. and Sun D. X. (2001) *Internet traffic tends toward Poisson and independent as the load increases*. Technical report, Bell Labs, http://www.stat.purdue.edu/~wsc/papers/lrd2poisson.pdf, accessed 1st June 2015.

CCITT (1988) *E.600: The International Telegraph and Telephone Consultative Committee (11/1988): Series E: Overall network operation, telephone service, service operation and human factors terms and definitions of traffic engineering*, https://www.itu.int/rec/dologin_pub.asp?lang=e&id=T-REC-E.600-198811-S!!PDF-E&type=items, accessed 1st June 2015.

Ferrant J.-L., Gilson M., Jobert S., et al. (2013) *Synchronous Ethernet and IEEE 1588 in Telecoms: Next generation synchronization networks*. Wiley-ISTE, New York.

GSM Association (2015) *Official document IR.92: IMS profile for voice and SMS: Version 9.0*, http://www.gsma.com/newsroom/wp-content/uploads/IR.92-v9.0.pdf, accessed 1st June 2015.

Horváth, G. (2001) Approximate waiting time analysis of priority queues. *Proceedings of the Fifth International Workshop on Performability Modeling of Computer and Communication Systems*, Erlangen, Germany, September 2001.

IEEE1588 (2008) *IEEE standard for a precision clock synchronization protocol for networked measurement and control systems*, July 2008, http://www.nist.gov/el/isd/ieee/ieee1588.cfm, accessed 1st June 2015.

IETF (1999) *RFC2544: Benchmarking methodology for network interconnect devices*, https://tools.ietf.org/html/rfc2544, accessed 1st June 2015.

ITU-T (2014) *G.8265.1/Y.1365.1: Precision time protocol telecom profile for frequency synchronization*, July 2014.

ITU-T (2013a) *G.8261: Timing and synchronization aspects in packet networks*, August 2013.

ITU-T (2013b) *G.8271.1: Network limits for time synchronization in packet networks*, pre-published, October 2013.

ITU-T (2012) *G.8032: Ethernet Ring Protection Switching*, February 2012.

ITU-T (2011) *Y.1564: Series Y: Global Information Infrastructure, Internet Protocol Aspects and Next-Generation Networks: Internet protocol aspects – Quality of service and network performance, Ethernet service activation test methodology, Recommendation.* March 2011, https://www.itu.int/rec/dologin_pub.asp?lang=e&id=T-REC-Y.1564-201103-I!!PDF-E&type=items, accessed 1st June 2015.

ITU-T (2008) *G.8264: Distribution of timing information through packet networks*, October 2008.

Kleinrock L. (1975) *Queueing Systems: Vol. I*, Wiley Interscience, New York.

Li X., Bigos W., Dulas D. et al. (2011) Dimensioning of the LTE access network for the transport network delay QoS. *Proceedings of the IEEE 73rd Vehicular Technology Conference*. Budapest, Hungary, May 2011.

Malhotra R. and van den Berg J. L. (2006) Flow level performance approximations for elastic traffic integrated with prioritized stream traffic. *Proceedings of the International Telecommunications Network Strategy and Planning Symposium*, doi: 10.1109/NETWKS.2006.300367.

Metsälä E. and Salmelin J. (eds) (2012) *Mobile Backhaul*, John Wiley & Sons, Ltd, Chichester, UK. doi: 10.1002/9781119941019.

NGMN Alliance (2011a) *Guidelines for LTE backhaul traffic estimation*, NGMN White Paper, https://www.ngmn.org/uploads/media/NGMN_White paper_Guideline_for_LTE_Backhaul_Traffic_Estimation.pdf, accessed 1st June 2015.

NGMN Alliance (2011b) *LTE backhauling deployment scenarios*. NGMN White Paper, https://www.ngmn.org/uploads/media/NGMN_Whitepaper_LTE_Backhauling_Deployment_Scenarios_01.pdf, accessed 1st June 2015.

Pietilainen A. and Virta T. (2013) Time synchronization with partial on-path support. *Proceedings of the Telcordia NIST-ATIS Workshop on Synchronisation in Telecommunication Systems*, 16th–18th April 2013, San Jose, CA.

RFC5357 (2008) *Two Way Active Measurement Protocol (TWAMP)*, https://tools.ietf.org/html/rfc5357, accessed 1st June 2015.

Riedl A., Perske M., Bauschert T. and Probst A (2000) Investigation of the M/G/R processor sharing model for dimensioning of IP access networks with elastic traffic. *Proceedings of the First Polish–German Teletraffic Symposium*, Dresden, Germany.

Sequeira L., Fernandez-Navajas J., Casadesus L. et al. (2013) *The Influence of the Buffer Size in Packet Loss for Competing Multimedia and Bursty Traffic*. Performance Evaluation of Computer and Telecommunication Systems (SPECTS), http://diec.unizar.es/~jsaldana/personal/influence_spects2013_in_proc.pdf, accessed 1st June 2015.

Stiliadis D. and Varma A. (1998) Latency-rate servers: A general model for analysis of traffic scheduling algorithms. *IEEE/ACM Transactions on Networking* 6 (5).

Szukowski M., Maciejewski H., Koonert M. and Chowanski B. (2011) Dimensioning of packet networks based on data-driven traffic profile modeling, *Proc. of the First European Teletraffic Seminar*, Poznań, Poland, 2011, http://www.zsk.ict.pwr.wroc.pl/zsk/repository/dydaktyka/air/ets_2011_dimensioning_of_packet_services.pdf, accessed 1st June 2015.

Vranken R., van der Mei R. D., Kooij R. E. and van den Berg J. L. (2002) Performance of TCP with multiple priority classes. *Proceedings of the International Seminar on Telecommunication Networks and Teletraffic Theory*, St. Petersburg, Russia, 29th January to 1st February: 78–87.

Walraevens J., Steyaert B. and Bruneel H. (2002) Delay characteristics in discrete-time GI-G-1 queues with non-preemptive priority queueing discipline. *Performance Evaluation* 50(1), http://smacs.ugent.be/intern/doc/smacs/j095.pdf, accessed 1st June 2015.

6

Design Examples

Jari Salo[1] and Esa Markus Metsälä[2]
[1]*Nokia Networks, Doha, Qatar*
[2]*Nokia Networks, Espoo, Finland*

6.1 Introduction

In this chapter, two design examples are given for the purpose of putting together design concepts discussed in Chapter 5. In the first example (Scenario 1) the operator has his own transport network, and microwave transport is deployed for the first few miles. The second example (Scenario 2) is similar to the first one, except that now the access transport between the cell site and the operator regional data center is based on leased line.

Table 6.1 presents the high-level design targets and boundary conditions assumed for both design examples of this chapter.

6.2 Scenario #1: Microwave

In this use case the transport for the first few miles from the eNB to the aggregation point is implemented by microwave links. The topology for the design example is that of Figure 6.1.

The following additional assumptions are made:

- Available microwave link bandwidth options are: 175 Mbps, 350 Mbps, 700 Mbps.
- Adaptive coding and modulation is used in microwave.

The pre-aggregation and aggregation networks are in this example assumed to be based on L3VPN over IP/MPLS. The pre-aggregation ring nodes function as provider edge (PE) routers having two VRFs, to be explained later in the section.

LTE Backhaul: Planning and Optimization, First Edition. Edited by Esa Markus Metsälä and Juha T.T. Salmelin.

Table 6.1 Input information for both scenarios

Type	High-level design inputs for access transport
IP plan	• 3000 base station sites • From IP planning viewpoint allow at least up to 9000 eNBs, or three eNB per site (future scalability requirement) • IPv4 used in access transport
Quality of service	• Average X2 user and control plane delay <5 ms • VoLTE packet delay variation due to transport network <5 ms (one-way peak) • Maximum 10% throughput degradation for best effort TCP data caused by transport network when using M/G/R-PS dimensioning
Availability	• Average eNB availability >99.99%. Availability requirement applies to the access and the pre-aggregation tiers of the transport network • Average eNB unavailability is evaluated as the sum of unavailable minutes over all eNBs divided by the product of the number of minutes within the observation interval and the number of eNBs • MTTR is assumed to be 8 hr for all transport nodes • MWR link unavailability due to rain and propagation anomalies can be excluded in the availability calculation
Synchronization	• Phase synchronization required • Aggregation and backbone tiers of the network do not support boundary clocks
Security	• IPsec required for user and control traffic • Management traffic protected at application layer
Geo-redundancy	• Each eNB is connected to two different MMEs and SP-GWs, which are located in two physically different core sites
User experience	• Network must support 20 MHz + 10 MHz LTE-A carrier aggregation 2 × 2 MIMO peak rates with FTP download
Traffic model (network launch plus one year)	• 50 Erlangs voice BH traffic per site • Site-level average BH data demand 40 Mbps
Network monitoring and capacity scaling	• Equipment performance counters should support efficient capacity management process • Network capacity scaling should be possible without hardware upgrades

6.2.1 Synchronization

The design requirement is that phase synchronization has to be supported. On the other hand, it is stated that pre-aggregation and aggregation tiers of the transport network do not support boundary clocks. This leaves very few design options:

- Option 1: Each site is equipped with GPS.
- Option 2: Each regional data center is required to be equipped with IEEE1588 (2008) grand-master and phase timing is delivered to the eNB sites over the PTPv2 protocol.

Option 2 would most likely be more economical since retrofitting all sites with GPS is likely to be costly. The main design question with option 2 is whether the access and pre-aggregation transport tiers can support the extremely strict delay variation requirements of packet phase synchronization (1 microsecond).

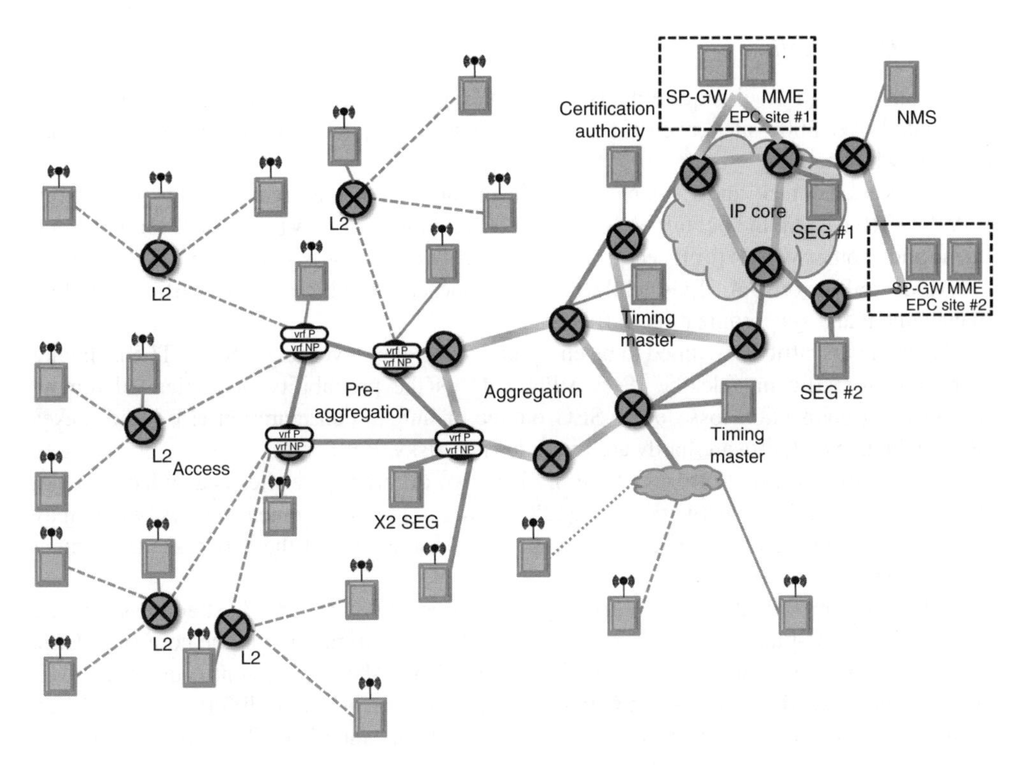

Figure 6.1 Topology for the design example

The sites are assumed to be not more than two microwave hops away from a pre-aggregation ring node, which is connected to the regional data center via high-speed fiber. If equipment performance results are not available, it is assumed that no more than two microwave links can be chained, as otherwise the delay asymmetry between UL and DL may be too large. Furthermore, some margin has to be left for the asymmetry introduced by the pre-aggregation fiber ring since the IEEE1588 (2008) grandmaster was assumed to be located in the regional core site.

The PTP packets should be mapped to strict priority queue e2e, in both UL and DL. Even with the strict priority mapping, the link loads should not be too high in either the pre-aggregation ring or in the microwave-based access network. The maximum tolerable load level depends on the link bandwidth and path delay asymmetry, which varies for each link and node. Some results are provided in Chapter 5. At the time of writing, there is only limited field experience on deployment of IEEE1588 (2008) based phase synchronization in production networks, and therefore performance monitoring should be carried out after deployment. This is possible either from performance counters collected by the PTP slave (implementation-dependent) or by utilizing active timing error measurements with special equipment. One option is to measure the air interface inter-cell frame timing offset with a GPS-synchronized RF scanner; this provides an e2e view of the radio frame timing quality in the network.

In this example, Option 2 is selected. For some remote sites, Option 1 (GPS) could also be used.

6.2.2 IP Planning

Based on the high-level requirements the IP plan has to possibly scale up to 9000 eNBs distributed over 3000 sites, even if the main option is to upgrade the single eNB for more capacity when needed.

IPsec is used for U- and C-plane traffic. In the absence of any requirements given from L3VPN VRF design, the default eNB VLAN configuration is one VLAN for IPsec-bypassed traffic and another VLAN for IPsec-protected traffic. In other words, one of the VLANs carries encrypted traffic to/from SEGs while the other one carries all bypassed traffic, including management and synchronization traffic.

Management traffic is assumed to be encrypted at a higher layer (e.g. SSH or TLS). Having management traffic outside the IPsec VPN reduces the probability of unplanned remote management connection loss due to SEG outage or human configuration error, and makes SEG maintenance (software/hardware upgrade) less risky. Further, the problem of having to split management traffic to two separate SEGs at two different core sites is avoided. Instead, the aggregation and core network will handle routing to the management servers, which may be distributed at several core sites. This makes geo-redundancy of the network management system easier to handle.

In the conventional site-to-site VPN implementation, IPsec requires host addresses at the eNB for U- and C-plane termination. To simplify policy configuration at the core site SEGs, the same address is configured for both U- and C-plane. There is no confusion at the eNB about traffic types since the U- and C-plane applications are bound to different protocol/port numbers. S- and M-plane are bound to the VLAN network interface IP address. To summarize, the IP addresses for one eNB are:

- One VLAN transport address for encrypted traffic. This is the address the remote SEG sees as the IPsec tunnel endpoint of eNB.
- One VLAN transport address for plaintext traffic. Management and synchronization traffic is terminated to this address.
- One eNB-internal host address (loopback) for user and control traffic.

The minimum number of application (host) addresses would be 2 ^ 14 = 16,384 where the 9000 eNBs were rounded up to the next smallest power-of-two. However, at network level it's easier to administrate and manage a 2 ^ 16 subnet, so one /16 subnet is allocated for eNB U- and C-plane addresses. This leaves some flexibility in U- and C-plane addressing albeit at the cost of increased address consumption. The subnetting has room for up to three eNBs per cell site as required in the design targets, whereas the initial deployment is one eNB per cell site.

Each VLAN accommodates a single eNB, which makes the design more robust against L2 problems as well as resulting in simpler performance management and fault isolation. For a single VLAN using a /29 subnet per eNB requires at least 9000 × 8 = 72k > 2 ^ 16 addresses. Hence /16 is not enough for future needs, so a /15 subnet supporting ~16k /29 VLANs is reserved. The subnetting for this example is assumed as:

- 10.0.0.0/15 for "P" VLAN
- 10.2.0.0/15 for "NP" VLAN
- 10.100.0.0/16 for U- and C-plane host addresses.

Table 6.2 Example eNB IP address allocation

eNB id	"P" VLAN	"NP" VLAN	U/C host address
260101	10.0.0.4/29	10.2.0.4/29	10.100.0.4/32
260102	10.0.0.12/29	10.2.0.12/29	10.100.0.12/32
260103	10.0.0.20/29	10.2.0.20/29	10.100.0.20/32
...	...	...	...

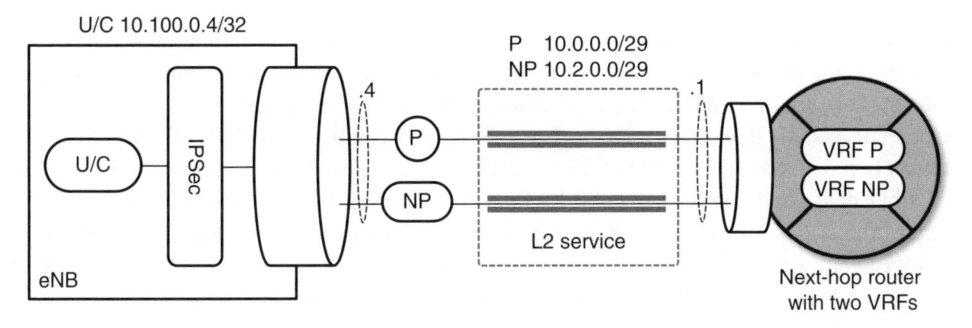

Figure 6.2 eNB addressing model for the planning example. The next-hop router is located in the pre-aggregation site. L2 service in this design example is realized by a microwave network

"P" and "NP" stand for protected and non-protected, respectively, and the protection here means cryptographic protection with IPsec. While this addressing may at first appear to excessively waste IP addresses, experience from practical networks has shown that it is almost always a better approach to overestimate the future address requirement.

As an example, the eNBs can be assigned addresses, as in Table 6.2 and shown in Figure 6.2.

The first three addresses in the VLAN subnets are reserved for VRRP-type router redundancy protocols. The allocation of address blocks should be further based on location, for example a contiguous block of /20 could be allocated to a regional data center serving up to ~500 eNBs. This subnet can be further subdivided into eNBs as the network evolves over time. Similarly, the eNB application address range for U- and C-plane could be divided to subranges based on location.

To satisfy the X2 one-way delay requirement of < 5 ms, it is not possible in all cases to route X2 traffic via centralized SEGs located at the core site. Along with phase synchronization, this presents one of the main problems in transport design. A small X2 delay is presumed a prerequisite for LTE-A radio features (e.g. CoMP). Having said that, at the time of writing these features are still in development phase and it is not clear whether the realized radio performance gain will justify the increased complexity in X2 transport. Also the actual X2 bandwidth requirements will be highly dependent on the CoMP implementation. In an extreme case it may be that a direct inter-eNB fiber connection is required. With these uncertainties in mind, there are several options on how to proceed with the design:

- Option 1: Estimate the number of sites that cannot satisfy the <5 ms X2 delay based on centralized solution where SEGs are located at the two core sites only. If the number of such base stations is very small—or if these sites are in low-traffic rural areas where the additional radio capacity brought by LTE-A radio features is unlikely to be needed in the

first place—then one may decide to neglect the X2 delay requirement for these base stations altogether. Thus, for the remote areas it can be, for the time being, assumed that no radio features requiring low X2 latency are needed.
- Option 2: Instead of two centralized gateways, distribute all SEGs to regional sites so that the X2 delay requirement can be satisfied for all base station sites. This solution increases both deployment and operational cost due to the increased number of SEGs that need to be operated.
- Option 3: Split the encrypted traffic to S1 and X2 traffic where the S1 gateways are at centralized core sites and the X2 gateways are at regional sites if needed. Like Option 2, this solution also increases maintenance costs and its implementation is highly complex in terms of IPsec and routing configuration. On the other hand, the resiliency requirements of the X2 gateways are not so stringent since gateway failure will cause only temporary capacity degradation, and maximum hardware redundancy will not be needed.
- Option 4: Assume that direct X2 tunnels can be established between eNBs. The feasibility of this option depends largely on the configuration management tools available. As an eNB in an urban area can easily have over 30 X2 links, the number of X2 IPsec tunnels will be in the tens of thousands for even a moderately sized network of 1000 eNBs. Even if the configuration problem were tackled, monitoring and troubleshooting of such a mesh network would remain an issue, not to mention the network behavior under transport network anomalies.

Option 1 would be simplest, but for the purpose of a design example also the least interesting. Therefore we will choose Option 3 here. With Option 3, remote regional hubs are deployed with their own SEGs handling only X2 traffic. In cases where the cell site is within about 500 km from the core site there is no need for regional X2 gateway[1] due to delay requirement. As mentioned, the availability of the remote X2 gateway does not need to be as high as that of a centralized gateway at the core site, since in the event of X2 failure S1 handover can be used temporarily. This allows for a longer MTTR or less redundant hardware. Further, with Option 3 the centralized S1 SEG can be taken to maintenance while still retaining X2 connectivity via a regional SEG. This would not be the case if X2 were routed via the SEG at the core site.[2]

Table 6.3 summarizes the main IP addresses needed for routing and IPsec traffic selector design.

Table 6.3 Summary of IP addresses in the design example

Application	Subnet	Notes
eNB "P" VLAN	10.0.0.0/15	Subdivided to regions
eNB "NP" VLAN	10.2.0.0/15	Subdivided to regions
eNB application	10.100.0.0/16	User and control plane. Subdivided to regions
S1-MME, S1-U	10.200.1.0/24	Subdivided to core sites
S1 security gw	10.4.0.0/24	Subdivided to core sites
X2 security gw	10.4.1.0/24	Subdivided to regional sites

[1] Velocity factor of optical fiber is about 0.7. One-way processing delay of a microwave link can be up to some hundreds of microseconds, depending on link speed and implementation.

[2] In the geo-redundant EPC solution, traffic can be offloaded by standard 3GPP mechanisms to one MME/SP-GW for the duration of a maintenance break. Redirecting protected X2 traffic from one core SEG is more complicated.

As an example for one eNB, the IPsec traffic selectors for protected traffic can then be selected as:

protect traffic from 10.100.0.4/32 to 10.200.1.0/25, via remote SEG 10.4.0.1 // security gateway at core site #1
protect traffic from 10.100.0.4/32 to 10.200.1.128/25, via remote SEG 10.4.0.129 // security gateway at core site #2
protect traffic from 10.100.0.4/32 to 10.100.0.0/16, via remote SEG 10.4.1.1 // X2 security gateway at regional site

The remote SEG address at core site #1 is 10.4.0.1 and at core site #2 it is 10.4.0.129. The regional SEG handling the delay-sensitive X2 traffic is 10.4.1.1.[3] Any protocol and port number is assumed to match the traffic selector. There are a total of three VPNs: two for geo-redundant S1 traffic and one for X2 traffic.

Additionally, a bypass policy is required for all bypassed traffic. A very loose bypass policy would be:

bypass traffic from 10.2.0.4/32 to 0.0.0.0

This policy bypasses all traffic to/from address 10.2.0.4. The default policy applied to a packet that does not match any of four traffic selectors is to discard the packet. The security policies in the example are fairly loose. Depending on the operator security principles, more stringent traffic selectors could be used.

From an eNB routing point of view the design is very simple. No static routes need to be defined at the next-hop router. In the UL direction, a default route and one static route are the minimum requirements. Since according to the subnet plan all remote SEGs are from 10.4.0.0/23 subnet, a minimum routing configuration for the example would be:

to 10.4.0.0/23	via 10.0.0.1	// "P" VLAN
default gateway	via 10.2.0.1	// "NP" VLAN

With this configuration, there is no need to update routing tables when new management domain servers, which need to communicate with the eNB, come online (since they use the default VLAN). Obviously, if the remote SEGs were not allocated from a contiguous subnet, more routing entries would be required. Careful design of IP subnetting can simplify radio network transport configuration, and facilitate easier configuration management of the network.

6.2.3 Availability

Basic availability calculation formulas are discussed in Chapter 5. The availability requirement given in the high-level design targets applies to the portion of the network shown in Figure 6.3. The allowed average eNB unavailability (due to transport) is 52 minutes per year, corresponding to four nines' availability. In this design example the problem is approached

[3] Traffic protection and routing between X2 SEGs need to be configured. Each SEG should handle the X2 traffic of a geographic region. This results in "X2 handover borders" that can be delay-optimized by considering transport and radio network topologies. If the eNB IP subnets are divided by geographic location, X2 routing becomes more manageable.

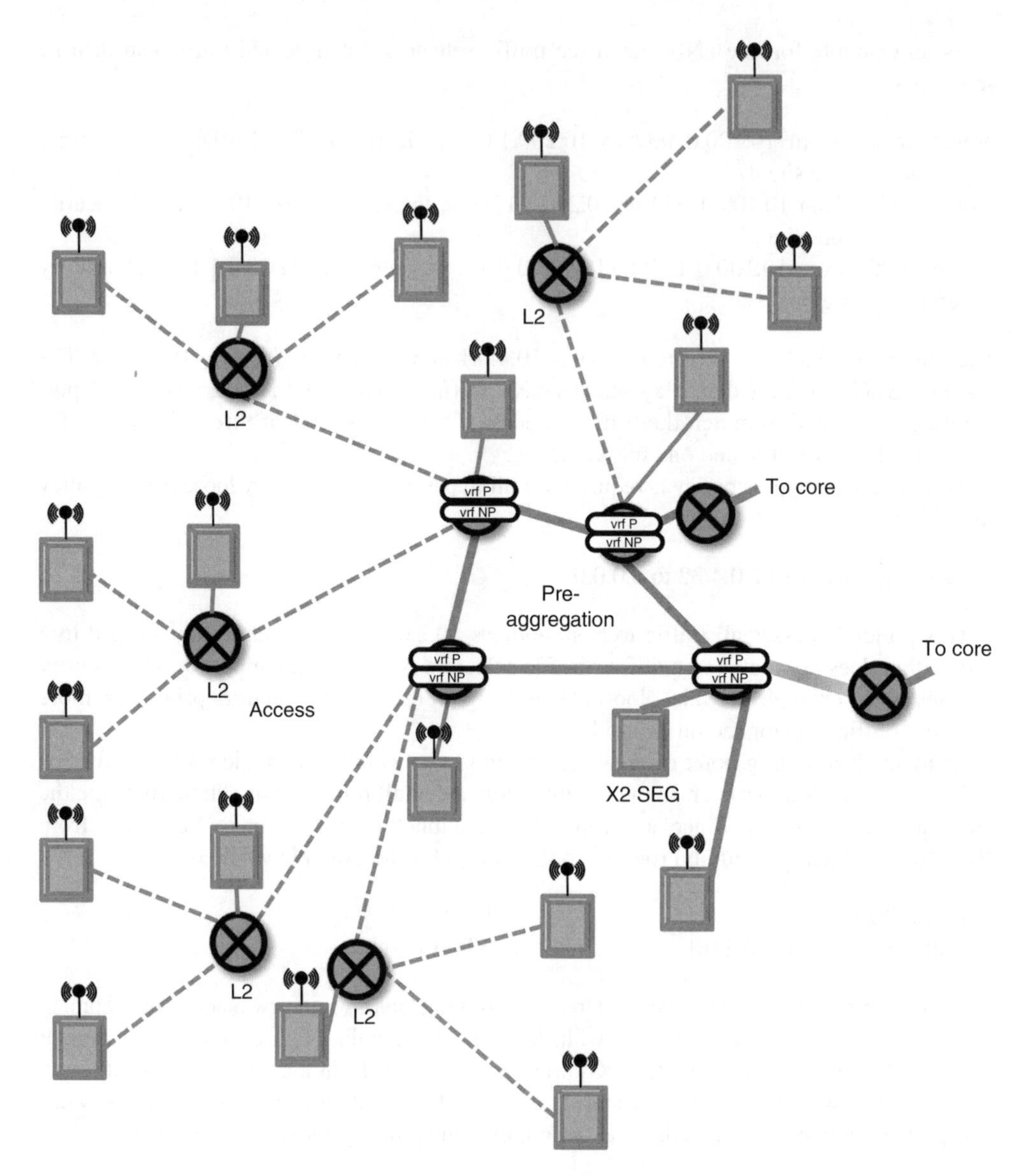

Figure 6.3 Portion of the network where the availability requirement applies

by first allocating the total unavailability target to different sections of the network, and then selecting a suitable resiliency technique for each network section to satisfy the chosen target.

When allocating the unavailability to different parts of the network, the number of eNBs being served by the network section under planning should be considered. This is because the failure of a transport link serving 100 eNBs will result in 100 times the unavailable minutes at network level than failure of a link serving one eNB. From Figure 6.3 it can be seen that the failure of a router node in the pre-aggregation ring results in service outage of up to seven eNBs. For this reason the availability requirement for the pre-aggregation ring should be higher than for the access links.

The pre-aggregation ring availability requirement is chosen here as > 99.999%, or a maximum of about five minutes of unavailability per year. The ring in the example topology has four nodes. The ring availability target is satisfied if each link in the ring has availability of at least 99.9%, which is a loose requirement. In this case the ring availability would be at least 1 – (1 – 0.999 ^ 2) × (1 – 0.999 ^ 2) = 99.9996%, or about two minutes per year.[4] This exemplifies the well-known fact that ring topology is theoretically very resilient against failures, assuming that the chosen ring protection works expectedly. For simplicity's sake, the link availability is assumed to include the equipment terminating the link at the pre-aggregation node (e.g. a router). The ring resiliency can be implemented, for example, with some fast rerouting mechanism at either the IP or MPLS layer.

The routing gateway of the eNB is located in the pre-aggregation node, and this equipment is assumed to be separate physical equipment from the ring transport so that its unavailability needs to be added to the e2e unavailability budget. Router failover can be controlled either by the router itself (e.g. VRRP) or by gateway selection, based on ICMP or BFD polling. With 8 hr of MTTR the required router, MTBF can be calculated from Equation 6.1:

$$A = \frac{MTBF}{MTBF + MTTR}. \tag{6.1}$$

Solving for MTBF from Equation 6.1 gives Equation 6.2:

$$MTBF = \frac{MTTR}{(1 - A)}. \tag{6.2}$$

With A = 99.999% and MTTR of 8 hr, the MTBF of a router should not be less than 91 years. To achieve this level of reliability the router configuration must be planned accordingly, with duplicated hardware and link aggregation groups divided to different line cards or different chassis. Implementing the availability target could be achieved with either a single highly redundant chassis or two separate router chassis. The availability of complex equipment such as a router depends not only on hardware redundancy but also on features such as software process modularity, decoupled C- and U-plane functions and online upgradeable software. Here it is assumed that the PE router by some means satisfies the required five nines' availability, but detailed analysis is omitted.

For remote regional hubs an X2 SEG may be deployed to reduce X2 delay. If the gateway becomes unavailable, this results in handovers being routed via S1 by eNB. Thus, the eNBs will be on air and the unavailability of the X2 IPsec gateway can be neglected here.

Based on the allocated pre-aggregation ring (unavailability ~2 minutes) and PE router (unavailability ~ 5 minutes), there remains 0.009% unavailability budget (or about 45 minutes a year) to allocate to the access microwave links. By strict interpretation of the given availability requirement this availability should be allocated to all the links so that even the tail eNB satisfies the availability requirement. Dividing unavailability evenly on the two-link

[4] The worst-case ring availability results by splitting the ring to two parallel serial systems with two links in each. Unavailability of a chain of two links is 1 – 0.999 ^ 2 and the parallel system of two such chains is unavailable only if both parallel chains are unavailable. See Chapter 5 for some basic availability formulas.

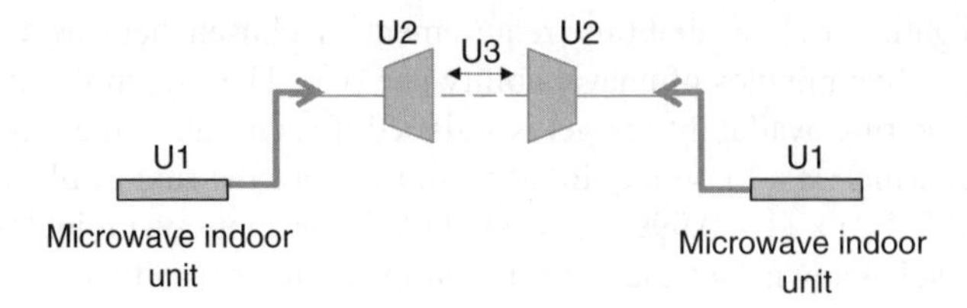

Figure 6.4 Microwave link hop availability model. Total unavailability including propagation (U3) is approximately $2U_1 + 2U_2 + U_3$

hops, the per-hop unavailability target is 0.0045%. The total link unavailability approximately is the sum of indoor (U1) and outdoor unit (U2) unavailability and the radio hop unavailability (U3; depends on rain intensity and propagation anomalies).

Based on the provided high-level design assumptions, the unavailability due to propagation (U3) may be excluded in this evaluation. It follows that the combined IU and ODU unavailability should not exceed 8 hr divided by 0.99996, which corresponds to an MTBF requirement of about 23 years per link. From this value and from reliability information from equipment vendor data sheets the link configuration satisfying the given target can be designed.

Example 6.1

Assuming an IDU and ODU MTBF of 75 and 25 years, respectively, it is clear that either of the units should be hardware redundant to satisfy the availability target of the link. Indeed, without any hardware redundancy the link unavailability is approximately two times $U_1 + U_2$, or ≈ 0.01% (52 minutes). See Figure 6.4. Protecting either IDU or ODU provides very high availability for the protected unit and in turn results in the unprotected unit dominating the total unavailability. Typically, physical access to ODUs is more difficult and hence it is chosen here to protect the ODU; this also gives larger improvement in availability than protecting the IU.

One link end becomes unavailable only if both ODUs fail simultaneously, and the theoretical probability of this happening is exceedingly small and so is assumed to be negligible. With hot-standby ODUs and 8 hr MTTR for IDUs the link unavailability is $\approx 2U_1$, or 0.0024%. This satisfies the link availability target of 0.0045%.

Summarizing the result of the calculations, the total unavailability due to pre-aggregation and access layer failures (topology of Figure 6.5) is given in Table 6.4, where less than 0.001% unavailability is rounded up to 0.001%. Tail eNBs have 36 minutes per year predicted unavailability, while a node connected to the pre-aggregation ring via fixed line has only 11 minutes of unavailability. Weighting the total unavailable minutes by the number of eNBs gives 5 × 36 + 23 + 11 = 214 unavailable minutes per year for the seven eNBs in total. However, the average eNB annual unavailability is 214/7 ≈ 31 minutes, which corresponds to 99.994% availability; this is better than the original design target. It can be seen that, especially in complex topologies, there is in theory some room for optimizing cost versus redundancy.

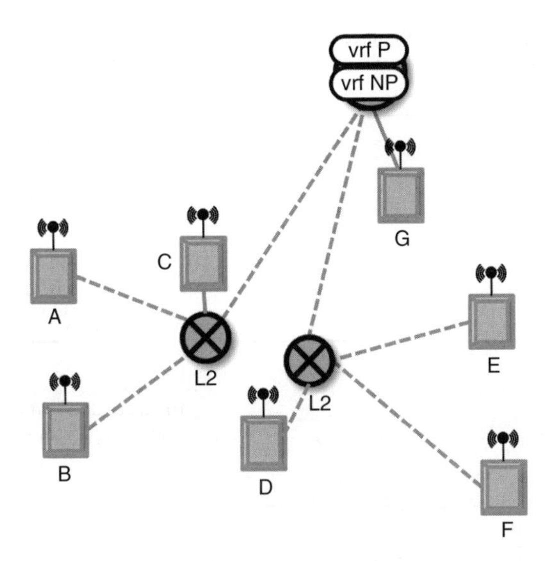

Figure 6.5 Topology considered

Table 6.4 Summary of availability calculations for eNB in the example topology

eNB	MWR unavailability	PE router unavailability	Pre-aggregation ring unavailability	Total unavailability	Total availability	Total unavailable minutes per year
A/B/D/E/F	0.00487%	0.001%	< 0.001%	< 0.00687%	> 99.993%	< 36
C	0.00244%	0.001%	< 0.001%	< 0.00444%	> 99.996%	< 23
G	0.0%	0.001%	< 0.001%	< 0.002%	> 99.998%	< 11

6.2.3.1 Quality of Service

The goal is to manage with as few service classes as possible. In practice this means that all traffic will be mapped to either three or four different quality classes. Although beyond the scope of this example (and the book), this includes 2G/3G traffic that also needs to be transported by the same network.

The mapping of traffic types to four service classes is assumed as shown in Table 6.5. In the table, voice is mapped to a lower priority than signaling. Here again, SPQ stands for strict priority queueing and WFQ for weighted fair queueing. The corresponding scheduling scheme chosen to implement the QoS for different traffic types is shown in Figure 6.6.

Two strict priority queues are used for delay-sensitive traffic and the w_2 weighted fair queue is used for BE data. Management traffic is mapped to the w_1 queue. The volume of LTE signaling traffic is expected to be in the order of 1% of LTE U-plane traffic, including IPsec headers. The signaling traffic volume of 2G and 3G is higher but the total signaling traffic per site is expected to be not more than a few megabits per second and hence signaling traffic cannot starve the lower priority traffic.

Table 6.5 Mapping of traffic types to classes

Traffic type	DSCP dec	VLAN priority	Queue	Notes
Signaling and network control, IMS signaling	46	6	SPQ1	S1AP, X2AP, NBAP, PTPv2, QCI5, IKE
Real-time packet, voice	34	5	SPQ2	2G/3G/LTE voice over IP
Base station management and high-priority data	10	1	WFQ1	O&M traffic, reserve for other important data, e.g. gaming or VIP data users
BE data	0	0	WFQ2	QCI9

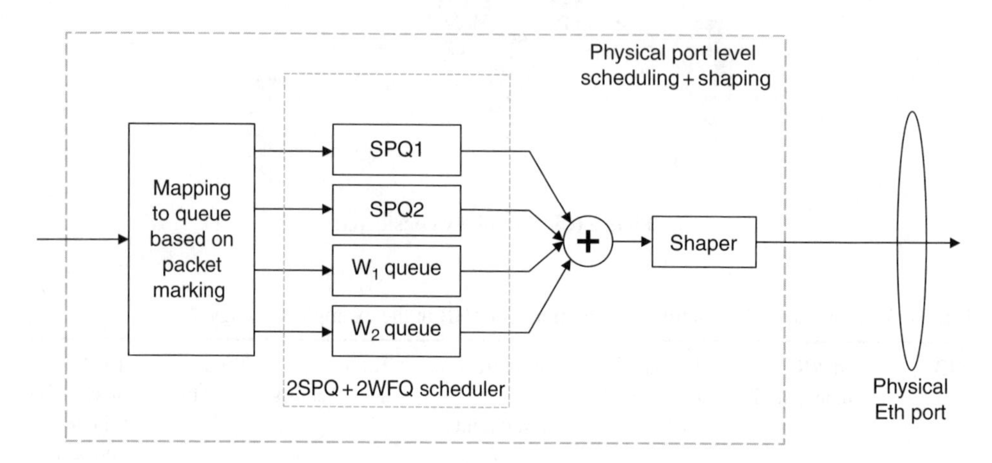

Figure 6.6 QoS scheme for the design example. Scheduling and shaping is at physical port level. The same scheme used in both uplink and downlink

On the other hand, 100 simultaneous LTE voice calls with narrow-band AMR codec and IPsec headers would consume 100 × 200 × 8 × 50 bits per second = 8 Mbps. The overall delay-sensitive signaling and voice traffic can be assumed less than 10% of the link bandwidth, assuming a 350-Mbps radio link. The IMS voice codec packet jitter requirements are tolerant enough so that it can be mapped to a second strict priority queue. The maximum queuing time for a 350-Mbps link is highly likely to be only some tens of microseconds since the higher-priority signaling traffic load is very low, and thus it is unlikely that there are many signaling packets ahead of the voice packet in the queue. Mapping voice to the same queue with signaling could also be used and would result in a very similar performance, with even simpler QoS implementation.

The bandwidth remaining over from SPQ queues is used by WFQ. To guarantee sufficient bandwidth for management traffic under congestion, a part of the WFQ bandwidth is allocated to the w_1 queue, say, 50%. Since most of the time management traffic is negligible, the WFQ1 bandwidth can be used by the w_2 queue carrying BE data. For 50% of bandwidth reservation for management traffic the queue weights would be $w_2 = w_1$. In principle, management traffic could be mapped to the same queue as the data traffic, but to guarantee smooth access to network elements under abnormal traffic conditions a dedicated queue is the approach favored here.

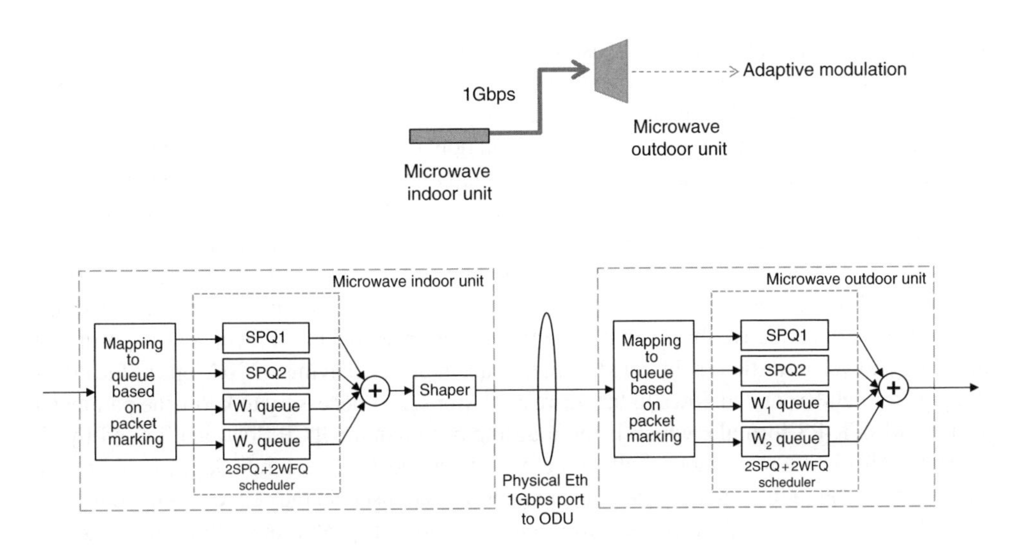

Figure 6.7 Scheduling and shaping for microwave links

The physical port level shaping bucket size is set equal to the size of the largest Ethernet frame that needs to be handled. This is IP MTU + Ethernet L1/L2 frame plus two VLAN headers (double tagging in microwave network). IP MTU is set to the largest value supported by the equipment to avoid fragmentation of the IPsec packets. This bucket size setting delays consecutive maximum size frames so that they are not sent back-to-back at a 1-Gbps line rate to the ODU and so risking buffer tail drop. The shaper average rate is set to the maximum microwave link speed with highest modulation order allowed by the adaptive modulation (Figure 6.7).

The microwave ODU scheduler parameters are set identically at what is described above. This will prioritize signaling and real-time traffic under adaptive modulation changes. Ingress policing is not used as it would drop packets instead of delaying them; and needlessly dropping their own packets is typically not in the best interests of an operator.

6.2.3.2 Bandwidth Dimensioning

The traffic model was given as:

- 50 Erlangs per site voice BH traffic
- site-level BH data demand 40 Mbps.

This, in terms of traffic information, is very minimal but on the other hand represents a typical traffic model available in the design phase. The input requirements also state that the network must be able to support a peak cell data rate of 20 MHz + 10 MHz carrier aggregation, which with 10% transport overhead corresponds to a peak rate of 250 Mbps per cell. From this "day one" requirement, it follows that the first mile microwave link should be 350 Mbps; with such high peak rate detailed first mile dimensioning becomes unnecessary since the site average bit rate of 40 Mbps is only a small fraction of this. The interesting question is mainly whether 350 Mbps is also enough for the second microwave hop that serves three eNBs.

Looking at the example topology for the second hop, the average traffic carried is 120 Mbps of data and 150 Erlangs of voice (three eNBs). Usually voice and data BHs do not coincide but for simplicity's sake both BHs are assumed here to take place simultaneously. For voice, the probability that the number of voice calls exceeds 170 is 1%, from the Erlang-B formula for 150 Erlangs of offered voice traffic. The bandwidth for 170 voice calls with IPsec headers is $\approx$ 15 Mbps. The total of signaling and voice is still less than 5% of the total link bandwidth.

To estimate the effective throughput reduction experienced by TCP download the portion of signaling and voice traffic is subtracted from the link capacity, leaving $\approx$ 330 Mbps as the effective pipe for data[5], rounded down to the nearest 10 Mbps. It is necessary to know the parameter R in the M/G/R-PS formula, which is the multiple of maximum TCP flow rate limited by air interface. This is often the most challenging part, since the TCP peak rate depends on several factors, including radio conditions and the number of other users in the cell. Cranking out some example numbers for a 10-Mbps, 30-Mbps and-50 Mbps UE peak rate (with rounding down, $R = 33, 10, 6$), the results shown in Figure 6.8 are obtained.

In design requirements it was stated that the transport network may degrade end-user performance (indicated by TCP throughput degradation) by a maximum of 10%. For excess

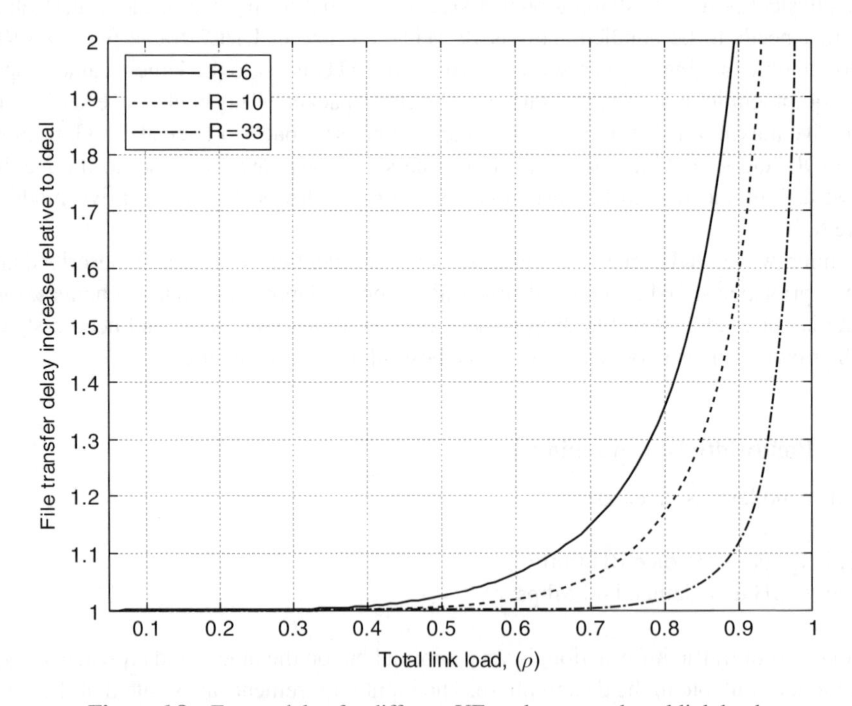

Figure 6.8 Excess delay for different UE peak rates and total link loads

[5]This is the simple quasi-static approximation from Chapter 5 where higher priority traffic is assumed to reserve a fixed portion of bandwidth.

file download delay of at most 1.1, the link load should be less than 65% and 90%, for R = 6 and R = 33 respectively. The result provides an idea of the sensitivity of the dimensioning result to the radio interface performance. The data traffic of 120 Mbps corresponds to a total effective data pipe load of 120/330, or 36%. From the graph it can be seen that even with the most optimistic of radio conditions (R = 6) the transfer delay with a 36% load is in the order of 1%. Hence for the given traffic model it can be concluded that 350 Mbps is enough link capacity for the second mile.

6.2.3.3 Network Capacity Management and Monitoring

Planning for capacity management and performance monitoring is often omitted in high-level design. However, it pays off to think ahead and consider how these tasks will be carried out after deployment, especially in terms of the performance-monitoring capabilities of different bits of equipment in the transport chain. The main performance areas to monitor include:

- Traffic volume and link load monitoring: In addition to 15 minutes or 60 minutes average, it would be helpful from a capacity-management perspective to have performance counters for measuring, e.g. 10-second or 1-minute peak traffic, per scheduler queue if possible.
- E2e QoS monitoring: It is important to select equipment that has built-in standard-compliant support for the TWAMP, which allows measurement of not only the basic packet loss and average packet delay statistics (per QoS class) but also one-way packet delay variation. Capability to measure e2e packet delay variation is increasing in importance because of PTP-based synchronization as well as voice over packet service.
- Path availability: Measurement of e2e path availability is especially important in the leased line use case. Even without leased lines, the e2e transport frequently consists of multiple network segments that are administered by different departments and network management systems within the operator organization. Having an active path availability measurement, based, for example, on UDP Echo or ICMP ping, helps in the centralized monitoring and troubleshooting of network availability issues.

With IPsec there is the added complexity that suitable security policies need to be in place when eNB is used as the monitoring endpoint. Otherwise, the default policy discards measurement protocol packets.

For the present design example, capacity monitoring of the DL user and control traffic can be done at the pre-aggregation nodes, since U- and C-plane traffic of an eNB is transported in a dedicated VLAN and each VLAN hosts a single eNB.

For e2e QoS monitoring both the eNB as well as the EPC core site router should support the TWAMP protocol. Otherwise, dedicated active measurement equipment needs to be installed at the core sites.[6]

The service availability could be monitored from core network SCTP and GTP path supervision alarms, or eNB alarms.

[6] Active QoS measurement need not be active all the time for all eNBs, but can be enabled periodically or per demand.

6.3 Scenario #2: Leased Line

6.3.1 Assumptions for the Use Case

In this example, the operator's own transport network is replaced with a regional leased line access transport, providing Ethernet service from the base station site to the operator's regional data center. Figure 6.9 illustrates the scenario. The high-level design boundary conditions are the same as for the first example, only the access and pre-aggregation cloud is now replaced with a black box leased line transport. The main difference to the MWR use case is how to match the QoS parameters to the SLA.

To make the example more interesting it is assumed that there are two candidate service providers with slightly different service offerings and pricing, (a summary of characteristics is given in Table 6.6). The main difference between the two is that the first provider offers a more fine-grained QoS offering with three service classes and at a higher price. The primary question for transport planning is to meet the design targets for the least possible cost.

6.3.2 Comparing Transport Providers

To support the required carrier aggregation cell air interface peak rates, about 240–250 Mbps of bandwidth is required; the exact requirement may vary a few percentage points in either

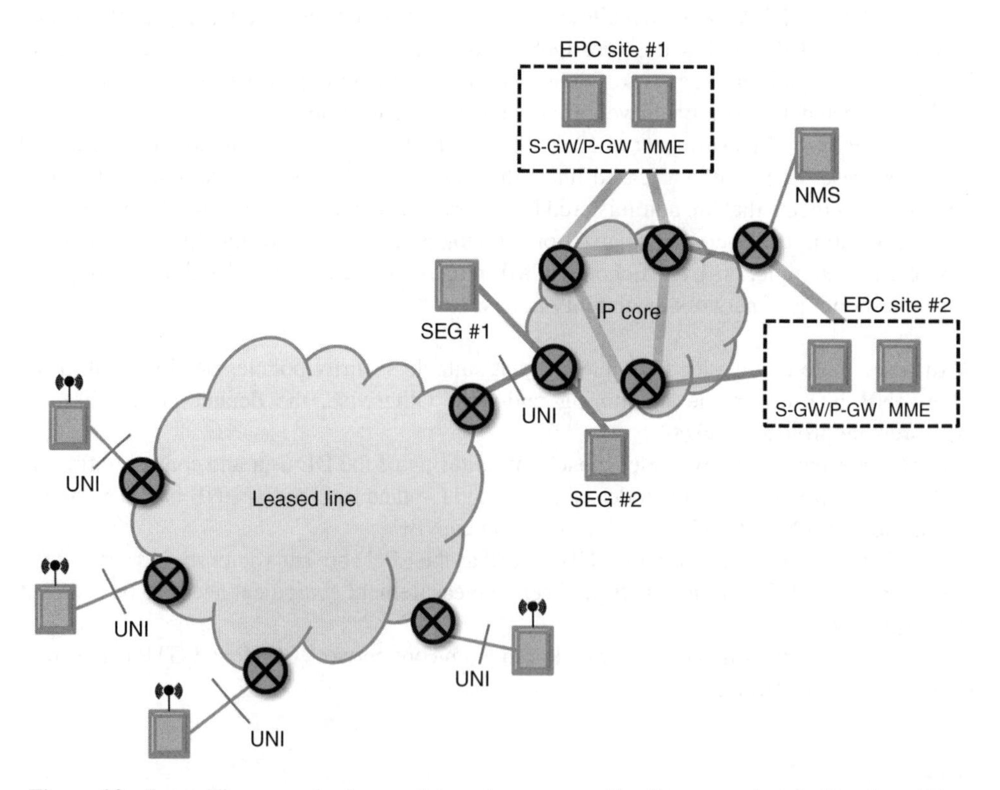

Figure 6.9 Leased line example. Aggregation and pre-aggregation tiers are replaced with a leased line

Table 6.6 Summary of provider service offering of the candidate service providers

Attribute	Provider #1	Provider #2
Round trip delay	Basic: <20 ms Priority data: <5 ms Real time: <5 ms	Basic: <20 ms Real time: <5 ms
Frame jitter (99% percentile)	Basic: n/a Priority: n/a Real time: 1 ms	Basic: n/a Real time: 3 ms
Frame loss rate	Basic: <0.1% Priority: <0.001% Real time: <0.1%	Basic: <0.1% Real time: <0.1%
Bandwidth profiles	CIR as fraction of bandwidth: 20% real-time, 20% priority, 60% basic EIR = 40% for basic, 0% for others	CIR as fraction of bandwidth: 10% real-time, 90% basic EIR = 0% for both service classes
Service availability	Better than 99.99%	Better than 99.99%
Service rates	100 to 1000M, 50M step	100 to 1000M, 100M step
Pricing[a]	$1000 per 100 Mbps per month	$700 per 100 Mbps per month
Class of service differentiation	VLAN ID or 802.1p based	VLAN ID based
Price distance dependency	None up to 500 km	None up to 500 km
Connection type	Point to point	Point to point
Synchronization	Phase sync not supported	Phase sync not supported

[a] The '$' notation is used here to indicate an arbitrary currency.

direction depending on the assumed packet size distribution, IP fragmentation and any other assumptions that the planner wishes to use. Making the most optimistic overhead assumptions possible, for provider #1 it could be possible to survive with a 250 Mbps bandwidth reservation. This would be just enough to satisfy the peak rate requirement. The voice traffic per site is 50 Erlangs, which corresponds to a not more than 5-Mbps bandwidth with 97% probability (from Erlang-B).

For provider #2, 300 Mbps reservation is required since the offered bandwidth granularity is 100 Mbps, and 200 Mbps capacity is not enough for the peak rate requirement. For provider #1 the basic traffic can use the excess information rate up to 40% only in the absence of real-time and high-priority traffic in the provider network.

This would set the price of transport per site to $2500 per month and $2100 per month, respectively, for the two providers. The next question concerns whether the remaining design targets can be satisfied with these minimum bandwidth requirements.

6.3.2.1 Synchronization

The design targets require that the base stations are phase synchronized. As the candidate providers do not support synchronization, the only way to meet this requirement is to install GPS at every site.

6.3.2.2 Availability

Both providers support the required availability target of 99.99%. The exact definition and calculation method of availability was not stated in the assumptions, so double-checking the contract details and constant SLA monitoring is recommended.

6.3.2.3 Quality of Service

Provider #1 supports three different traffic classes while Provider #2 supports two. Also the QoS guarantees for priority data and real-time are better for Provider #1. For IMS voice the e2e packet jitter requirements from 3GPP in the order of 30…50 ms (depending on the test profile) and therefore 3GPP-compliant IMS voice codecs are expected to be fairly robust against packet delay variations. On the other hand, it should be kept in mind that LTE radio interface may introduce delay variation in the order of 10…20 ms in poor radio conditions (peak delays can be up to 50 ms) so it is not generally recommended to consume e2e jitter budget in transport, if this can be reasonably avoided. The limiting factor comes from VoLTE–VoLTE calls where both ends generate air interface jitter. Regardless, from an IMS voice quality perspective there is not much practical difference between the two providers, even if Provider #1 guarantees a slightly smaller packet jitter.[7]

Provider #2 does not support any service class with a low packet loss rate. This could be a problem for signaling traffic, which should be delivered with high reliability. For example, a signaling procedure involving five S1 messages would suffer a packet loss with about 0.5% probability, which could lead to KPI degradation. With Provider #1 signaling traffic could be mapped to the priority data service class which offers a low packet loss rate. With Provider #2 it would be necessary to tune SCTP retransmission timers to smallest possible values to mitigate any delay in S1 signaling procedures due to packet loss.

Another point to consider is the TCP throughput degradation from packet loss. The basic TCP throughput formula, discussed in Chapter 4, predicts that throughput is inversely proportional to the square root of the TCP segment loss probability. Therefore, reducing packet loss by a factor of 10 would increase TCP throughput by a factor of about three. While this relationship may not hold exactly for the latest TCP variants, it is still worth noting that even if committed bandwidth is allocated an up to > 200-Mbps peak rate, this throughput may not be achievable with applications transported over TCP. This is because TCP throughput degrades severely from packet loss (see Chapter 4 for further discussion). For the basic service class, both providers guarantee a less than 0.1% packet loss. The high probability of packet loss lets the providers operate the basic service with small transmit buffers or use early packet discard (e.g. RED). In practice, it remains an open question whether the highest LTE and LTE-A peak rates can be achieved with TCP over the candidate networks, due to the 0.1% packet loss rate stated.

---

[7] If IEEE1588-2008 frequency synchronization was required in the design targets the smaller jitter of Provider #1 would be advantageous. Since the phase synchronization packet delay requirements are in the order of microseconds, the only way they can be satisfied is if the transport provider offers a dedicated phase timing service to the base station sites (instead of black box transport).

6.3.2.4 IP Planning

Provider #2 requires that basic and real-time service classes are separated in different VLANs. Since the requirement is to encrypt both user and control traffic (which have different QoS requirements), some means of routing these traffic types to different VLANs is needed, and the split to protected and non-protected VLANs from the MWR use case is not applicable (assuming that only destination-based routing is supported by the eNB).

Further complication is that U-plane traffic comes in BE and voice varieties which have different QoS requirements; voice and IMS signaling traffic should go to the real-time VLAN, while BE QCI9 traffic should go to the basic VLAN. One way to circumvent the problem is to configure two SEG addresses at the remote side of the leased line transport: one for carrying basic QoS VPNs and one for carrying real-time VPNs (signaling, voice). In this case the number of IPsec VPNs is doubled compared to the previous MWR use case (Scenario 1). The eNB should be able to send voice and data user traffic in different VPNs, which may prove to be difficult if only basic traffic selectors based on source/destination address and protocol/port are supported by the eNB. Correspondingly, in the DL direction the SEG should be able to route voice/signaling and BE data to different VPNs and VLANs. Depending on the routing features of the SEG, this could require different U-plane host addresses for voice and data traffic at the eNB. Overall, implementing QoS with Provider #2 requires non-standard solutions regarding IPsec traffic selector and routing implementation.

With Provider #1 the problem of splitting encrypted U- and C-plane traffic to different VLANs because of QoS requirements can be solved if the eNB supports mapping outer (IPsec) IP header DSCP bits to 802.1p bits in the VLAN header, which is typically supported by the eNB. In this case one VLAN per eNB is enough, and protected and non-protected traffic can be sent in the same VLAN, since service differentiation is based on the VLAN header 802.1p bits.

6.3.2.5 X2 Delay

One-way delay requirement for the X2 interface is 5 ms or less. Neither provider satisfies this requirement for any service class, since only point-to-point connections are offered between base station sites and the regional operator data center. The guaranteed one-way X2 delay would be <10 ms. This could prove to be a bottleneck for deploying radio features requiring low X2 latency in the future.

6.3.2.6 IPsec

Both providers offer point-to-point connection from base station site to the operator core site. The SEG could be placed at the core site co-located with MME and SP-GW, or it could be placed at the regional core site. From an X2 delay perspective, it is beneficial to have the SEG at the regional site. On the other hand, the number of required gateways increases and so does the configuration and maintenance effort. From a resiliency and maintenance angle, it is beneficial to have several small SEGs rather than two big ones. Hardware-redundant SEG configuration (e.g. active-standby dual-chassis) will have to be used in either case since SEG is a single point failure that could risk service outage in large service areas. If the

SEG is located at the regional site, the path from the SEG to both core sites will have to be protected if there is a risk that an attacker could obtain physical access to the transported signal.

6.3.3 The Solution Summary

The required QoS can be satisfied with either provider. From a purely (simplified, short-term) economics point of view, Provider #2 is the winning choice due to the higher bandwidth offered for less cost per site. Cell peak rate requirement is the most critical item when targeting cost efficiency. Relaxing the peak rate requirement would result in large transport cost savings. Since peak rate is often an important competitive factor in the business, it may be that the marketing department of the operator is not willing to relax this requirement, even though cost savings would be achievable in transport.

The problem with the Provider #2 solution arises from the combination of QoS and IPsec. Since the service level differentiation is based on VLANs, there should be a way to send encrypted signaling/voice and BE data in different VLANs, in both the UL and the DL direction. If only the basic destination address-based routing is supported, this requires more complex IP configuration, and is dependent on the available eNB features (e.g. how to separate voice and data traffic to different VLANs).

While Provider #2 offers a more cost-efficient choice of transport from the bandwidth point of view, QoS differentiation presents some technical problems that either make network configuration complex or not possible at all. For this reason Provider #1 would be the primary candidate.

Finally, the obvious should be stated: there are numerous other decision criteria involved in the selection of a transport provider than the mere cost per site or a list of technical characteristics considered in this design exercise. However, as it is hard to quantify such immaterial non-engineering factors in terms of hard numbers, they are considered beyond the scope of this design example.

Reference

IEEE1588 (2008) *IEEE standard for a precision clock synchronization protocol for networked measurement and control systems*, July 2008, http://www.nist.gov/el/isd/ieee/ieee1588.cfm, accessed 1st June 2015.

7

Network Management

Raimo Kangas[1] and Esa Markus Metsälä[2]
[1] *Nokia Networks, Tampere, Finland*
[2] *Nokia Networks, Espoo, Finland*

7.1 Introduction

Requirements toward network management of the mobile backhaul network are increasing due to the vast growth of network capacity and due to the increasing complexity of the mobile network. This chapter looks more specifically at how the planning and optimization functions affect network management. The final part of the chapter, Section 7.9, examines network management planning.

Network management consists of element management systems (EMS) and actual network management systems (NMS). In the widest definition network management also includes service management and customer care. This widest definition is typically called "operations support systems" (OSS). The OSS includes all the necessary systems used by the telecommunications service providers to manage their networks. Either the operations and management (O&M) interface connects the managed network elements directly to the NMS or the O&M interface is provided by the EMS.

Network management serves many user groups in the network operator organization, for example those involved in the planning, optimization, configuring, monitoring and troubleshooting of the mobile backhaul network. Network management provides a centralized focal point to collect information about the managed mobile backhaul network and enables network optimization either manually or automatically. The optimized configuration is deployed to the mobile backhaul network elements by using NMS—the system provides one centralized location where one can rapidly change network element configurations when necessary. Part of the network element configurations can be aligned with operator-wide policies with the help of the system. The collected and maintained information includes network topology and up-to-date configuration, enabling new planning cycles, simpler troubleshooting and the isolation of faults.

NMS saves effort and reduces errors, compared to EMS. The vast amount of measurement data, configuration data and different observation data requires a functioning, well-designed

LTE Backhaul: Planning and Optimization, First Edition. Edited by Esa Markus Metsälä and Juha T.T. Salmelin.

NMS that is capable of storing at least a short-term history of performance data required for statistical KPI calculation, as one example.

The typical NMS provides the following functionality: fault management, performance management, CM, optimization and, lately, SON). These management areas are discussed in the following sections.

7.2 NMS Architecture

The high-level network management architecture when looking at the external interfaces includes connectivity to planning and optimization tools, trouble ticket systems, inventory systems and EMS. The O&M interface between the EMS and network elements is also called a "southbound interface." The interface between planning and other external systems is called a "northbound interface," as shown in Figure 7.1.

Many operators organize network management to have a centralized or a nation/country-level system in addition to regional NMS systems, creating a layered NMS configuration (Figure 7.2). The centralized NMS provides multiple benefits, for example the possibility to manage border areas between regions due to mobility management, and enable a single location to provide 24/7 support instead of multiple ones.

One of the key issues and challenges is that the transport network management is separate from the actual mobile network management. The LTE radio and core elements have a dual role in the network, since in addition the actual LTE radio and core functions, the elements, include transport functions. These elements have two aspects within a single network element: an integrated mobile application part and a transport part that provide a single O&M interface with the EMS or that is directly connected to the NMS. In Figure 7.3, the LTE network elements (such as eNBs, S-GWs, etc.) are marked "LTE NEs" and the backhaul network elements are marked "Backhaul NEs" (such as switches, routers, microwave devices). The shaded part in LTE NEs indicates that some level of transport functionality is integrated into

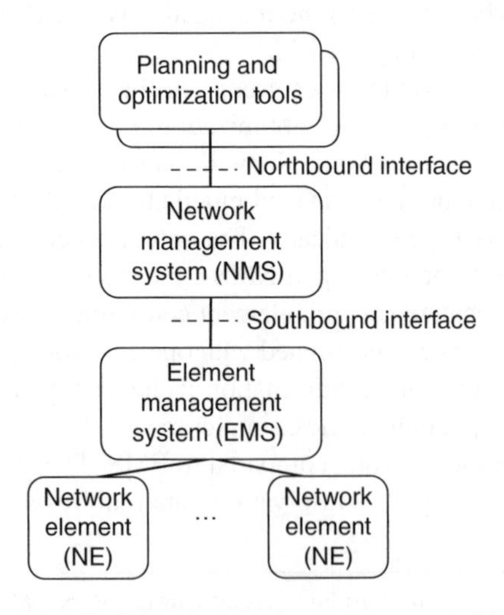

Figure 7.1 High-level network management architecture

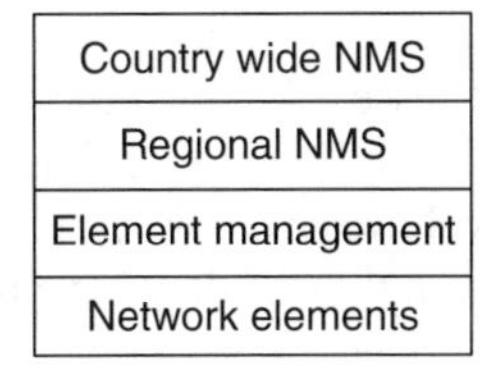

Figure 7.2 Layered NMS: Centralized/regional/EMS levels

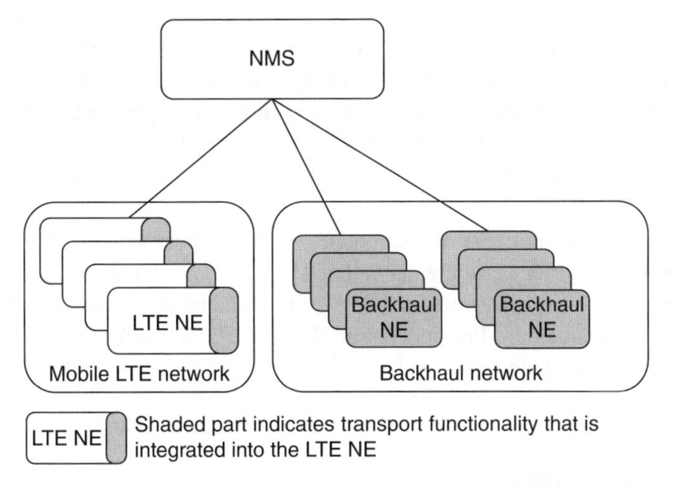

Figure 7.3 Common NMS for mobile LTE network and the backhaul network

each of the LTE NEs. At the very least, this would be a network interface with the protocol stack being that defined by 3GPP, since without it the LTE NE would not be able to connect to other LTE elements.

The main advantage of having both the backhaul and the mobile radio together in the same NMS, as in Figure 7.3, is being able to provide a combined view of the network. This is necessary to manage relationships and dependencies between these two logically separate networks. The combined network management is also an advantage because the mobile network elements include a transport function, which shares a common O&M function with the mobile radio part. Separating the radio and the transport functions under different NMSs/EMSs is not possible in the case of mobile network elements, due to the fact that the element usually provides a common O&M interface, a common info model. etc.

The following high-level requirements and expectations are typically brought up towards the NMS: redundancy, high availability, security, real-time performance (to react faster to changes) and automation to help users and save opex. Operators are unwilling, because of the costs involved, to increase the number of operating personnel even to match the growth of the managed network or to keep abreast of mobile technology evolution and new radio generations, like LTE, and this means that NMS is a critical part of the business's success.

The network growth and network technology evolution steps set requirements for scalability in addition to the increasing amount of mobile network services to be managed.

Customer-specific extensions are supported in order to help adaptation to operator- and operator-group-specific business processes. An operator may have a huge number of tools and tool chains with which NMS needs to be integrated. This number could be more than a hundred when taking into account the evolution of mobile radio technology starting with GSM (or even before that). GSM networks, even though they have been operating for decades, are still part of the managed and operated services.

Multivendor capability has a high priority due to the fact that operators have multiple mobile network infrastructure vendors. The ability of a multivendor-capable NMS to manage a common radio network consisting of equipment from multiple vendors reduces the amount of tools and allows users to concentrate on a single tool usage, which translates to higher efficiency. A reduced number of tools reduces opex as well due to smaller licensing costs.

From standardization viewpoint basic network management concepts are defined in ITU-T (2000). 3GPP (2015) defines more concepts and general principles. The most accurate definition covers the northbound interface that can be utilized by the centralized management system. The northbound interface can be provided either by NMS or by the network element manager (i.e. EMS).

The IETF approaches network management from the perspective of IP networks, and defines management interfaces for elements that are used to implement IP networks. The most commonly used IETF specifications for network management are the ones defining SNMP (Simple Network Management Protocol) and NETCONF (Network Configuration Protocol). These are both described below.

7.3 Fault Management

The purpose of network fault management is to detect faults in the network and make a notification of the fault. Fault notification typically means an alarm event is emitted and sent to the NMS. The alarm contains a standardized set of attributes defined by ITU-T and 3GPP, such as the location of the fault and its type, severity, probable cause and time stamp.

The growth of the network size and its increasing complexity have set new requirements for the fault management when it looks to avoid overloading network management users.

The network element system and EMS should reduce the amount of alarms, for example by preventing repeating alarms (toggling alarms). Fault correlation combines alarms when they are triggered by the same fault. For example, a break in backhaul connectivity may cause several fault notifications to be emitted from the mobile network side. Correlation is possible at the NMS level, where alarms originating from the backhaul network and from the mobile radio network can be consolidated.

Fault identification is the first step where the system is checking the nature of the alarm and trying to make an automatic fix or recovery. Fault clearance can take place automatically or with the help of the user. Fault management can make fault clearance proposals based on a specific fault's history or on its knowledge base of network faults generally. The network element stores information of the latest alarms to prepare for situations like O&M connection breaks. After connectivity is restored, the EMS/NMS have functions that target synchronizing alarm information.

Fault management provides a mechanism for O&M connection supervision in the form of a heartbeat signal. The "unavailable status" indication is emitted in case the heartbeat signal is not registered after a certain period.

The NMS has an alarm-forwarding capability that can transfer alarms to an external system, for example to the centralized NMS. The alarm-forwarding function includes filtering and mapping functions.

Alarm events can trigger automatic recovery actions and optimization algorithms. The nature of the alarm can be a threshold that indicates non-optimal (abnormal) functioning of the network, suggesting that tuning and optimization are required.

7.4 Performance Management

Network performance management in EMS/NMS collects performance indicators emitted by the network elements. Performance data monitoring, to ensure the network works as planned and in an optimal way, is a vital part of network management. The network performance data form the basis of the network capacity extension forecasts.

The amount of collected performance data is reduced with the help of performance data aggregation. The aggregation applies statistical functions, like averaging, summarizing, to the raw performance data over a predefined measurement period. This measurement period can typically range from, for example, 1 minute, 15 minutes, 1 hr to 6 hr, and so on.

Besides the collection, performance management includes administration of measurements (AOM) within network elements. The AOM function enables the user to activate/inactivate the performance measurement and adjust measurement periods.

In addition to collecting them, NMS can permanently store performance data, as well as perform data cleaning, reporting, analysis, forwarding and export.

Data reporting typically presents the data as graphs and plots.

Performance data analysis includes a functionality to perform automatic actions based on predefined thresholds and profiles. A more advanced functionality is network optimization, which uses performance data as input to calculate optimized configuration settings for the network elements.

The data-forwarding function can automatically send data to an external system, for example to the centralized NMS. The data sending may include filtering and additional aggregation. NMS provides an interface for export measurement data with an external system, which could then generate related reports. Example performance measurements in the backhaul network are utilization rates, QoS levels and resource availability.

In summary, combined radio and backhaul measurement reporting helps to locate bottlenecks and enables accurate and efficient optimization and planning cycles to be set up for the network.

7.5 Configuration Management (CM)

Network configuration management (CM) provides functionality to monitor and control network element configurations. Configuration management is a central planning and optimization function, especially in the detailed network design phase, since with it network element configurations and configuration changes can be provisioned to the network elements as planned.

The CM in NMS has storage for all of the configuration data provided by the network elements. The configuration data transfer from the elements to the NMS can be initiated from the NMS or from the network element independently.

The modern NMS and modern network elements maintain synchronization of the configuration data almost in real time, based on events or notifications sent by the network elements.

Configuration management supports single and mass provisioning of configuration data, CM data export/import, CM data consistency checking, CM back-up and restore, CM history, CM reporting, templates and configuration profiles, audit logs, editors, configuration data copy, network element configuration compare, workflow automation and network configuration discovery.

The network element and service lifecycle can be divided into the following steps:

1. installation and service deployment
2. network and network element operation
3. network and network element upgrade.

The network capacity and network coverage extensions include new network element installations. Network planning produces configuration data for the new network elements before the actual installation. The initial network element configuration is transferred from the planning system to the network management with the CM data import functionality. The CM provisioning function is used after the O&M connection to the element is established. The O&M connection from network element to the NMS can be established automatically or based on user actions (manual network integration to NMS), as is discussed more specifically in Section 7.7.

Effective single and mass-provisioning functions are necessary in order to modify either a single network element configuration or multiple network element configurations at the same time. This provisioning is essential when deploying new services and features to the network.

The CM data consistency within the network element and across the network elements is one of the essential CM functions in order to ensure that the configuration of the element follows predefined rules and policies.

NMS provides multiple interfaces for integrating the NMS as part of the operator tool chain, and in the case of CM the most important interface is the network-planning interface.

7.5.1 Maintaining an Up-to-Date Picture of the Network

Working with the latest data is a precondition for exploiting the full potential of any network. NMS retrieves configuration data from network elements whenever desired, and saves this information to allow operators to create an archive of actual network configuration, giving the NMS users vital insight into the network configuration, which subsequently enables further optimization of the network. Built-in northbound interface capability allows for the export of data to external systems, such as the centralized NMS.

7.5.2 Configuration History

Configuration history enables operators and NMS users to browse and report configuration data changes. This feature makes it possible to view a network configuration at a given date and time. The recorded configuration history data include actual network configuration changes, reasons for the changes and changes of the network reference configurations.

Configuration history provides parameter history (fully managed object parameter set at the time of change), enabling change supervision, evaluating change rate, detecting repetitions and keeping track of performed optimizations.

Configuration history enables users to view the delta between the current actual and a historic configuration. The delta between two historic configurations is useful to check for modifications during a certain performance measurement report period. Obviously, viewing and browsing all changes (all configurations) in the past for selected managed objects is necessary to be able to indicate the history of and pending changes in the NMS.

Correlation of the configuration history data with performance data history helps user to detect the impact of a configuration change to KPIs.

7.5.3 Configuring Network

Network complexity is the main obstacle to achieving the best possible configuration. Configuration function provides offline-planning capabilities to prevent any mistakes being made on a running network. Current and planned configurations can be compared to identify the best solution. Plausibility checks prevent errors from the start. Import functionality enables one to process network plans which have been prepared outside of NMS in the operator's planning environment.

In the configuration phase all three NMS components work hand in hand:

- Configuration visualization and modification are used for displaying configuration data and to facilitate data editing data when creating a network plan.
- Validation is used for use-case-specific running of consistency checks.
- Configuration management is used for uploading network element data and for provisioning network plans.

Once a new configuration has been defined, the NMS provides all necessary changes in the relevant network-element-specific data formats. The automation of all download tasks ensures network consistency and minimizes operational effort. The result is complete transparency of all planning, without delay, and with a minimized margin of error. All provisioning steps are controlled by the configuration function.

7.5.4 Policy-Based Configuration Management

Policy-based configuration management, as shown in Figure 7.4, introduces a capability to add or change a configuration with templates.

Network audits can be scheduled, and policies can be changed for individual parameters or objects in the network. In addition, exception management allows permanent and temporary exceptions to the parameter policy.

The majority of the parameters can be defined and changed with policy templates. Without policies, millions of parameters need to be managed individually as per managed object in the network.

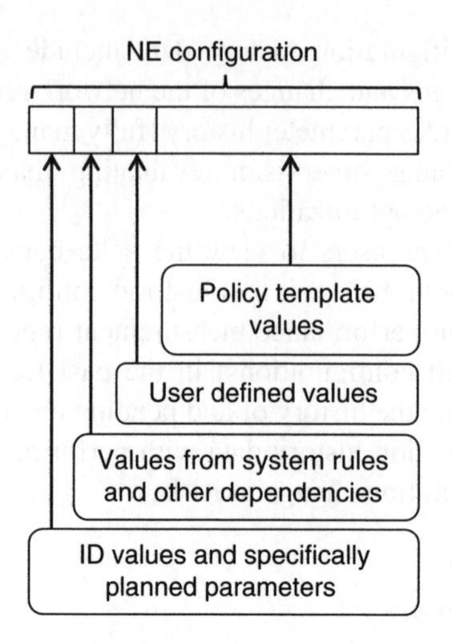

Figure 7.4 Policy based configuration management

Policies provide the means to abstract the desired behavior of the network elements. With policies, different parameter settings can be defined for groups of similar elements in the network. Grouping criteria can be location-related or feature- or layer-based in the network. Policies also enable settings for different services. For some parameters, permanent exceptional values or temporary changes can be defined. With policy audits, a fast, reliable and continuous control of an operator's parameter policy can be applied which also considers the necessary temporary and permanent parameter exceptions.

Defining and maintaining policies is one planning task necessary, for example, for when a new network feature is introduced.

7.5.5 *Planning Interfaces*

In accordance with open policy, the NMS provides a set of well-defined open interfaces for the upper level systems like network-wide systems and other external systems the operator has needed to integrate with the NMS. Open application interfaces provide easy integration and data exchange with external third-party tools.

Examples of interfaces typically used are 3GPP northbound interface, XML (Extensible Markup Language) interface, CSV (comma-separated values) interface, Excel interface and Web services.

7.5.5.1 XML Interface

The XML interface for CM data enables a seamless exchange of radio and core network configuration data between the NMS and external systems, making it possible to import and export planned configuration data as well as export the actual network configuration. This interface also provides import and export potential for templates.

7.5.5.2 CSV Interface

The CSV interface for CM data enables CSV-formatted file exchange between external planning/reporting systems and the NMS for radio and core network data. The users can import planned configuration and export actual configuration.

7.5.5.3 Web Service

A Web service interface is based on SOAP (Simple Object Oriented Access Protocol) used for exchanging structured data in the form of XML. The SOAP protocol uses HTTP or SMTP (Simple Mail Transfer Protocol) underneath.

The interface is compliant with concepts of service-oriented architecture (SOA) and can be easily integrated with an operator's external workflows and tool chain. This interface is suitable for automation and guarantees security for remote access.

7.5.5.4 Excel

A native Excel interface enables easy and smooth data transfer back and forth with Microsoft Excel. This interface is suitable when using Microsoft Excel as planning tool.

7.5.6 Network Configuration Discovery

The network configuration discovery function automatically locates and builds logical/ physical network configuration, for example for MPLS and VPNs.

Network configuration discovery saves time when installing a new NMS on top of an existing network, by building the backhaul network topology automatically. The discovery can detect new installed network elements and merges the added connections and topology with the existing network configuration.

7.5.7 Configuration Management of Backhaul Network

Today, transport-specific backhaul elements such as transport switches and routers are configured either by vendor-specific EMSs or by using command line interfaces (CLI). Commands provided by the CLI are often collected together in scripts and run as a sequence. These scripts are a major cause of network misconfigurations, due to missing consistency checks.

NMS-based CM brings a number of benefits: up-to-date inventory information, cross-domain consistency checking between the radio and the backhaul, auditing leading to less parameter errors, efficient operations due to mass modification possibilities, reducing unnecessary variation in configurations by utilizing policy-based CM, quicker troubleshooting and combined configuration changes with a radio network.

All the above advantages lead to increased service availability and reduced opex.

The challenge is the available O&M protocol that is suitable for CM purposes. The backhaul network elements and network element vendors do not necessarily provide a proper interface that can be utilized to build machine-to-machine interworking between the NMS and the network element.

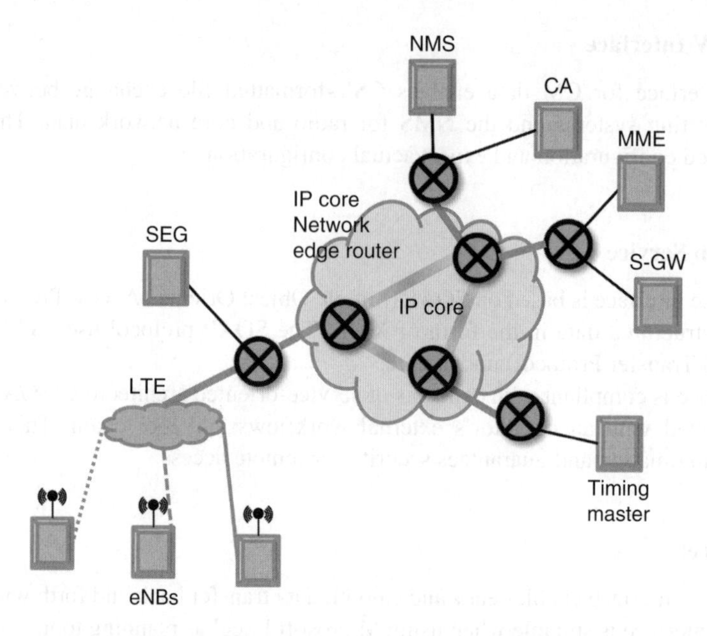

Figure 7.5 Example of combined LTE radio and backhaul configuration management

> **Example 7.1**
>
> Consider a task of configuring IPsec services that protect the LTE backhaul. There are two (peer) elements in this case: the base station (eNB) with its integrated IPsec functionality and the security gateway (SEG), which terminates the IPsec tunnels from a number of LTE eNBs.
>
> Both elements need to have a matching configuration of a number of parameters (detailed in Table 5.14 of Chapter 5), such as traffic selectors, definition of cryptographic algorithms and protocol variants.
>
> When the configuration support exists in the same NMS for both of these elements, significant configuration effort savings can be achieved as well as quality improvement, since errors due to incompatible configurations can be avoided. Errors easily happen when separate tools are used, such as CLI or other element-specific management systems.

In Figure 7.5, a single NMS supports CM functions for both the LTE radio network and the LTE backhaul, including elements such as base stations, core network elements, switches, routers, SEGs, certification servers and timing systems. Clearly, the more the NMS supports CM over various elements, the bigger the benefits achievable in terms of opex and quality.

7.6 Optimization

Network management includes a function to optimize network configurations and settings.

The optimization algorithm mainly uses measurement data as input. The optimization algorithm is run either automatically, triggered by some threshold or initiated manually by

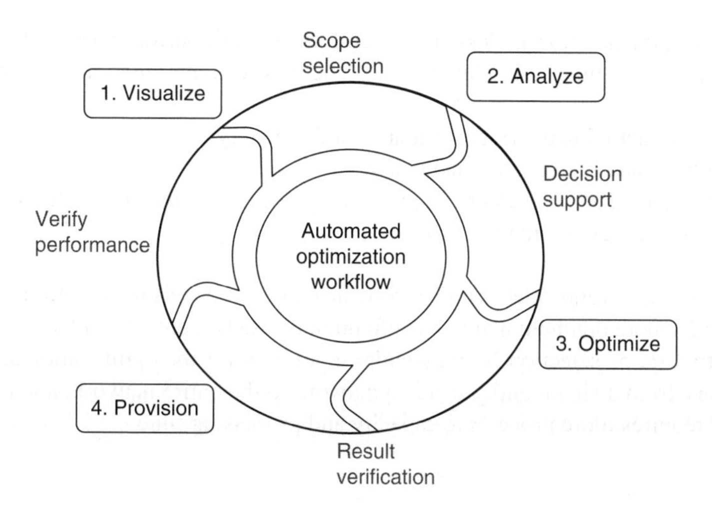

Figure 7.6 Optimization process steps

the NMS user. The resulting configuration is deployed to network elements by using the CM function introduced in the previous section. The related process steps are depicted in Figure 7.6.

The optimization algorithms include programmed logic which improves performance measurement results by changing the values of the configuration parameters. The logic has an in-built knowledge of how and which configuration parameter improves network performance and quality. The end-user effort spent in optimization is minimized when letting the optimization algorithms analyze the network and propose improved configurations for network elements. The more optimal configuration can be found faster because the automated optimization can be run many times faster than manual optimization can.

For tuning and optimizing the configuration, actual measurements from the network are more accurate than a prediction-based approach. The benefit of accurate results can be seen in improved network performance and the capacity utilization rate.

Beside measurements, other input data for optimization are online tracing data, network observations and alarms.

The seamless integration of the optimization function and other NMS functionality found from the NMS, enables easier and more efficient automation, for example for data transfers. This leads to further savings in effort and increases the speed of the optimization process. The optimization function can also be invoked automatically as part of other NMS workflows. Optimization delivers back the modified configuration including the optimal configuration parameter changes.

The optimization function retrieves the actual network configuration data via the CM function explained earlier. An optimization algorithm typically requires settings derived from company policies and quality targets. Optimization results are automatically calculated by the optimization algorithm, and results are deployed to the network via the CM function.

The optimization function can also be activated based on the KPI threshold value or based on alarms. The optimization function shares the KPI formulas with the other NMS functions, like the performance management function. This ensures consistent performance analysis in all of the tools. Operators can define their own KPI formulas to suit their network and business needs.

The optimization function is designed to collect network performance and configuration data, to display it on a map and in table views. The following modules are essential for this:

- visualization of actual network performance and topology
- graphical object and parameter management
- interface to import custom KPIs from preferred performance management tool
- integration with other automation processes.

The closed loop optimization means fully automatic optimization where the optimization function iterates toward a more optimum network configuration based on predefined KPI targets.

Optimization scope selection is an essential input item for the optimization algorithm; the scope can vary from a single entity inside a network to the entire network, and obviously the larger scope requires more processing capacity and processing time.

7.7 Self-Organizing Network (SON)

The 4G LTE network architecture included a new functionality area called a "self-organizing network" (SON). The self-organization function is mainly meant to reduce the cost of operations through increased efficiency and better operational quality. The configuration changes are done by the network element independently without the NMS user's intervention. The cost savings are due to the reduced manual work required by the network operator personnel. The need and demand for efficiency and automation is due to the increased size of network required to deliver the higher data rate and much higher capacity.

The initial target of SON was to add more logic to the network and network elements to perform self-optimization, self-configuration and self-healing. After that, SON has got many other forms and usage targets. The SON approach is also applied to existing mobile technologies like W-CDMA and GSM.

SON is classified as "distributed SON" and "centralized SON," depending on where the logic is running. In the distributed SON approach, the intelligence is in the network element; in centralized SON, it is in the NMS, or another separate centralized system.

The SON functionality also allows the operator to react to more real-time changes and to enjoy a more optimal network configuration compared to manual or semi-manual methods in network planning and optimization. The logic includes, for example, an optimization algorithm improving backhaul QoS parameters. The SON algorithms improve over time, taking into account more diverse conditions. The other improvement target is performance in order to make algorithms perform as close to real time as possible while at the same time avoiding hysteresis. The network quality in the future depends on how good and efficient the SON algorithms are.

SON, especially the centralized SON function provided by NMS, is multivendor-capable. For example, the MVI (multi-vendor interface) capable self-optimization (automated network optimization) uses multivendor CM and multivendor performance data collection.

The LTE eNodeB creation, one of the centralized SON functions controlled by the NMS, automates site integration and site configuration by reducing manual steps. The site integration includes multiple steps (as shown in Figure 7.7). The first step is auto-connection, in which the new site is opening the O&M connectivity to the EMS and network management. The second

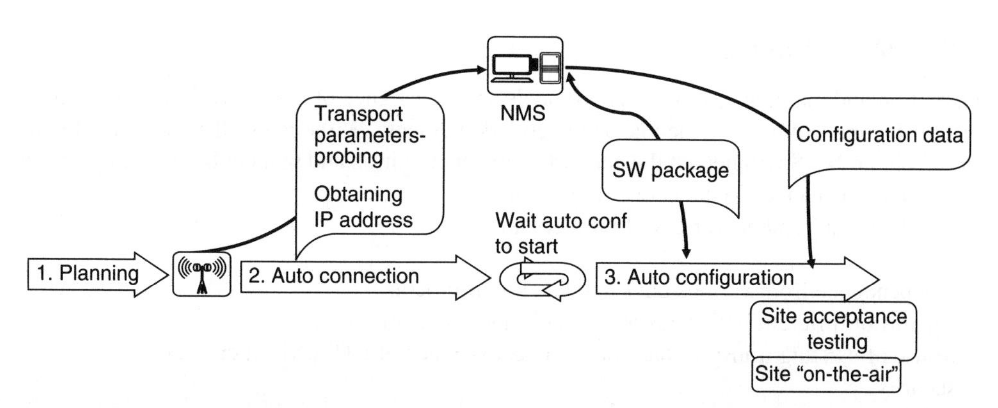

Figure 7.7 New LTE site creation by using an NMS-controlled SON function

step is auto-configuration, where the site fetches configuration data and performs self-configuration. The configuration data include transport configuration, radio configuration and software packages.

Example 7.2

Question: How can the auto-connection phase be arranged in practice?

Answer: Basic enabler for connectivity to the central NMS is that of obtaining an IP address for the eNodeB itself, and obtaining the address of this central NMS. DHCP (Dynamic Host Configuration Protocol) can be used to get these IP addresses as well as other parameters that are needed by the client (the eNodeB).

With DHCP, the client first sends a DHCP broadcast message to the subnet it resides in, trying to find a DHCP server which is managing IP addresses. If the DHCP server is located in the same subnet (VLAN), it responds directly. Otherwise, a DHCP relay agent (e.g. a service provided by a router) has to be configured to the subnet to forward the message to a remote DHCP server. The DHCP server then allocates an IP address to the client.

To communicate with the configuration server (e.g. NMS), eNodeB needs to know the IP address of this server. This address, as well as the address of the SEG, can be provided by the DHCP server. When eNodeB has IP layer connectivity to the server, further configuration data can be downloaded from the central configuration server.

The NMS functionality related to SON is the monitoring possibility for example concerning automatic configuration actions taking place independently in the network and in the network elements.

The SON algorithms in the network element typically include configuration parameters, such as parameters to disable/enable the algorithm and thresholds to tune the function of the algorithm. These configuration parameters are managed by using the NMS CM function.

7.8 O&M Protocols

Operation and maintenance (O&M) protocols are the essence of the network management to establish connectivity with the managed network elements. Without a well-functioning O&M protocol, the NMS will not get the network element condition and will not be able to provision configuration changes to the network elements.

The key requirements for O&M protocol are:

- independence from various physical data link protocols
- support of different higher-layer protocols and applications
- re-use of existing transport facilities (e.g. coexistence of LTE eNB interface)
- security
- reliability
- robustness
- low cost
- low-resource consumption (especially the consumption of processing and memory resources in the network elements).

The communication between the network management and the network can be divided into the following layers:

- application
- O&M protocol
- networking protocol
- physical level or data link level.

The application layer has logic of the specific O&M area, for example fault management, CM etc.

O&M protocol provides services to transport operation and maintenance data and operations between the management system and the network element.

The next two levels, networking and physical levels, can be mapped, for example, to TCP/IP in the Internet protocol suite.

Security, reliability and robustness are essential characteristics of the O&M communication between the network management and network element. Loss of communications may mean a severe situation where NMS does not have an up-to-date picture of the alarms, measurements, states and configuration. From a planning and optimization point of view, the loss of the current state could affect automated optimization. For the SON function the impacts are more related to loss of the network status and loss of information of changes due to the SON function itself.

The reliability could be improved at the link level by implementing redundancy, but that means higher costs. Network-level redundancy is not necessarily sufficient due to the challenge of covering redundancy e2e. Robustness of the O&M protocol itself varies, for example SNMP doesn't ensure communication is delivered. The best option is to build robustness into the network management application layer by using the O&M protocol. One of the techniques used is the "heartbeat approach," where the network element periodically sends a message telling the EMS and NMS that the connectivity works and the element itself is working as explained in the fault management part.

Secure and protected communications are mandatory and in practice implemented in every network. An example protocol used is TLS. TLS may be used on its own to protect the O&M applications, or it may be used in combination with IPsec, as IPsec is the 3GPP-standardized method for LTE backhaul for the U- and C-planes. For the O&M, there is no 3GPP-standardized protection method, but solutions are vendor- and network-specific.

The O&M interface is either a direct interface from the network management or provided by the EMS. The O&M communication between the network management and the element is connectionless due to the fact that it is not possible to maintain simultaneous connections to elements.

Typical protocols provided by the network elements and vice versa supported by the NMSs and the backhaul elements are SNMP, NETCONF, vendor-specific protocols and CLI. The mobile network elements provide many other options, for example Common Object Request Broker Architecture (CORBA), HTTP and XML.

7.8.1 SNMP

SNMP (Simple Network Management Protocol) is the most common protocol specifically for IP networks and is available in nearly all backhaul network elements as one option. SNMP is defined by the IETF in *RFC 3411–RFC 3418* (Harrington et al., 2002; Presuhn et al., 2002). The latest version is SNMPv3 providing, for example, improved security. The IETF claims SNMPv3 to be the full Internet standard having the highest maturity level for an RFC.

The network elements' vendors adopt new SNMP versions with some delay, meaning that the NMS is expected to support more than one SNMP version.

The SNMP protocol is widely used due to its simplicity and its low processing cost. SNMP uses the UDP and the IP. SNMP is mostly used for monitoring and performance management. The improved security enables SNMP to be used for CM.

The SNMP architecture consists of two main parts: agent software running on network elements and manager software running on the NMS or the EMS.

The managed element measurements, configuration and alarming are modeled as objects. The objects are organized into a hierarchical structure. The hierarchical structure containing objects and type definitions are described by a management information base (MIB).

Objects within the MIB are defined by using a subset of Abstract Syntax Notation One (**ASN.1**), defined in McCloghrie et al. (1999). The objects are organized into hierarchies and each object is identified by an object identifier (OID). Each OID identifies an object that can be read or set via SNMP.

IETF has defined multiple MIBs, and network element vendors provide MIBs in addition to the SNMP agent software included in the network element.

The SNMP protocol includes the following operations:

GetRequest
  A *manager-to-agent* request to retrieve the value of an object or list of objects.
SetRequest
  A *manager-to-agent* request to change the value of an object or list of objects.
GetNextRequest
  A *manager-to-agent* request to discover available object and their values.

GetBulkRequest
Optimized version of *GetNextRequest.* A *manager-to-agent* request for multiple iterations of *GetNextRequest. GetBulkRequest* was introduced in SNMPv2.

Response to the requests returns object bindings and acknowledgment from agent to manager for GetRequest, SetRequest, GetNextRequest, GetBulkRequest and InformRequest. Error reporting is provided by error-status and error-index fields.

"Trap" is an asynchronous notification from an agent to a manager. SNMP traps enable an agent in a network element to notify the NMS of some significant event by generating a trap type of SNMP message. The format of the trap message was changed in SNMPv2 and the protocol data unit (PDU) was renamed SNMPv2-Trap.

Acknowledged asynchronous notification was introduced in SNMPv2 and was originally defined as manager-to-manager communication. Acknowledge is sent as InformRequest.

SNMPv3 made no changes to the protocol except the addition of security features: encryption of packets to prevent snooping by an unauthorized source, message integrity protection mechanism and authentication.

The identification of SNMP entities ensure that communication takes place with known SNMP entities. Each SNMP entity has an identifier called SNMPEngineID. SNMP communication is possible only if an SNMP entity knows the identity of its peer. Communication with traps and notifications is an exception to this rule.

SNMPv3 improves the large-scale deployment of SNMP like accounting and fault management and remote CM because of the improved security.

Performance management application with network management using SNMP protocol utilizes polling for obtaining measurement counter values from the network elements. The polling faces challenges from possible communication errors, etc. The standard MIBs used on most backhaul elements include a cyclic counter, for example the number of bytes transmitted and received, on each of its interfaces. The basic traffic statistics for a single network element and even the entire network can be collected by having a SNMP poller application that periodically stores the counters. However, it should be noted that SNMP counters (on the network element) do not count the number of packets per interval, but only give a running total. In order to compute packets per interval, the polling of each network element should take place at precise times. Quite often, data for a particular period are lost due to unreliable UDP transport or when a poller in the NMS crashes.

A typical polling interval for SNMP is five minutes.

Backhaul elements generate traps to notify the network management, for example, of exceeded threshold values or equipment faults. Fault management within the NMS is the application used to collect these traps.

Example 7.3

Question: What are the mechanisms SNMP provides for reliable alarm notifications?

Answer: With SNMPv2, traps are unacknowledged, so they may get lost. Agents may send a trap many times, which of course still does not guarantee it will be correctly received. The other end (SNMP manager) may periodically retrieve certain objects and by this way collect information.

SNMPv3 provides for acknowledged alarm notifications with Informs. With the acknowledgement, the manager lets the agent know the alarm is received.

7.8.2 NETCONF

The Network Configuration Protocol (NETCONF) is a newer protocol which tries to address the increasing complexity of larger communication networks spanning large countries. NETCONF is defined by the IETF (Enns, 2006; Renns et al., 2011). SNMP has challenges as those already mentioned, the PM polling type of inaccuracy and also limitations in scalability when a great many tasks need to be processed simultaneously. NETCONF has been designed to overcome the limitations encountered with SNMP.

NETCONF is used to perform management functions, mainly targeted at configuration data provisioning but capable of monitoring certain network element configuration and network element operational state information. The protocol is based on client-server architecture. The client applications are located within the NMS and multiple servers are deployed in the network, embedded within network elements. The server side issues remote procedure calls responded by the client side agents. NETCONF operations are used for CM functions, for example for provisioning configuration changes.

A network management transaction consists of a request and a corresponding response to that request. A single transaction may be composed of several protocol messages because of protocol message size limits.

Network element configuration is modeled as managed objects and parameters which can be configured by using the *edit-config* NETCONF operation.

NETCONF protocol provides more efficient way to exchange data between NMS and network element than SNMP, when the number of network elements increases.

The NETCONF manager and agent require a common agreement of the network element configuration. The modeling language YANG (Björklund, 2002) is employed to describe the common agreement. YANG is used to describe the network element configuration syntax and operational characteristics. YANG models configuration as a structured tree with attribute declarations that include type definitions. The supported data types include simple data types—for example string and integer—and complex data types—for example list and structured list.

7.9 Planning of Network Management System

The planning of NMS is a part of the network operation and maintenance function and part of the whole mobile network planning. The planning process mostly follows planning of large IT system planning processes (Mangalaraj, 2014). The initial planning phase is followed by enhancement and extension synchronized with the managed growth of the mobile network. The introduction of new mobile network generations, like LTE, means extensions to the NMS as well. The LTE network, as mentioned in the earlier part of this chapter, poses new requirements for NMS. Therefore, LTE means more than an ordinary extension, for example due to the automation requirement discussed above.

Looking at the planning of the whole mobile network and all the related entities in a stepwise approach, the network management should be planned and installed first. Otherwise, the network rollout—including new network element installations, network monitoring and other functions—requires an intermediate solution typically meaning more manual tasks and incomplete integration, due to the missing NMS. The intermediate solution may be difficult to be changed during the course of the network rollout project with a tight schedule and a large project team usually consisting of subcontractors, etc.

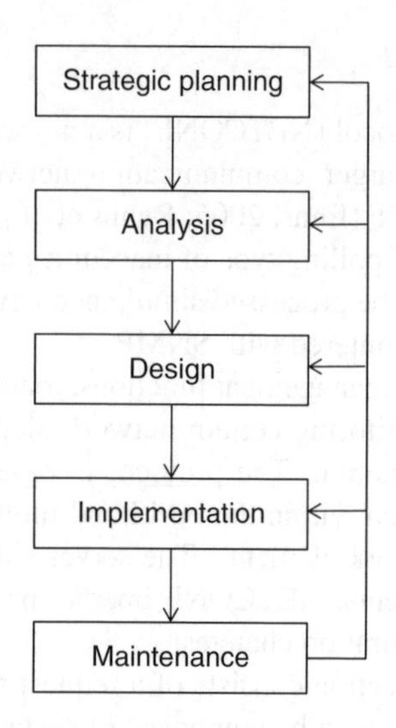

Figure 7.8 NMS planning process

Planning can use all sorts of methodologies, including analysis, design and implementation phases.

The typical steps, as shown in Figure 7.8, are:

1. Strategic planning
2. Analysis
3. Design
4. Implementation
5. Maintenance.

7.9.1 Strategic Planning

The strategic planning of the NMS is derived and related to the operator business strategy. The key facts are the planned network size as network elements, amount of users, network partitioning as regions and estimated amount of data. The data mostly consist of measurement, alarm, CM and tracing data. In addition, the business strategy or the company's IT strategy may set requirements for network managements, for example concerning security.

Strategic planning includes big decisions, such as whether the network management is outsourced partially or completely. One option to consider is outsourcing the NMS, with NMS as a service (NaaS) provided by an NMS vendor. The other big decision affecting the NMS strategy is a full or partial outsourcing of the network management service. This, however, does not remove the need to design and implement the NMS.

7.9.2 Analysis

The analysis of an operator's business processes related to NMS ensures alignment between the NMS and the operator's business processes. An analysis of network management processes produces requirements (e.g. for functionality and interworking). The analysis phase includes tendering and vendor selection of the systems, in addition to those delivered by the network element infrastructure vendor.

7.9.3 Design

The design phase includes, for example, dimensioning and capacity planning; network management applications and tools planning; interworking planning, including interface definitions, contingency redundancy planning (resiliency), security and an internal DCN plan; O&M network planning; and O&M protocol definitions.

Dimensioning and capacity planning is based on projections concerning the mobile network size and the number of network elements. A countrywide mobile network is typically divided into regions in order to split the huge network into more manageable pieces even though there would be a centralized NMS as well. It is possible that some of the network management and operations functions are distributed to the regions and some are centralized. This regional design is an input to the NMS capacity and dimensioning planning, producing information like how many regional NMSs are necessary.

The network management tools and applications design looks into the processes related to network rollout, performance monitoring, fault monitoring, mobile service management, etc., and produces a plan of the NMS application content and high-level requirements for the applications. NMS is connected to many other systems, for example to the radio and transport planning systems, trouble ticket systems and hardware inventory. The interworking between the regional and centralized NMSs is designed based on the responsibility split between regional and central functions. The amount of data transferred from the regional system to the centralized system should be limited and reduced in order to not overload the central system. For example, PM data aggregation and summarizing practices are designed to reduce PM data and the same is done with the fault management data.

Network operation contingency redundancy meaning resilience of NMS is a mandatory requirement to prepare for disasters like hardware failures, environmental issues and power outages. The virtualization vendors and storage vendors provide disaster recovery products. Preparedness for the bigger environmental issues can be solved by distributing NMS geographically. Virtualization products and storage systems enable building a stretched cluster where the computing hardware and storage systems are distributed geographically. The distribution poses bandwidth and latency requirements for the connectivity solution between the locations.

Network management security and network security is of paramount importance for asset protection. Security has become far too broad as a single subject since it includes many aspects.

Security planning includes NMS system isolation from the public Internet with firewalls. The security system requires continuous maintenance and updates in order to have the latest technology. Secure communications ensure protection between distributed communication links of the distributed software entities and between users and the NMS system.

The commonly available technologies to secure communication are SSL/TLS and IPsec (encryption of IP Packets at the IP network layer). The technology used in the network management domain is SSL/TLS defined in IETF as a standard secure protocol.

The CA is typically an independent network element. The CA itself uses certificates and either guarantees its own identity or relies on an additional CA to verify its identity. The CA systems form a hierarchy having a root server that guarantees its own identity.

The network management DCN is partitioned into subnetworks by VLANs . The network management DCN requires the DNS that can serve mobile network elements. DNS and VLAN configurations are designed as part of the DCN design.

The O&M connectivity between the NMS and the mobile network element is typically built by IP technology. IP address planning is a crucial and common task between NMS planning and network planning. the DHCP server can be used to distribute the IP addresses, as discussed in Example 7.2.

7.9.4 Implementation

Implementation includes all actions to execute the created NMS strategy and perform the planned actions to establish the NMS. Network management is a large and complex system having a lot of functionality and a large database, with links to many external systems. The system acceptance testing is a time-consuming task, but an essential step to ensure the functioning of the system. Training and documentation are vital parts of the implementation.

7.9.5 Maintenance

Maintenance includes system maintenance actions, such as continuous improvement rounds leading to complete or partial new planning cycles with possible extensions where the previous process steps are revisited.

References

3GPP (2015) *3GPP specification series*, http://www.3gpp.org/DynaReport/32-series.htm, accessed 1st June 2015.

Björklund M. (2010) *RFC 6020: YANG: A data modeling language for Network Configuration Protocol (NETCONF)*. IETF, Reston, VA.

Enns R. (2006) *RFC 4741: NETCONF Configuration Protocol*, http://www.ietf.org/rfc/rfc4741.txt, accessed 1st June 2015.

Enns R., Björklund M., Schoenwaelder J. and Bierman A. (2011) *RFC 6241: Network Configuration Protocol (NETCONF)*, http://www.ietf.org/rfc/rfc6241.txt, accessed 1st June 2015.

Harrington D., Presuhn R. and Wijnen B. (2002) *RFC 3411: An Architecture for Describing Simple Network Management Protocol (SNMP) Management Frameworks*. IETF, Reston, VA.

ITU-T (2000) *Principles for a telecommunications management network: ITU-T recommendation M.3010*, https://www.itu.int/rec/dologin_pub.asp?lang=e&id=T-REC-M.3010-200002-I!!PDF-E&type=items, accessed 1st June 2015.

Mangalaraj G. (2014) Strategic information systems planning: A literature review. *MWAIS 2014 Proceedings* 17, http://aisel.aisnet.org/mwais2014/17, accessed 1st June 2015. IETF, April 1999.

McCloghrie K., Perkins D. and Schoenwaelder J. (1999) *RFC 2578: Structure of Management Information Version 2 (SMIv2). RFC 2578*, IETF, Reston, VA.

Presuhn R., Case J., Rose M. and Waldbusser S. (2002) *RFC 3418: Management Information Base (MIB) for the Simple Network Management Protocol (SNMP)*. IETF, Reston, VA.

8

Summary

Esa Markus Metsälä and Juha T.T. Salmelin
Nokia Networks, Espoo, Finland

Normally, in mobile system descriptions backhaul is drawn as a narrow line connecting one eNB to the core cloud. That is, of course, not at all the truth: the backhaul is a very complex system especially when thousands of eNBs are connected via packet-based backhaul to the network. How to plan, deploy and then optimize and manage the backhaul efficiently for LTE is a major exercise, with major differences from previous mobile network generations, in areas like synchronization, QoS and security.

Before planning, the mobile operator needs to carefully analyze the strategy and the big picture issues surrounding backhaul. If there are no cost limits, or if own fibers are easily available, it may be possible to build a backhaul which does not limit capacity or create large delays, and is futureproof for even higher requirements than the next year's needs. On the other hand, if there is no own fiber installed or a great amount of money to be spent then more detailed planning and optimization and a compromise between costs and performance are needed.

This book has presented the basic guidelines for the planning and optimization of the backhaul based on the operator's own strategy and cost structure. The various operators and the different regions have their own particular needs and resources when setting up their backhaul. The amount of end-user activity differs from area to area and the used services vary, based on the needs of individuals, so the traffic load in terms of volume and mix of services is always different.

In addition to the discussion and presentation of LTE backhaul planning and optimization topics in standard textbook format, many examples, questions and problems with model answers have been included to help readers understand the main issues of each topic. There is no "one size fits all" solution to backhaul optimization, for no two networks can ever be the same, but it is hoped that the examples, cases and guidelines in this book will help planners tackle the many issues found when planning and optimizing a backhaul network or when working with service providers.

LTE Backhaul: Planning and Optimization, First Edition. Edited by Esa Markus Metsälä and Juha T.T. Salmelin.

After reading this book, one should get the impression that there is a lot to do before LTE backhaul reaches an acceptable level in terms of performance and quality. Yes, LTE backhaul is much more than a narrow line in a mobile network description.

As stated in the introduction, a good knowledge of the LTE network and IP networking is useful for getting the most out of this book and for this the reader is referred to Metsälä and Salmelin's book *Mobile Backhaul* (John Wiley & Sons, Ltd, Chichester, UK, 2012), which offers further information and references.

First the basic LTE backhaul technologies were presented in Chapter 2, including requirements that originate from the LTE radio. Then economical modeling was presented to understand the cost-versus-technology selections in Chapter 3. In Chapter 4 the theory of dimensioning the backhaul was discussed, and this was followed in Chapter 5 with practical planning and optimizing guidelines. Chapter 6 was dedicated to two main network case examples, own backhaul and leased backhaul. Finally, the network management overview was given in Chapter 7.

It is hoped that this book will make it easier to plan and optimize the LTE backhaul, meeting performance targets and reaching the quality and security levels that will be required over the next couple of years, while complying with reasonable capex and opex budgets as set by the organization. Of course new features will be introduced in 3GPP and some of these will likely impose new requirements on the backhaul. Topics under discussion were reviewed as part of Chapter 2 with their potential impact on the backhaul. This outlines the different scenarios. Furthermore, it allows steering the present design toward the direction that future LTE radio features anyway demand.

Index

LTE Backhaul: Planning and Optimization, First Edition. Edited by Esa Markus Metsälä and Juha T.T. Salmelin.